Springer Series on Atomic, Optical, and Plasma Physics

Volume 108

The Springer Series on Atomic, Optical, and Plasma Physics covers in a comprehensive manner theory and experiment in the entire field of atoms and molecules and their interaction with electromagnetic radiation. Books in the series provide a rich source of new ideas and techniques with wide applications in fields such as chemistry, materials science, astrophysics, surface science, plasma technology, advanced optics, aeronomy, and engineering. Laser physics is a particular connecting theme that has provided much of the continuing impetus for new developments in the field, such as quantum computation and Bose-Einstein condensation. The purpose of the series is to cover the gap between standard undergraduate textbooks and the research literature with emphasis on the fundamental ideas, methods, techniques, and results in the field.

More information about this series at http://www.springer.com/series/411

Vladimir Krainov · Boris M. Smirnov

Atomic and Molecular Radiative Processes

With Applications to Modern Spectroscopy and the Greenhouse Effect

Springer

Vladimir Krainov
Moscow Institute of Physics and Technology
Dolgoprudny, Moscow, Russia

Boris M. Smirnov
Joint Institute for High Temperatures
Russian Academy of Sciences
Izhorskaya, Moscow, Russia

ISSN 1615-5653 ISSN 2197-6791 (electronic)
Springer Series on Atomic, Optical, and Plasma Physics
ISBN 978-3-030-21954-3 ISBN 978-3-030-21955-0 (eBook)
https://doi.org/10.1007/978-3-030-21955-0

This Springer imprint is published by the registered company Springer Nature Switzerland AG
The registered company address is: Gewerbestrasse 11, 6330 Cham, Switzerland

Vladimir Krainov

Preface

Starting from Newton, the spectroscopy played an important role in the development of various directions of physics. Spectroscopy data were a basis of foundation of quantum physics, and the theory of radiative transitions was the first complete theory of quantum mechanics. This book contains the description of quantum theory of radiative transitions involving atomic particles (atoms and molecules) and atomic systems (nanoclusters and microparticles) under simple conditions. Namely, an electromagnetic field interaction with atomic particles and systems is weak, the interaction has a nonrelativistic character, and we are restricted by the strongest, so-called dipole radiative transitions. An important part of the contemporary spectroscopy consists in some phenomena which result from the interaction of laser interaction with atomic particles, and data banks which collect a rich information about radiative transitions. The goal of this book is also to establish the connection between the classical theory of radiative transitions in atomic particles or systems and contemporary branches of spectroscopy.

Dolgoprudny, Russia
Moscow, Russia

Vladimir Krainov
Boris M. Smirnov

Contents

Introduction

This book is devoted to radiative processes involving atomic particles and atomic systems, and at the beginning, the understanding of these problems was connected at a great extent with spectroscopy. The history of spectroscopy starts from I. Newton who used the word "spectrum" in order to describe various colors of light and decomposed white light in rainbow of colors with a prism. The understanding of these problems is contained in his treatise "Optics" (1704). The development of spectroscopy includes many separate steps, and we indicate a small part of those which are of importance for radiative atomic properties. Namely, J.von Fraunhofer (1815) replaced a prism with a diffraction grating to decompose white light that allows him to observe dark spectral lines in the solar spectrum and to find their wavelengths. G. R. Kirchhoff and R. Bunsen (1859) proved the identities of emission and absorption spectral lines and concluded that each element has specific properties as regards the light it emits. This allows one to identify an atom of each element by a combination of corresponding spectral lines which are responsible for radiative transitions of these atoms. Since the width of a separate spectral line is relatively small, this fact becomes a fine instrument to measure parameters of atomic particles. This instrument was of decisive importance in creation the quantum mechanics and in the understanding the physics of atomic particles.

This book contains three lines in the analysis of radiative atomic processes. The first one gives the apparatus of quantum mechanics for the description of the properties of atomic particles and systems, as well as processes with their participation. The second line is the analysis of radiative processes in gases or plasmas due to atomic particles from which they contain. The third line of this book relates to applications of this area. Indeed, spectroscopy allows to measure some parameters of atoms with a high accuracy, and this property may be used in precise devices; in the first place, this consists in the detection of atoms of a certain element in various applications, starting from metallurgy. Radiative processes involving atomic particles are the basis of lasers, detectors of infrared radiation, optical magnetometers, etc. In addition, such phenomena, as optical alignment of atoms, cooling of an atom ensemble up to low temperatures are explained by the interaction of radiation with atoms under certain conditions. This will be considered in this book.

The material of the first part of the book, which is devoted to the quantum theory of radiative transitions in atoms and atomic particles, is analogous to that description of this quantum apparatus for radiative processes in the classical books [1, 2]. In order to simplify this apparatus, we consider the strongest radiative transitions which correspond to the dipole approximation. We combine spectroscopy of atoms and radiation transitions in atoms with the atomic structure and principles of atom construction [3–17]. We consider also the structure of simple molecules, their spectroscopy and radiative transitions in simple molecular systems and nanoclusters, which are presented in [18–27].

Because we are guided by the active readers, who make some evaluations and estimations about certain atomic systems, we include some reference data in the book which refer to spread information for atomic systems and spectra [28–30], including Grotrian diagrams [31–34] and the rates of radiative transitions in atomic systems [35–43]. The second part of the book uses statistical properties of radiation and emission of gases and plasmas [44–48]. The latter includes both hot or ionized gases [46, 47, 49–52] and planet atmospheres [53– 61]. Interactions of radiation with atomic particles in gases and plasmas lead to broadening of spectral lines [62, 63] that influences on the character of radiation for these atomic systems. The third part of the book which is connected with applications of radiative processes includes fine spectroscopy of gases [64–66] and uses polarized light and atoms [67–69] and behavior of radiative atoms in magnetic fields.

An essential part of applications is connected with highly excited atoms [70, 71]. The peculiarity of contemporary applications of fine spectroscopy is a narrow spectral beam of radiation compared with a natural width of spectral lines for atomic states [64, 66, 72] that leads to various effects in the behavior of atomic systems. The above list of references which includes only books shows a wide circle of problems which are based on contemporary spectroscopy. Our goal is to join these problems such that the last parts of the book follow from the previous ones. In addition, we try to make the analysis as simple as it is possible without the loss of its rigidity. Therefore, this book is best suited for graduate students and researchers in the field of quantum mechanics, atomic and molecular physics, plasma physics and laser physics. It should be noted that the material of the previous books of the authors [73–75] is used partially in this book which contains a more wide analysis of the structure of atomic particles, radiative processes in gases and plasmas, and special applications of the fine spectroscopy.

References

1. E.U. Condon, G.H. Shortley, *Theory of Atomic Spectra* (Cambridge University Press, Cambridge, 1949)
2. L.I. Sobelman, *Atomic Spectra and Radiative Transitions* (Springer, Berlin, 1979)
3. F. Hund, *Linienspectren and periodisches System der Elemente* (Springer, Berlin, 1927)
4. H.A. Bethe, Quantenmechanik der Ein- und Zweielectronenprobleme, in *Handbuch der Physik*, vol. 24-1 (Springer, Berlin, 1933)

5. G. Herzberg, *Atomic Spectra and Atomic Structure* (Dover, New York, 1944)
6. H.A. Bethe, E.E. Salpeter, *Quantum Mechanics of One and Two-Electron Atoms* (Springer, Berlin, 1957)
7. H.A. Bethe, *Intermediate Quantum Mechanics* (Benjamin Inc., New York, 1964)
8. H.G. Kuhn, *Atomic Spectra* (Longmans, London, 1969)
9. N. Bohr, *The Correspondence Principle* (Elsevier, Amsterdam, 1976)
10. L.D. Landau, E.M. Lifshitz, *Quantum Mechanics* (Pergamon Press, Oxford, 1965)
11. M.G. Veselov, L.N. Labtzovskij, *Theory of the Atom Structure of Electron Shells* (Nauka, Moscow, 1986; in Russian)
12. H. Haken, A.C. Wolf, *The Physics of Atoms and Quanta* (Springer, Berlin, 2000)
13. B.M. Smirnov, *Physics of Atoms and Ions* (Springer NY, New York, 2003)
14. J. Barrett, *Atomic Structure and Periodicity* (Wiley, Berlin, 2003)
15. B.H. Bransden, C.J. Joachain, *Physics of Atoms and Molecules* (Prentice Hall, Harlow, 2003)
16. H. Haken, H.C. Wolf, *The Physics of Atoms and Quanta* (Springer, Berlin, 2004)
17. C. Food, *Atomic Physics* (Oxford University Press, Oxford, 2004)
18. W. Demdroter, *Atoms, Molecules and Photons* (Springer, Berlin, 2006)
19. B.M. Smirnov, Phys. Usp. **44**, 221(2001)
20. I.G. Kaplan, *The Pauli Exclusion Principle* (Wiley, Chichester, 2017)
21. C. Herzberg, *Molecular Spectra and Molecular Structure,* vol. 1–4 (van Nostrand, New York, 1964–1966)
22. K.P. Huber, G. Herzberg, *Molecular Spectra and Molecular Structure. Constants of Diatomic Molecules* (Van Nostrand, New York, 1979)
23. J.D. Graybell, *Molecular Spectroscopy* (McGrawHill, New York, 1988)
24. S.V. Khristenko, A.I. Maslov, V.P. Shevelko, *Molecules and Their Spectroscopic Properties,* (Springer, Berlin, 1998)
25. J.M. Brown, *Molecular Spectroscopy* (Oxford University Press, Oxford, 1998)
26. J.L. McHale, *Molecular Spectroscopy* (Prentice Hall, New Jersey, 1999)
27. P.F. Bernath, *Spectroscopy of Atoms and Molecules* (Oxford University Press, Oxford, 2005)
28. B.M. Smirnov, *Clusters and Small Particles Processes in Gases and Plasmas* (Springer NY, New York, 1999)
29. B.M. Smirnov, *Cluster Processes in Gases and Plasmas* (Wiley, Berlin, 2010)
30. S. Svanberg, *Atomic and Molecular Spectroscopy* (Springer, Berlin, 2012)
31. R. Kakkar, *Atomic and Molecular Spectroscopy: Basic Concepts and Applications* (Cambridge University Press, Delhi, 2015)
32. A.A. Radzig, B.M. Smirnov, *Reference Data on Atoms, Molecules and Ions* (Springer, Berlin, 1985)
33. B.M. Smirnov, *Reference Data on Atomic Physics and Atomic Processes* (Springer, Heidelberg, 2008)
34. W.M.Haynes (ed.), *CRC Handbook of Chemistry and Physics*, 97th edn. (CRC Press, London, 2016–2017)
35. S. Bashkin, J. Stoner, *Atomic Energy Levels and Grotrian Diagrams,* vol. 1–4 (North Holland, Amsterdam, 1975–1982)
36. A.S. Yatzenko, *Grotrian Diagrams of Neutral Atoms* (Nauka, Novosibirsk, 1993; in Russian)
37. T. Shirai, J. Sugar, W. Wiese e.a., *Spectral Data and Grotrian Diagrams for Highly Ionized Atoms* (AIP, New York, 2000)
38. G.G. Telegin, A.S. Yatzenko, *Optical Spectra of Atmospheric Gases* (Nauka, Novosibirsk, 2000; in Russian)
39. W.L. Wiese, M.W. Smith, B.M. Glennon, *Atomic Transition Probabilities—H Through Ne*, vol. 1. Nat. Stand. Ref. Data. Ser. Nat. Bur. Stand. **4** (1966)
40. W.L. Wiese, M.W. Smith, B.M. Miles, *Atomic Transition Probabilities—Na Through Ca*, vol. 2. Nat. Stand. Ref. Data. Ser. Nat. Bur. Stand. **22** (1969)
41. W.L. Wiese, B.M.Glennon, Atomic transition probabilities, in *American Institute of Physics Handbook*, ed. by D.E.Gray (McGrow Hill, New York, 1972). Chap. 7, pp. 200–263

42. W.L. Wiese, G.A. Martin, Transition probabilities, in *Wavelengths and Transition Probabilities for Atoms and Atomic Ions*. Nat. Stand. Ref. Data. Ser. Nat. Bur. Stand. **68**, Part II, 359–406 (1980)
43. L.J. Curtis, *Atomic Structure and Lifetimes* (Cambridge University Press, Cambridge, 2003)
44. http://physics.nist.gov/PhysRefData/ASD1
45. https://www.nist.gov./srd
46. https://www.nist.gov./pml/productservices/physical-reference-data
47. https://www.nist.gov./pml/atomic-spectra-database
48. H.R. Grim, *Plasma Spectroscopy* (Cambridge University Press, New York, 1964)
49. F. Reif, *Statistical and Thermal Physics* (McGrow Hill, Boston, 1965)
50. E. Lindholm, Ark. Mat. Astron. Fys. **32**A, 1(1946)
51. H.M. Foley, Phys. Rev. **69**, 616(1946)
52. Ya.B. Zel'dovich, Yu.P. Raizer, *Physics of Shock Waves and High-temperature Hydrodynamic Phenomena* (Academic Press, New York, 1966)
53. L.D. Landau, E.M. Lifshitz, *Statistical Physics*, vol. 1 (Pergamon Press, Oxford, 1980)
54. C. Kittel, H. Kroemer, *Thermal Physics* (Wiley, New York, 1980)
55. M. Capitelli, C.M. Ferreira, B.F. Gordiets, A.I. Osipov, *Plasma Kinetics in Atmospheric Gases* (Springer, Berlin, 2000)
56. V.N. Ochkin, *Spectroscopy of Low Temperature Plasma* (Wiley, Berlin, 2009)
57. B.M. Smirnov, *Fundamental of Ionized Gases: Basic Topics of Ionized Gases* (Wiley, Weinheim, 2012)
58. B.M. Smirnov, *Theory of Gas Discharge Plasma* (Springer, Heidelberd, 2014)
59. K.Ya. Kondrat'ev, *Radiation in the Atmosphere* (Academic Press, New York, 1969)
60. E.J. McCartney, *Absorption and Emission by Atmospheric Gases* (Wiley, New York, 1983)
61. M. Iqbal, *An Introduction to Solar Radiation* (Academic Press, New York, 1983)
62. R.M. Goody, Y.L.Yung, *Atmospheric Radiation* (Oxford University Press, Oxford, 1995)
63. K.N. Liou, *An Introduction to Atmospheric Radiation* (Academic Press, Amsterdam, 2002)
64. G.W. Petry, *A First Course in Atmospheric Radiation* (Sunlog Publishing, Madison, 2006)
65. W. Zdunkowski, T. Trautmann, A. Bott, *Radiation in the Atmosphere* (Cambridge University Press, Cambridge, 2007)
66. M. Wendisch, P. Yang, *Theory of Atmospheric Radiative Transfer* (Wiley, Singapore, 2012)
67. B.M. Smirnov, *Microphysics of Atmospheric Phenomena* (Springer, Switzerland, 2017)
68. H.R. Griem, *Spectral Line Broadening by Plasmas* (Academic Press, New York, 1974)
69. I.I. Sobelman, L.A. Vainstein, E.A. Ukov, *Excitation of Atoms and Broadening of Spectral Lines* (Springer, Berlin, 1998)
70. V.S. Letokhov, *Laserspektroskopie* (Vieweg, Braunschweig, 1977)
71. V.N. Ochkin, N.G. Preobrazhensky, N.Va. Shaparev, *Optphalvanic Effect in Ionized Gas* (Gordon and Breach, London, 1998)
72. W. Demdroter, *Laser Spectroscopy: Basic Principles* (Springer, Berlin, 2008)
73. E. Collet, *Polarized Light: Fundamentals and Applications* (Dekker, New York, 1993)
74. D.S. Klinger, J.W. Lewis, C.E. Randull, *Polarized Light in Optics and Spectroscopy* (Academic Press, Boston, 1997)
75. S. Huard, G. Vacca, *Polarization of Light* (Wiley, Chichester, 1997)

Chapter 1
Single-Photon Transitions of Atomic Particles

Abstract Radiative transitions in atoms and atomic particles result from interaction between the radiation field and atomic electrons. Representing an atom as a system of electrons located in the Coulomb motionless center, we have that bound atomic states has a discrete character and bound energy states are described by certain quantum numbers. A weakness of interaction between an atom and radiation field follows from a small velocity of bound electrons compared to light speed. Therefore, parameters of radiative transitions in atoms may be determined within the framework of the perturbed theory of quantum mechanics. This approach exhibits that dipole transitions are the strongest one and allows one to analyze the character of atom spectroscopy which including the selection rule and the sum rules. Some applications of atom spectroscopy are based on a narrow width of spectral lines for transitions between discrete states of atomic particles.

1.1 Principles of Atomic and Molecular Structure

1.1.1 One and Two-Electron Atoms

An atom is the simplest object; we consider its interaction with the radiation. Therefore, first we analyze general properties of atoms. As a physical object, an atom consists of a positively charged nucleus and electrons. Because the nucleus mass is large compared to that of electrons, one can represent an atom as a motionless Coulomb center and electrons which interact with this charged center and with each other. Being guided by this standpoint, we consider general properties of atoms as quantum systems. A more detailed description of atoms and atomic systems is given in specific books, for example, [1–7].

We first consider properties of the hydrogen atom or hydrogenlike ions [8, 9] which consist of a nucleus with the charge Ze (Z is an integer, e is the electron charge) and one electron. It is convenient to describe this system applying to it the apparatus of quantum mechanics [10]. The Hamiltonian of this system has the form

V. Krainov and B. M. Smirnov, *Atomic and Molecular Radiative Processes*, Springer Series on Atomic, Optical, and Plasma Physics 108,
https://doi.org/10.1007/978-3-030-21955-0_1

$$\hat{H} = -\frac{\hbar^2}{2m_e}\Delta - \frac{Ze^2}{r}, \tag{1.1.1}$$

where m is the electron mass, $\hbar$ is the Planck constant, r is a distance between an electron and nucleus. The atom energy levels may be determined as the eigenvalues ε_n of the Schrödinger equation for the electron wave function Ψ_n, that has the form

$$\hat{H}\Psi_n = \varepsilon_n \Psi_n \tag{1.1.2}$$

Simultaneously, this allows one to determine quantum numbers which describe the atom state. In the case of the hydrogen atom, the electron quantum numbers are the electron principal quantum number n, the orbital momentum of the bound electron $l \le n-1$, the momentum projection m onto a given direction $m = -l .. +l$ and the projection of electron spin onto this direction $\sigma = \pm 1/2$. Thus, an electron state in the Coulomb field is characterized by quantum numbers

$$n, l, m, \sigma$$

The energy of the bound state with these quantum numbers is

$$\varepsilon_n = -\frac{Ry}{n^2}, \ Ry = \frac{m_e e^4}{2\hbar^2} = 13.6 \text{ eV} \tag{1.1.3}$$

This binding electron energy characterizes the atom ionization potential J_n for this state, i.e. $J_n = -\varepsilon_n$. As is seen, the electron energy levels of the hydrogen atom are degenerated with respect to the quantum numbers l, m and σ. Therefore $2n^2$ states correspond to each electron level. In addition, $n = 1$ corresponds to the ground state. In the case of the hydrogenlike ion with a charge Z of the nucleus we have

$$\varepsilon_n = -\frac{Z^2 Ry}{n^2} \tag{1.1.4}$$

Note that the Hamiltonian (1.1.1) describing the hydrogen atom does not contain the electron spin. This means that the space and spin electron coordinates are separated, and the electron wave function Ψ is the product of the electron $\varphi(\mathbf{r})$ and spin χ_σ wave functions

$$\Psi = \psi(\mathbf{r})\chi_\sigma, \tag{1.1.5}$$

where $\mathbf{r}$ is the electron coordinate, and $\sigma = \pm 1/2$ is the spin projection onto a given direction. Correspondingly, the spatial wave function which describes the electron state which quantum numbers are nlm, is

$$\psi(\mathbf{r}) = R_{nl}(r)Y_{lm}(\theta, \varphi), \tag{1.1.6}$$

where r, θ, φ are electron spherical coordinates, $Y_{lm}(\theta, \varphi)$ is the spherical function.

For an atom containing several electrons we use the one-electron approximation, where the atom wave function is the combination of products of one-electron wave functions. This combination must satisfy to the Pauli exclusion principle [11–13] according to which location of two electrons in identical electron states is prohibited. This means that in the one-electron approach two electrons cannot be characterized by identical quantum numbers. If the electron state is described by its space coordinate and spin, this means that two electrons with identical spin direction cannot be located at the same coordinate. The Pauli exclusion principle requires a certain symmetry of the wave function, namely, the total wave function of electrons changes a sign as a result of permutation of two electrons.

The Pauli exclusion principle causes a specific interaction between atomic electrons and influences on the atomic structure. It is convenient to demonstrate this for the helium atom where two electrons are characterized by the same quantum numbers, namely, the principal quantum number is $n = 1$ and the orbital momentum is $l = 0$. In this case the spin state of the helium atom is separated from the spatial one, and there are two spin states of this atom, with the total spin S which may be equal 0 and 1. The spin wave function for the singlet spin state ($S = 0$) is antisymmetric with respect to permutation of two electrons, and the spin wave function for the triplet spin state ($S = 1$) is symmetric at this operation. Since the total wave function is the product of the spatial wave function of electrons and their spin function, from the Pauli exclusion principle follows that in the first case ($S = 0$) the spatial wave function is symmetric with respect to exchange of spatial electron coordinates, and for the triplet spin state ($S = 1$) the spatial wave function of electrons is antisymmetric with respect to electron exchange. In other words, the spatial wave function of electrons for these states is given by

$$\Psi(\mathbf{r}_1, \mathbf{r}_2) = \frac{1}{\sqrt{2}}[\psi(\mathbf{r}_1)\varphi(\mathbf{r}_2) \pm \psi(\mathbf{r}_2)\varphi(\mathbf{r}_1)], \tag{1.1.7}$$

where sign plus corresponds to the singlet state ($S = 0$) and sign minus refers to the triplet state ($S = 1$). Next, $\mathbf{r}_1$, $\mathbf{r}_2$ are spatial coordinates of two electrons, the spatial wave functions $\psi(\mathbf{r})$ and $\varphi(\mathbf{r})$ correspond to states with $n = 1$ and $n = 2$ respectively.

One can define the atom energy E as

$$E = \langle\Psi|\hat{H}|\Psi\rangle, \tag{1.1.8}$$

where Ψ is the wave function of electrons for a given state, and $\hat{H}$ is the electron Hamiltonian. It is seen, that the energies E of singlet and triplet states are different even in the approximation under consideration where the wave function of individual electrons are identical. Then the difference of energies between singlet and triplet states ΔE are equal

$$\Delta = 2\langle\psi(r_1)\varphi(r_2)]\hat{H}|\psi(r_2)\varphi(r_1)\rangle, \tag{1.1.9}$$

It follows from this consideration that the energies are different for states with the various total spins. It is seen, that this effect is determined by the symmetry of the spatial wave function, rather than by the magnetic interaction of spins. The interaction due to the symmetry of the electron wave function, or the exchange interaction, is of importance for atomic properties.

1.1.2 Light Atoms

We now, following to [14], analyze the quantum numbers of the light atom with several electrons neglecting the relativistic effects. The Hamiltonian of the system of atomic electrons in a light atom is given by

$$\hat{H} = -\frac{\hbar^2}{2m_e}\sum_i \Delta_i - \sum_i \frac{Ze^2}{r_i} + \sum_{i \neq j} \frac{e^2}{|\mathbf{r}_i - \mathbf{r}_j|}, \tag{1.1.10}$$

where $\mathbf{r}_i$ is a spatial coordinate of i-th electron. In this case the interaction between electrons is important, unlike the case of the hydrogen atom. Our goal is to determine quantum numbers of a light atom. According to general principles of quantum mechanics, an operator corresponding to a given quantum number commutates with the Hamiltonian (1.1.3).

Let us introduce operators of a total atomic spin $\hat{\mathbf{S}}$ and of a total atomic orbital momentum $\hat{\mathbf{L}}$ as

$$\hat{\mathbf{S}} = \sum_i \hat{\mathbf{s}}_i; \ \hat{\mathbf{L}} = \sum_i \hat{\mathbf{l}}_i, \tag{1.1.11}$$

where $\hat{\mathbf{s}}_i$ is the spin of i-th electron, $\hat{\mathbf{l}}_i$ is the orbital momentum of the i-th electron.

Since in this approach the Hamiltonian does not depend on electron spins, one can obtain that the commutator $\left[\hat{\mathbf{S}}, \hat{H}\right] = 0$, and hence the total electron spin S and its projection M_S onto a given direction are the atomic quantum numbers. One can find also that the commutator $\left[\hat{\mathbf{L}}, \hat{H}\right] = 0$. Hence, the total orbital momentum L and its projection M_L onto a given direction are also atomic quantum numbers. Thus, the quantities L, S, M_L, M_S are quantum numbers of a light atom, if we neglect relativistic effects. Each atomic level is degenerated, including $(2M_L + 1)\,(2M_S + 1)$ states.

The atomic spectroscopy allows us to determine energies of atomic levels with a high accuracy. Therefore below we consider the most significant relativistic interactions. In the case of light atoms, this is spin-orbital interaction, i.e., interaction between the spin and the orbital magnetic moment. As a result, an atom is characterized by the total atomic angular momentum J which is the sum of spin and orbital momenta. Then quantum numbers J, L, S, M_J determine an atomic state. Here M_J is the projection of the atomic total momentum onto a given direction.

Table 1.1 Electron shells of atoms with valence p-electrons

LS-shell	LS-term	Total momentum J
p	^2P	1/2
p	^2P	3/2
p^2	^3P	0
p^2	^3P	1
p^2	^3P	2
p^2	^1D	2
p^2	^1S	0
p^3	^4S	3/2
p^3	^2D	3/2
p^3	^2D	5/2
p^3	^2P	1/2
p^3	^2P	3/2
p^4	^3P	2
p^4	^3P	0
p^4	^3P	1
p^4	^1D	2
p^4	^1S	0
p^5	^2P	3/2
p^5	^2P	1/2
p^6	^1S	0

The degeneration power, i.e. the number of states with the same energy, is equal to $(2J + 1)$.

The notations for atomic orbital momenta $L = 0, 1, 2, 3, 4, 5...$ are used S, P, D, F, G..., respectively. In the notation of the atomic state, the value $(2S + 1)$ is presented as the left upper index of the notation for orbital momentum L. The total angular momentum J is presented as the right lower index. For example, the notation $^2P_{3/2}$ means that the total spin is $S = 1/2$, the total orbital momentum is $L = 1$ and the atomic total angular momentum is $J = 3/2$.

We now construct the atomic structure by subsequent addition of electrons to the Coulomb field of a nucleus; we use the model that the Coulomb interaction potential of an electron with the charged nucleus is large compared to the interaction potential with other electrons. Taking into account the Pauli principle and constructing the lowest bound state, one can put one electron into one state of the hydrogenlike system. As a result, the lowest bound state has the atomic shell structure, i.e., electrons are located in the nucleus Coulomb field and in a self-consisting field of other electrons occupying the lower energy states of the hydrogenlike ion.

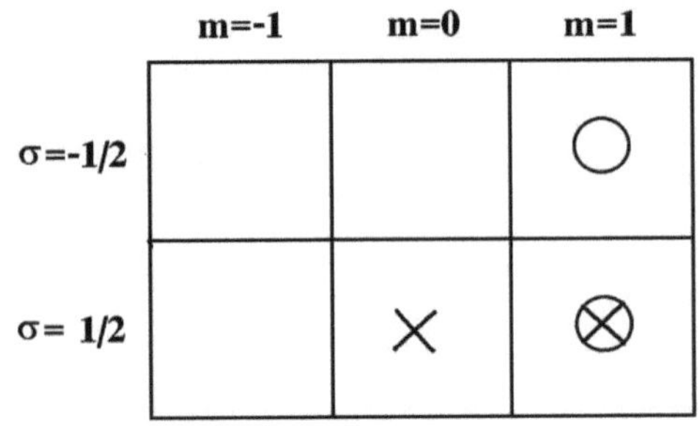

Fig. 1.1 Filling of the electron shell of the nitrogen atom. Six cells of the box take into account the electron states, where m is the projection of electron momentum onto a given state, and σ is the projection of the spin onto this direction. Filling of electron shell of the nitrogen atom for the state with the maximum spin $S = 3/2$ is denoted by crosses, and the atom state with the maximum atom orbital momentum $L = 1$ is denoted by circles

The state of an individual electron in the self-consisting field is characterized by quantum numbers n, l, m, σ. Notations s, p, d, f, g... denote orbital momenta $l = 0, 1, 2, 3, 4$... of the corresponding electron. In the notation of an atomic state, the number of electrons with a given value of quantum numbers nl is presented as right upper index after l. Thus, electrons are distributed on atomic shells, so that each atomic shell corresponds to the fixed value of nl. For example, the electron shells of oxygen atom with 8 electrons are denoted as $1s^2 2s^2 2p^4$. If we distribute electrons over atomic shells, states with the lower energy correspond to lower values of the principal quantum number n. At the fixed value of n, states with lower energy correspond to lower values of the orbital momentum l. This empirical rule is valid, at least, for small values of quantum numbers, i.e. for light atoms. However, it is violated if s and d electron shells are filled. In fact, an electron of s-shell with higher value of the principal quantum number n is filled before than electron d-shell is occupied with the lower value of n. In particular, the electron structure of the ground state in the vanadium atom with $Z = 23$ is $V(3d^3 4s^2)$, while the electron structure of the ground state in the chromium atom with $Z = 24$ is $Cr(3d^5 4s)$. However, this empirical rule is valid for filling of p-shells.

States of a given electron shell can have different energies. Indeed, atomic energy levels are eigenvalues of the Schrödinger equation (1.1.8) and according to the Pauli principle the wave function of atomic electrons is antisymmetric with respect to exchange of any two electrons. Therefore, the splitting of atomic levels follows from the exchange interaction inside the electron system. Due to the exchange nature of this splitting, dipole radiative transitions are forbidden between these states (Table 1.1).

Taking into account that atomic states are determined by quantum numbers $JLSM_J$, we consider as an example the nitrogen atom containing unperturbed atomic shell $N(2p^3)$ (we do not take into account inner atomic shells). In order to solve this problem, we distribute these electrons over states which differ by orbital momentum projection m onto a given direction and by the spin projection σ onto this direction, as it is given in Fig. 1.1. As a result, three p-electrons are distributed over six states. Because of the identity of electrons, there are $C_6^3 = 20$ possibilities of this distribution. Thus, we have 20 different states for electron shell $2p^3$. States with maximum values 3/2 of the total spin determine the electron term $^4S_{3/2}$ including $2J + 1 = 4$ states with different momentum projections M_S. The states with the maximum orbital momentum form the level $^2D_{3/2,5/2}$ which includes 10 states. Hence, an additional electron term $^2P_{1/2,3/2}$ contains 6 states.

Thus, the nitrogen atom with the electron shell $2p^3$ is characterized by three electron terms which are placed according to energy growth. From the first empirical Hund rule [16] it follows that the lowest state should have the maximum spin. According to the second empirical Hund rule [16], if spins of two states coincide with each other the lower one has maximum orbital momentum. Hence, the ground state of nitrogen atom is $^4S_{3/2}$, the next two excited states are $^2D_{3/2,5/2}$ and the higher excited states are $^2P_{1/2,3/2}$.

In the case of D states of the nitrogen atom, 3 electrons are distributed over 10 states. According to the third Hund rule [16], if the number of electrons is less that a half of corresponding states (this case), the lower state of fine structure should have a larger total momentum. Therefore the energy of state $^2D_{5/2}$ is lower than the energy of the state $^2D_{3/2}$. In the case of P-states of the nitrogen atom, 3 electrons are distributed over 6 states. According to the third Hund rule [16], if the number of electrons is more or equal to a half of corresponding states, the lower state of fine structure should have a less total momentum. Therefore the energy of state $^2P_{1/2}$ is

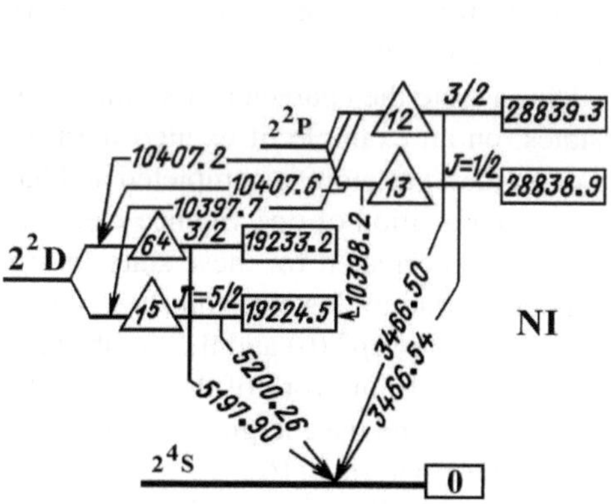

Fig. 1.2 Energy levels of lower states of the nitrogen atom with the electron shell $2p^3$ of electron shells [15]. Excitation energies presented inside rectangles are expressed in cm^{-1}, the radiative lifetimes are located inside triangles and are given in s, the wavelengths of radiative transitions are presented near arrows in Å, energies of fine splitting of levels are given inside squares in cm^{-1}. The total atomic momentum is shown above arrows

lower than the energy of the state ${}^2P_{3/2}$. In the same manner, one can analyze the electron states of other atoms with an incomplete electron shells containing valence p-electrons. Figure 1.2 [15] contains energies of lower states of the nitrogen atom and presents also the wavelengths of radiative transitions between these states

1.1.3 LS- and jj-Coupling Schemes

Though relativistic interactions are small for light atoms, they determine the addition rules of electron momenta in the total atomic momentum. Having in mind light atoms, we analyze how to construct the atomic structure. We above consider the LS-coupling scheme. It is based on the assumption that the exchange interaction inside the atom is large compared to the spin-orbit interaction that determines the fine structure of atomic levels. Note that within the framework of the atomic shell structure, internal electrons fill complete electron shells, so that atomic states are determined by valence electrons. Next, ignoring the spin-orbit interaction, we sum separately orbital momenta of valence electrons into the atomic orbital momentum L, and spins of valence electrons into the total atomic spin, as it was shown above from commutation properties of momenta for light atoms. Hence, in this scheme (so called LS coupling scheme) quantum numbers of the atomic electron system are LSM_LM_S, where M_L and M_S are the projections of the total orbital momentum of the atom and the total atomic spin onto a given direction.

These atomic momenta L and S are added into the total atomic momentum J, and atomic quantum numbers are $LSJM_J$, where M_J is the projection of the total angular momentum of the atom J onto a given direction. Another limiting case corresponds to the other character of momentum summation, so that at first the orbital momentum and spin of an individual electron are added in the total electron momentum j, and then total momenta of individual electrons are added into the total atomic momentum.

It is convenient to demonstrate the character of summation of momenta, as well as notations for the states, on an example of excited inert gas atoms. Indeed, the valence electron shell of an inert gas atom is completed and has the form np^6; lower excited states correspond to transition of one valence electron from the state np to the state $(n+1)s$, and the electron shell for these states is $np^5(n+1)s$. There are 12 excited states which correspond to this electron shell; this number is the product of the number of atomic core states np^5 (6) and the number of states for the valence electron $(n+1)s$ (2). Within the framework of the LS coupling scheme, the states ${}^3P_2$, ${}^3P_1$, ${}^3P_0$, ${}^1P_1$ are realized, and the energies of their excitation from the ground state increase according to the Hund rule [16].

Within the framework of jj-coupling scheme, this sequence of states is as follows:

$$s\left[\frac{3}{2}\right]_2,\; s\left[\frac{3}{2}\right]_1,\; s'\left[\frac{1}{2}\right]_0,\; s'\left[\frac{1}{2}\right]_1 \qquad (1.1.12)$$

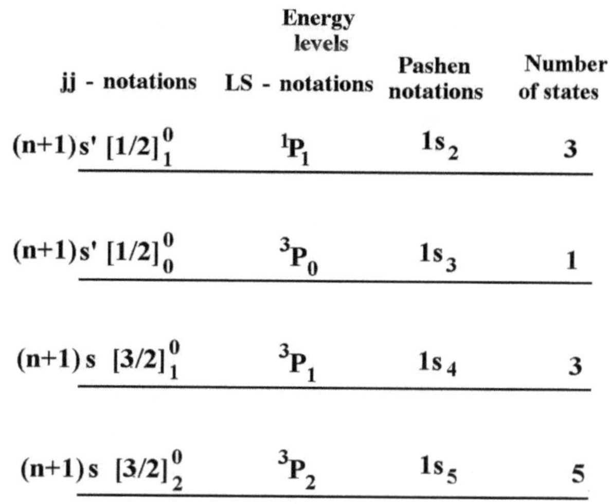

Fig. 1.3 Energy levels of low excited states of argon atom with the electron shell $3p^54s$. Parameters of radiative transitions to the ground state are given. Electron energies are counted off the first excited state. Notations of Paschen are given for these levels. Notations within the frames of LS- and jj-coupling schemes are shown to the right of energy levels

The notations are such that the total momentum of the atomic core is given inside square brackets, and the low index is the total atomic momentum. Figure 1.3 presents notations of electron terms of an excited atom of inert gases. In addition, so called Paschen notations are used often for excited atoms of inert gases due to their simplicity. The sequence of states under consideration at the increasing of the excitation energy is denoted as $1s_5$, $1s_4$, $1s_3$ and $1s_2$. Below we take the energy of the state $1s_5$ to be zero and denote the difference of excitation energies of the other states and this state as ε_4, ε_3 and ε_2. Table 1.2 contains some parameters of lowest excited states for atoms of inert gases [17, 18]. The energies of excitation of these states demonstrate the character of interaction between electron shells for a given atom.

One can describe this interaction numerically by extraction the principal interactions in these atoms from the Hamiltonian

$$\hat{H} = -a\hat{\mathbf{l}}\hat{\mathbf{s}}_1 - b\hat{\mathbf{s}}_1\hat{\mathbf{s}}_2, \tag{1.1.13}$$

where $\hat{\mathbf{l}}$ is the operator of the orbital momentum of the atomic core, $\hat{\mathbf{s}}_1$ is the spin operator of the core, $\hat{\mathbf{s}}_2$ is the spin operator of the excited s-electron. The parameter a of this formula is responsible for the fine splitting of the ion levels; in absence of the excited s-electron we have $a = 2\Delta_f/3$, where Δ_f is the energy difference for $^2P_{3/2}$ and $^2P_{1/2}$ ion states. The other parameter of this formula b is responsible for exchange interaction between the valence electron and the atomic core.

From solution of the Schrödinger equation

$$\hat{H}\Psi_k = \varepsilon_k\Psi_k$$

with the Hamiltonian (1.1.13) one can find the energies of states. Taking $\varepsilon_5 = 0$, one can determine the eigen energies of the Schrödinger equation, which are [19]

Table 1.2 Energetic parameters of the first excited states of inert gas atoms. All the energetic parameters are expressed in cm^{-1} [19]

Atom	Δ_f	ε_3	b	b/a	$\varepsilon_2 - \varepsilon_4$	$\sqrt{9a^2/4 + b^2 - ab}$
Ne	780	777	1488	2.9	1430	1430
Ar	1432	1410	1453	1.5	1649	1653
Kr	5370	5220	1600	0.46	4930	4923
Xe	10537	9129	1966	0.32	9140	8674

$$\varepsilon_{2,4} = \frac{3}{4}a + \frac{1}{2} \pm \frac{1}{4}\sqrt{9a^2 - 4ab + 4b^2}, \ \varepsilon_3 = \frac{3}{2}a \tag{1.1.14}$$

These values and their components are given in Table 1.3, where $a = 2\Delta_f/3$ and $b = \varepsilon_4 + \varepsilon_2 - \varepsilon_3$. It follows from data of Table 1.2, that the exchange interaction potential depends slowly on the atomic number. We also give in the Table 1.2 the value $\sqrt{9a^2/4 + b^2 - ab}$, that is, according to the obtained formulas equals to the difference $\varepsilon_2 - \varepsilon_4$. From the data of this Table one can conclude that the Hamiltonian (1.1.13) describes enough well excited states of the inert gas atoms at various relations between parameters a and b.

Evidently, one can expect that in the limiting case $b \gg a$, the LS scheme of the momentum coupling takes place, while in the other limiting case $b \ll a$ the jj scheme of momentum coupling is realized. It is seen, that in the argon and krypton cases we have an intermediate coupling, and then the Hamiltonian (1.1.13) describes well the positions of energy levels for these atoms.

Analyzing the results given in Table 1.2, one can find that the relativistic interaction is determined mostly by the spin-orbit interaction. Even in the krypton case the ion fine splitting and the difference of energies for levels $1s_3$ and $1s_5$ differ by 2.8%, whereas in the xenon case this difference is about 13%. Next, under assumptions used, the relative position of levels of $1s_2$ and $1s_4$ states differs from its experimental value by 0.14% in the krypton case, while in the xenon case this difference is above 5%. We conclude that the higher is the accuracy of the approach under consideration, the less are the relativistic corrections to the atom energy, and the error of this model increases nonlinearly with respect to relativistic interactions. Hence, additional relativistic interactions are remarkable when the $j - j$ coupling scheme is valid. In addition, the interaction of shells of $1s$ and $2p$ in Paschen notations, that is stronger for xenon than for other inert gas atoms, influences on the accuracy of the model under consideration also. Note that this interaction between shells, which mixes shells $np^5(n+1)s$ and $np^5(n+1)p$ as a result of exchange interaction between electrons, influences on the parameter b which is responsible for exchange interaction in the atom. Because the $(n+1)p$ excited electron interacts weakly with internal electrons, it does not influence on the fine splitting of levels of the atomic core; thus, the interaction between shells does not influence on the spin-orbit interaction of the atomic core. Therefore, deviation of the difference $\varepsilon_3 - \varepsilon_5$ from the fine splitting of ion levels for xenon can be produced by additional relativistic effects.

1.1.4 Parentage Scheme of Atom

We consider an atom or atomic ion within the framework of one-electron approximation where the total wave function of electrons is the combination of products of one-electron wave functions; on the basis of the Pauli principle, the total wave function of electrons change its sign at the permutation of any two electrons. Next, these electrons are distributed over the shells, and usually internal atom shells are not excited, i.e. they are filled. If the shell of valence electrons is not filled, there are different atomic states in the same electron shell. These combinations can be taken on the basis of total wave function of electrons which are eigenfunctions of the electron Hamiltonian. In particular, ignoring relativistic interactions, we have that quantum numbers LSM_LM_S are quantum numbers of a light atom.

In the analysis of transitions of an atom to an excited state, it is convenient to apply one-electron approximation where one valence electron varies its state. For description of this process we extract the transferring electron and represent the wave function of valence electrons as the product of the wave function of the transferring electron and the wave function of an atomic core. This approach is called the parentage scheme of the atom. On the basis of the total symmetry of the atomic electron system, the total wave function of n valence electrons is given by

$$\Psi_{LSM_LM_S}(1, 2, \ldots\ldots, n) = \frac{1}{\sqrt{n}}\hat{P} \sum_{L'M'_L S'M'_S m\sigma} G^{LS}_{L'S'}(l, n)$$

$$\begin{bmatrix} l & L' & L \\ m & M'_L & M_L \end{bmatrix} \begin{bmatrix} \frac{1}{2} & S' & S \\ \sigma & M'_S & M_S \end{bmatrix} \psi_{l\frac{1}{2}m\sigma}(1) \cdot \Psi_{L'S'M'_LM'_S}(2, \ldots\ldots, n), \qquad (1.1.15)$$

Here the operator $\hat{P}$ transposes positions and spins of a test electron which is described by the argument 1 and atomic electrons; LSM_LM_S are quantum numbers of the atom, $L'S'M'_LM'_S$ are the quantum numbers of the atomic core, $l, 1/2, m, \sigma$; are quantum numbers (the orbital momentum, the spin, and their projections onto a given direction) of an extracted valence electron, $G^{LS}_{L'S'}(l, n)$ is so called the fractional parentage coefficient or the Racah coefficient [20, 21] which is responsible for binding of an electron and atomic core to form the atom. Note that there are a finite number of states of an atomic core at the removal of the valence electron from an atom. Table 1.3 lists the values of fractional parentage coefficients for s and p-electron shells. In the case of d- and f-electrons removal of one valence electron can lead to various states of the atom with the same values L and S. Then the atom state is described by the additional quantum number v that is the seniority of state.

Fractional parentage coefficients satisfy to some relations. In particular, it follows from the condition of normalization of the wave function

$$\sum_{L'S'v} \left[G^{LS}_{L'S'}(l, n, v)\right]^2 = 1$$

Table 1.3 Fractional parentage (Racah) coefficients for valence s- and p-electron shells. The electron shell and the state term for the atom and atomic core are indicated

Atom	Atomic rest	$G_{L'S'}^{LS}$	Atom	Atomic rest	$G_{L'S'}^{LS}$
$s(^2S)$	(^1S)	1	$p^3(^2P)$	$p^2(^1S)$	$\sqrt{2/3}$
$s^2(^1S)$	$s(^2S)$	1	$p^4(^3P)$	$p^3(^4S)$	$-1/\sqrt{3}$
$p(^2P)$	(^1S)	1		$p^3(^2D)$	$\sqrt{5/12}$
$p^2(^3P)$	$p(^2P)$	1		$p^3(^2P)$	$-1/2$
$p^2(^1D)$	$p(^2P)$	1	$p^4(^1D)$	$p^3(^4S)$	0
$p^2(^1S)$	$p(^2P)$	1		$p^3(^2D)$	$\sqrt{3/4}$
$p^3(^4S)$	$p^2(^3P)$	1		$p^3(^2P)$	$-1/2$
	$p^2(^1D)$	0	$p^4(^1S)$	$p^3(^4S)$	0
	$p^2(^1S)$	0		$p^3(^2D)$	0
$p^3(^2D)$	$p^2(^3P)$	$1/\sqrt{2}$		$p^3(^2P)$	1
	$p^2(^1D)$	$-1/\sqrt{2}$	$p^5(^2P)$	$p^4(^3P)$	$\sqrt{3/5}$
	$p^2(^1S)$	0		$p^4(^1D)$	$1/\sqrt{3}$
$p^3(^2P)$	$p^2(^3P)$	$-1/\sqrt{2}$		$p^4(^1S)$	$1/\sqrt{15}$
	$p^2(^1D)$	$-\sqrt{5/18}$	$p^6(^1S)$	$p^5(^2P)$	1

1.2 Single-Photon Transitions

1.2.1 Rates of Single-Photon Transitions for Nonrelativistic System

Radiative transitions result from interaction between the radiation field and atomic systems. If we deal with radiative transitions in atoms, which are transitions between bound states of valence electrons, they are governed by a small parameter that is the ratio of a typical electron velocity to the light speed c. A typical velocity of atom valence electrons is $e^2/\hbar$, and in this case a small parameter of the theory is the fine structure constant $e^2/\hbar c = 1/137$. This small parameter provides low rates of radiative transitions compared to typical atomic rates, and therefore the theory of radiative transitions is the perturbation theory of quantum mechanics. Below we represent the apparatus of quantum mechanics for radiative transitions which is based on the above small parameter. In addition, because typical times of radiative transitions are large, spectral lines for these transitions are narrow. This fact is the basis of spectroscopy.

In considering radiative transitions between two discrete states, we use below notations given in Fig. 1.4. The radiative transition under consideration takes place between states i and j, and a photon which partake in this transition has the energy $\hbar\omega = E_j - E_i$, where E_j and E_i are the energies of states between them the transition proceeds. We below represent the nonrelativistic theory of radiative transitions, where

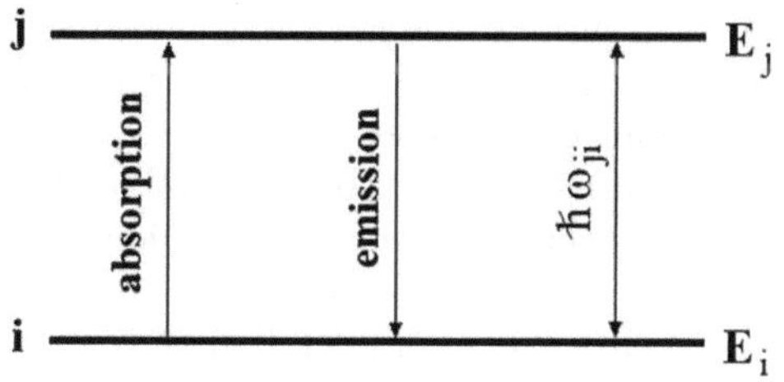

Fig. 1.4 Radiative transitions between two discrete states and used notations

the operator of interaction between the electromagnetic field and the atomic system is given by

$$\hat{V} = -\frac{e}{mc}\left(\hat{\mathbf{p}}\mathbf{A} + \mathbf{A}\hat{\mathbf{p}}\right) + \frac{e^2}{2m_e c^2}\mathbf{A}^2 \tag{1.2.1}$$

where $\mathbf{A}\,(\mathbf{r}, t)$ is the vector potential of the electromagnetic radiation) radiation field, and $\hat{\mathbf{p}}$ is the momentum operator of an atomic electron that participates in the radiation transition. We presume that the radiation wavelength is large compared to a typical atomic dimension (i.e. the Bohr radius). This statement can be written as

$$\hat{\mathbf{S}} = \sum_i \hat{\mathbf{s}}_i;\ \hat{\mathbf{L}} = \sum_i \hat{\mathbf{l}}_i,\ \frac{c}{\omega} \gg \frac{\hbar^2}{m_e e^2} \tag{1.2.2}$$

where ω is the radiation frequency. This inequality can be rearranged to read $e^2/(\hbar c) \ll 1$ since the radiation frequency ω is of the order of the frequency of an atomic transition $m_e e^4/\hbar^3$. Then we see that the dipole approximation is applicable as the consequence of the condition $v/c \ll 1$. The interaction Hamiltonian can be written in the simpler form

$$\hat{V} = -\mathbf{E}\cdot\mathbf{D} \tag{1.2.3}$$

Here $\mathbf{E}$ is the electric field strength of the radiation field being considered, and $\mathbf{D}$ is the dipole moment operator of the atomic particle.

Through most of this book, we shall also assume that the energy of interaction between the radiation field and the atomic system is small compared to the transition energy. This justifies the description in the framework of the perturbation theory for emission and absorption processes involving photons of the radiation field. The perturbation theory is applicable if the electric field strength E is small compared to a typical internal electric field of the atom. From the dimensionality analysis we can take this internal field to be of the order of $m_e^2 e^5/\hbar^4$. Thus, we assume that

$$E \gg E_o;\quad E_o = \frac{m_e^2 e^5}{\hbar^4} = 5.14\cdot 10^9\ \mathrm{V/cm} \tag{1.2.4}$$

If this criterion holds true, one can use the operator (1.2.1) for the analysis of radiative transitions between states of an atomic particle and use a small parameter to construct the perturbation theory.

We use the interaction potential (1.2.1) as a perturbation for radiative processes under consideration. Additional terms of expansion over a small parameter $v/c,\ v \sim e^2/\hbar$ correspond to weak processes. Within the framework of the perturbation theory under consideration, the electric field strength of the radiation field may be represented in the form of a sum of those for monochromatic waves

$$\mathbf{E} = \sum_{\omega} \mathrm{Re}\left[\mathbf{E}_{\omega} \exp(i\omega t)\right], \tag{1.2.5}$$

where $\mathbf{E}_{\omega}$ is the complex strength of the electric field of a frequency ω. Let us denote by $\delta\omega$ the difference between neighboring wave frequencies and assume the differences of neighboring wave frequencies to be nearby. Denoting the spectrum width by $\Delta\omega$, we assume that the value $\mathbf{E}_{\omega}$ varies slightly within this width, i.e., the criterion $\Delta\omega \gg \delta\omega$ is fulfilled. Thus, many frequency components partake in the radiative transitions.

If the phases of the amplitudes $\mathbf{E}_{\omega}$ are statistically independent, one can neglect an interaction between waves. Then basing on the time-dependent perturbation theory for radiative transitions, we employ the first-order perturbation theory which leads to transition of one photon in the absorption or emission processes. Then the transition lifetime $1/w$, or the residence time in the upper state of transition is large compared to a typical atomic time that is of the order of $\hbar^3/(m_e e^4)$.

Since the interaction between an atomic electron and each monochromatic component of the electromagnetic wave can be considered separately, one can represent the operator of interaction of an atomic electron and a monochromatic wave of the radiation field in the form

$$\hat{V} = -\mathbf{D} \cdot \mathrm{Re}\left[\mathbf{E}_{\omega} \exp\left(i\omega t\right)\right] \tag{1.2.6}$$

The Hamiltonian of an atomic system, i.e. a system of bound electrons, in absence of the electromagnetic field is $\hat{H}_o$ and describes the behavior of valence electrons or vibrations of atoms for a molecular system. Let us denote by ψ_k the system of eigenfunctions of the Hamiltonian $\hat{H}_o$, where ε_k are the corresponding energy eigenvalues. These wave functions are solutions of the time-independent Schrödinger equation

$$\hat{H}_0 \psi_k = \varepsilon_k \psi_k \tag{1.2.7}$$

The wave function of the system in terms of the eigenfunctions ψ_k can be presented in the form

$$\Psi = \sum_{k} c_k \psi_k\left(\mathbf{r}\right) \exp\left(-\frac{i\varepsilon_k t}{\hbar}\right) \tag{1.2.8}$$

Here **r** is a set of coordinates of atomic electrons, or the coordinates that determine vibrational or rotational molecular states. Let us the expansion (1.2.8) for the wave function of atomic electrons in the Schrödinger equation

$$i\hbar\frac{\partial\Psi}{\partial t} = \hat{H}\Psi, \tag{1.2.9}$$

where $\hat{H} = \hat{H}_o + \hat{V}$ is the total Hamiltonian of the system. Then multiplying the resulting equation by ψ_m^*, we integrate over coordinates of the atomic electron and use the orthogonality conditions for the eigenfunctions ψ_k of the unperturbed atomic electron. We thus obtain for the coefficients c_k the equation

$$i\hbar\frac{\partial c_m}{\partial t} = \sum_k V_{mk}c_k \exp\left(i\omega_{mk}t\right) \tag{1.2.10}$$

Here

$$\omega_{mk} = \frac{\varepsilon_m - \varepsilon_k}{\hbar},$$

V_{mk} is the matrix element of the time-dependent interaction operator, and this equation includes the states of atomic electrons. Using first-order perturbation theory, we suppose that the perturbation is absent at time $t = 0$, before which the electron is in a state i. Therefore in zero approximation we have $c_{ji}^{(0)} = \delta_{ji}$, where δ_{ji} is the Kronecker function.

Let us take the radiation field to be linearly polarized, that means that $\mathbf{E}_\omega$ is a real value. Subsequently we replace $\mathbf{E}_\omega$ by $\mathbf{E}$. This replacement does influence the result. On the basis of (1.2.6) we have

$$\hat{V} = -\mathbf{D}\cdot\mathbf{E}\cos\omega t, \tag{1.2.11}$$

and in the first approximation the set of equations (1.2.9) gives

$$c_j^{(1)} = -\frac{\mathbf{D}_{ji}\cdot\mathbf{E}}{2\hbar}\left\{\frac{1-\exp\left[i\left(\omega_{ji}-\omega\right)t\right]}{\omega_{ji}-\omega} + \frac{1-\exp\left[i\left(\omega_{ji}+\omega\right)t\right]}{\omega_{ji}+\omega}\right\} \tag{1.2.12}$$

As it follows from (1.2.12), the strongest transitions correspond to such states j for which the resonance condition $\omega_{ji} = \omega$ is fulfilled. In this case, one can conserve only the first term in (1.2.12). The probability for transition to the state j is

$$\left|c_j^{(1)}\right|^2 = \frac{E^2\left(\mathbf{D}_{ji}\cdot\mathbf{s}\right)^2}{\hbar^2}\cdot\frac{\sin^2[(\omega_{ji}-\omega)t/2]}{(\omega_{ji}-\omega)^2}, \tag{1.2.13}$$

where **s** is the polarization vector, that is the unit vector directed along the electric field. In the case of small a and large t, the factor of formula (1.2.13) $\sin^2(at)/a^2$ may be replaced by $\pi t\delta(a)$, where $\delta(a)$ is the Dirac delta function. This delta function is $\delta(a) = 0$, if $a \neq 0$, and $\delta(a) = \infty$ at $a = 0$, and

$$\int_{-\infty}^{\infty} \delta(a)\mathrm{d}a = 1 \tag{1.2.14}$$

The function $\sin^2(at)\left(\left(\pi a^2 t\right)\right)$ has properties of delta-function $\delta(a)$, and just this expression is used often for definition of delta function. Note that the condition $t \to \infty$ is required for this. Assuming this requirement to be fulfilled, we have for the transition rate

$$w = \frac{1}{t}\left|c_j^{(1)}\right|^2 = \frac{E^2}{2\hbar^2}\left(\mathbf{D}_{ji}\cdot\mathbf{s}\right)^2\delta\left(\omega_{ji}-\omega\right) \tag{1.2.15}$$

This formula is valid if t is not extremely large, as long as $wt \ll 1$ and the perturbation theory is applicable. From another standpoint, the relationship $\omega_{ji}t \gg 1$ should be fulfilled in order the delta function could be formed.

We now take into account that the radiation field has many modes. Then one can replace by integration the summation over frequencies in the following manner

$$\sum_{\omega} \to \frac{1}{\delta\omega}\int \mathrm{d}\omega,$$

where $\delta\omega$ is the frequency difference of neighboring modes (see above), and the quantity $1/\delta\omega$ is the energy density of the states. Next, let us express the rate of transition w through the number of photons n_ω located in one state with a frequency ω, instead of the electric field strength $\mathbf{E}$. Then the quantity

$$\frac{1}{2}\left(\frac{E^2+H^2}{8\pi}\right) = \frac{E^2}{8\pi} \tag{1.2.16}$$

is the average energy density of the electromagnetic field in a single mode of the field. Arriving this result, we replace the time average value of $\cos^2\omega t$ by $1/2$, and take into account that the electric field strength $|\mathbf{E}|$ for an electromagnet plane wave is equal to the magnetic field strength $|\mathbf{H}|$. We then have for the energy density of modes in the frequency range from ω up $\omega + \mathrm{d}\omega$ as $E^2/(8\pi)\cdot \mathrm{d}\omega/\delta\omega$. From another point of view, this quantity is also given by

$$\frac{\hbar\omega n_\omega \cdot 2d\mathbf{k}}{(2\pi)^3} = \frac{\hbar n_\omega \omega^3 \mathrm{d}\omega}{\pi^2 c^3}, \tag{1.2.17}$$

where $\mathbf{k}$ is the photon wave number, $d\mathbf{k}/(2\pi)^3$ is the number of states per unit volume and unit frequency ω, n_ω is the number of photons located in one state, and the factor 2 takes into account two photon polarizations. The wave number k is connected with the photon frequency ω by the dispersion relation $\omega = ck$, where c is the light speed. As a result, we find the connection between the electric field strength E and the number of photons n_ω located in one state

$$w = \frac{4\omega^3 n_\omega}{\hbar c^3} \left(\mathbf{D}_{ji} \cdot \mathbf{s}\right)^2, \ \omega_{ji} = \omega \qquad (1.2.18)$$

The integration takes into account the property (1.2.14) of the delta function, that is correct, if the frequency width $\Delta\omega$ of the transition is large compared with typical values of the difference $|\omega_{ji} - \omega| \gg t^{-1} \gg w$, i.e. $\Delta\omega \gg w$. Thus, formula (1.2.18) is valid in the case of a weak field strength with a broad frequency spectrum and at moderate times of observation. It is necessary to require also the incoherence of waves with different frequencies.

Formula (1.2.18) is derived for the case $n_\omega \gg 1$. Indeed, we assume in its derivation that the electric field strength does not change at the radiation transition, whereas the number of photons in this state decreases by one as a result of the absorption process. Nevertheless, formula (1.2.18) is correct at any values of n_ω. Namely, from the nature of the photon absorption, it follows that the absorption probability is proportional to the number of photons in this state, and this formula is valid also at $n_\omega \sim 1$.

One can average the expression (1.2.18) for the rate of the radiative transition over the polarization direction **s** and integrate over the solid angle $\langle \cos^2\theta \rangle = 1/3$, where θ is the angle between vectors **D** and **s**. As a result, we obtain the transition rate

$$w_{ij} = w\left(0, n_\omega \to k, n_\omega - 1\right) = \frac{4\omega^3}{3\hbar c^3} |\mathbf{D}_{ji}|^2 g_j n_\omega \qquad (1.2.19)$$

Here we introduce the factor g_j which is the statistical weight of the final state and equals to the number of states of the final level. In contrast to the expression (1.2.18), formula (1.2.19) does not depend on the field polarization.

The above expressions (1.2.18) and (1.2.19) relate to the process of photon absorption. In the same manner, one can find the rate of the photon emission. For a certain photon polarization described as before by the unit vector **s** this rate by analogy with formula (1.2.18) takes the form

$$w_{ji} = \frac{4\omega^3}{\hbar c^3} \left(\mathbf{D}_{ji} \cdot \mathbf{s}\right)^2 g_i \cdot (n_\omega + 1), \qquad (1.2.20)$$

where g_i is the statistical weight of the lower state of transition.

The value $1/w$ is the time between two subsequent events of the photon absorption. But a time of absorption of one photon is much less; it is determined by the uncertainty principle, which gives that this value is of the order of $1/\omega_{ji}$. Next, from formula (1.2.12) it follows that at very small times the transition rate is proportional to t^2. However, this rate is so small that it is experimentally unobservable. On the other hand, the result for the transition rate is incorrect at large times, where $w_{ji}t > 1$. In this limit, the rate of transition differs from the linear dependence.

In contrast to the case of photon absorption, photon emission from an excited atomic state includes two parts. The first one, so-called spontaneous transition rate, describes emission in the absence of an external electromagnetic field, so that the spontaneous radiation occurs due to the interaction of a charged atomic particle with the electromagnetic vacuum field. The second part takes place under the action of an external electromagnetic field, corresponds to production of photons stimulated by the external field, and is called induced radiation. As a result, the total rate of emission from the excited state j may be represented in the form

$$w_{ji} = w\left(j, n_\omega \to 0, n_\omega + 1\right) = A_{ji} + B_{ji} n_\omega, \tag{1.2.21}$$

where the quantities A_{ji} and B_{ji} are called the Einstein coefficients. Subsequently we present the connection between Einstein coefficients and the spontaneous lifetime τ of an excited state considering the system of atoms and photons which are found in thermodynamic equilibrium.

1.2.2 *Intensity of Radiative Transitions*

Let us introduce the intensity I of the absorption process as the energy change per unit time due to the photon absorption. Then multiplying formula (1.2.18) by the photon energy $\hbar\omega$, we obtain the emission intensity

$$I = \frac{4\omega^4 n_\omega}{c^3} \left(\mathbf{D}_{ji} \cdot \mathbf{s}\right)^2 \tag{1.2.22}$$

Although this expression does not contain the Planck constant, it becomes classical only in the limit $n_\omega \gg 1$. In order to transfer to the classical limit, it is necessary to replace the matrix element of the dipole moment of an atomic system, $\mathbf{D}_{ji}$, by the Fourier component of the time-dependent dipole moment with respect to the transition frequency $\omega_{ji} = \left(\varepsilon_j - \varepsilon_i\right)/\hbar$. This is an example of the Bohr correspondence principle [22].

We now determine the intensity of the spontaneous radiation process as the energy radiated per unit time per one atomic particle or system which is described by the dipole operator $\mathbf{D}$. Characterizing the photon polarization by a unit vector $\mathbf{s}$, one can obtain from expression (1.2.20) the intensity of radiative transition from an upper state j to a lower state m as

$$\hbar\omega_{jm} w_{jm} = \frac{4\omega_{jm}^4}{c^3} \left(\mathbf{D}_{jm} \cdot \mathbf{s}\right)^2 g_m, \tag{1.2.23}$$

where $\hbar\omega_{jm}$ is the photon energy or the energy difference between states j and m. Averaging over polarizations of and emitted photon and taking, for simplicity, $g_m = 1$, one obtain the total intensity of spontaneous radiation

$$I_{jm} = \frac{4\omega_{jm}^4}{3c^3}|\mathbf{D}_{jm}|^2 \tag{1.2.24}$$

Transiting to the classical limit, we summarize the partial intensity I_{jm} over all states m with the energy below that of the state j. This gives for the radiative intensity from a state j in the classical limit

$$I_j = \sum_{m,\omega_{jm}>0} I_{jm} \tag{1.2.25}$$

Using general principles of quantum mechanics, we find that the matrix element $\mathbf{D}_{jm}$ is proportional to $\exp\left(i\omega_{jm}t\right)$, that gives

$$\frac{\mathrm{d}^2\mathbf{D}_{jm}}{\mathrm{d}t^2} = -\omega_{jm}^2\mathbf{D}_{jm}. \tag{1.2.26}$$

From this relation we obtain

$$I_j = \frac{4}{3c^3}\sum_{m,\omega_{jm}>0}|\ddot{\mathbf{D}}_{jm}|^2 = \frac{4}{3c^3}\sum_{m,\omega_{jm}>0}|\omega_{jm}^2\mathbf{D}_{jm}|^2 \tag{1.2.27}$$

It is important that the summation in formula (1.2.27) over all the states gives

$$\sum_m |\ddot{\mathbf{D}}_{jm}|^2 = \left(\ddot{\mathbf{D}}^2\right)_{jj}. \tag{1.2.28}$$

Note that the main contribution in the quasi-classical sum (1.2.27) is produced by states m which are nearby by energy to the state j. In fact, the values of the matrix elements $\ddot{\mathbf{D}}_{jm}$ calculated with quasi-classical wave functions decrease quickly with an increasing energy difference between states j and m, because of oscillations in the integrand, which grow sharply with the increasing of the energy difference. Indeed, matrix elements of the dipole moment operator decrease exponentially, while square of energy difference increases much slower. Now we take into account that states m, which are close to the state j by energy, are distributed symmetrical with respect to j. This gives

$$\sum_{m,\omega_{jm}>0}|\ddot{\mathbf{D}}_{jm}|^2 = \sum_{m,\omega_{jm}<0}|\ddot{\mathbf{D}}_{jm}|^2 = \frac{1}{2}\left(\ddot{\mathbf{D}}^2\right)_{jj} \tag{1.2.29}$$

In the classical limit, the diagonal matrix element of a quantum operator for a given physical quantity coincides with the average value of this operator. This average value, in turn, yields the classical value for the physical quantity. Thus from formula (1.2.27) we obtain the total radiation intensity of an atom in the classical limit

$$I(t) = \frac{2}{3c^3}\left[\ddot{\mathbf{D}}(t)\right]^2 \tag{1.2.30}$$

The classical dipole moment $\mathbf{D}$ in this expression should be regarded as a given function of time. It follows from the above analysis, that the radiative lifetime of a given state is determined by spontaneous transitions to nearby states in the classical limit.

1.2.3 Selection Rules for Single-Photon Radiative Transitions

The rate of radiative dipole transitions under consideration according to formula (1.2.15) is proportional to the square of the matrix element from the dipole moment operator. We first formulate the selection rule for an individual electron located in a structureless central field, and its state is described by quantum numbers n, l, m, σ. The selection rule requires a nonzero value of the matrix element of the dipole moment operator. Let us consider the matrix element of the dipole moment operator $-e\mathbf{r}$ of an individual electron using the wave function (1.1.6). The factor that includes the angle dependence of the wave functions is expressed through the Clebsch-Gordan coefficients and is equal

$$\int Y^*_{l'm'_l} Y_{1q} Y_{lm_l} d\Omega = \sqrt{\frac{3\,(2l+1)}{4\pi\,(2l'+1)}}\left\langle l1, m_l q | l'm'_l \right\rangle \left\langle l1, 00 | l'0 \right\rangle$$

Here the spherical functions Y_{lm_l} and $Y_{l'm'_l}$ correspond to initial and final angular wave functions, respectively, the index q characterizes a photon polarization. The Clebsch-Gordan coefficient $\left\langle l1, m_l q | l'm'_l \right\rangle$ is nonzero only when $m'_l = m_l,\ m_l \pm 1$ since $q = -1, 0, +1$.

Thus, selection rules for one-electron transitions are of the form

$$l - l' = \pm 1;\ s - s' = 0;\ m_l - m'_l = 0, \pm 1;\ m_s - m'_s = 0 \tag{1.2.31}$$

In the case of several identical electrons located in the unfilled electron shell, the above selection rule must be generalized. Since the operator of the dipole moment is the orbital vector, then matrix elements of all its components can be nonzero only for transitions in which the orbital momentum of the electron shell changes no more than by one, i.e. $L \to L, L \pm 1$. In addition, the selection rule forbids transitions between any two states with $L = 0$. This rule follows from the value of the Clebsh-

Gordan coefficients [23, 24]; in this case because of the spherical symmetry of states with $L = 0$, there is no the vector for direction of the matrix element of the dipole moment.

Since the operator of the dipole moment

$$\mathbf{D} = e\sum_i \mathbf{r}_i$$

is odd function at the reflection of all electrons with respect to plane which contains the atomic center, matrix elements of this operator between states of the same parity are zero. This is so called Laporte rule [25]. This rule forbids transitions with $L = L'$ in single-electron atoms since in this case the orbital quantum number L determines the parity of the state $(-1)^L$. On the other hand, there is no direct connection between the total orbital momentum of the multi-electron atom and its parity; thus, the condition $L = L'$ does not contradict the Laporte rule.

We consider a light atom which undergoes a radiative single-electron transition from a state with orbital angular momentum L and spin S to a state with orbital angular momentum L' and spin S'. The fine structure of the atom in the initial state is determined by the total angular momentum $\mathbf{J} = \mathbf{L} + \mathbf{S}$. We designate by M the projection of the total angular momentum J on a fixed direction. Find the relative probability of the atom to be in the final state J' after single-electron transition. We presume LS-coupling to hold true. The states are characterized by the following quantum numbers: the orbital momentum L, the spin momentum S and the total momentum J. Since the dipole moment D_q is is the orbital vector, its matrix elements are diagonal with respect to S.

On the basis of formula (1.2.20) we find the total rate of emission from the excited state to a given final state that is proportional to

$$w \sim \sum_{q,M'} \left|\langle JMLS \left| D_q \right| J'M'L'S\rangle\right|^2,$$

where $q = -1, 0, 1$ and J, M_J, L, S are, respectively, the total angular momentum, its projection onto a given direction, the orbital and spin angular momentum quantum numbers of the atom in its initial state. We omit all other quantum numbers necessary to describe the initial and final states.

One can present atomic spectra and radiative transitions in atoms within the framework of the Grotrian diagram. Examples of the Grotrian diagrams are given in Fig. 1.5 for the sodium atom and in Fig. 1.6 for the nitrogen atom [18]. The sodium atom has one valence electron above filled shells; atomic levels and radiative transitions relate to a single valence electron only. In the case of the nitrogen atom there are three valence electrons, and their coupling corresponds to the LS-scheme of addition of electron momenta.

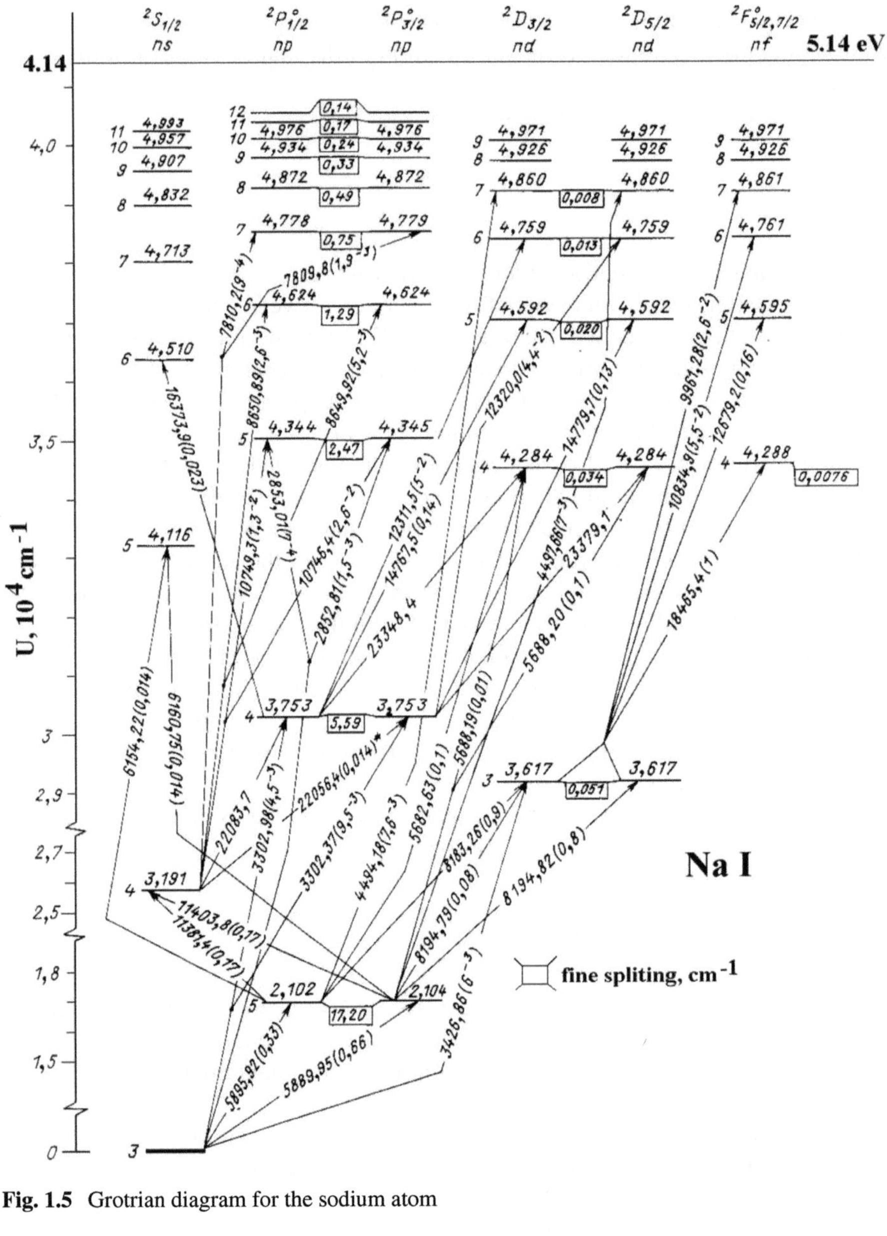

Fig. 1.5 Grotrian diagram for the sodium atom

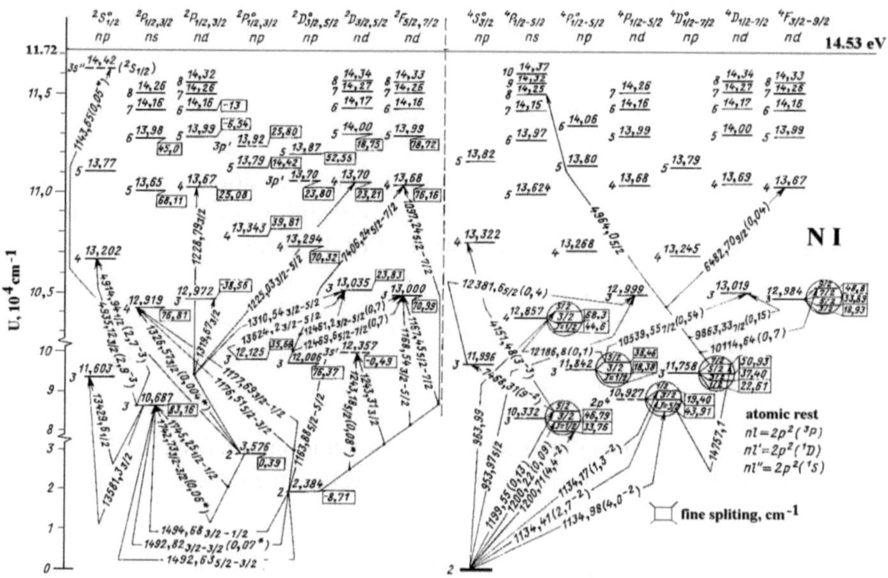

Fig. 1.6 Grotrian diagram for the nitrogen atom

The matrix element of the dipole moment operator for the LS-scheme of momentum addition for the initial and final states can be expressed through the Wigner $6j$-symbols

$$w \sim (2L'+1)(2J'+1)\sum_{q,M'} \langle L'1,00|L0\rangle^2 \langle 1J',qM'|JM\rangle^2 \begin{Bmatrix} L' & 1 & L \\ J & S & J' \end{Bmatrix}^2 = \tag{1.2.32}$$

$$= (2L'+1)(2J'+1)\langle L'1,00|L0\rangle^2 \begin{Bmatrix} L' & 1 & L \\ J & S & J' \end{Bmatrix}^2$$

Selection rules on $L \to L'$ follow from the selection rules (1.2.31) for the one-electron atom. The same rules follow also from the Clebsch-Gordan coefficient $\langle L'1,00|L0\rangle$ of formula (1.2.32).

The mathematical formalism based on usage of Clebsch-Gordan coefficients and the fractional parentage coefficients possesses the central place in the atomic theory. This formalism allows one to take into account the symmetry of atomic particles in the analysis of their properties. In particular, the selection rule for radiative transitions is based on this formalism. The rate of radiative transition between two atomic states is proportional to the square of the matrix element of the dipole moment operator between transition states. Within the framework of the scheme for the light atom, this transition is possible between states with $\Delta S = 0$ and $\Delta L = 0, \pm 1$.

In the same manner, selection rules for J follow from the properties of the Wigner $6j$-symbol

$$\begin{Bmatrix} L' & 1 & L \\ J & S & J' \end{Bmatrix}$$

This symbol is nonzero if $J - J' = 0, \pm 1$ (except the transition $J = 0 \rightarrow J' = 0$).

According to formula (1.2.32), the relative probability for the transition to the state with a given total angular momentum J' is of the form

$$w \sim (2J'+1) \begin{Bmatrix} L' & 1 & L \\ J & S & J' \end{Bmatrix}^2 \sum_{J'} (2J'+1) \begin{Bmatrix} L' & 1 & L \\ J & S & J' \end{Bmatrix}^2 = (2L+1))\,(2J'+1) \begin{Bmatrix} L' & 1 & L \\ J & S & J' \end{Bmatrix}^2$$

This sum is derived using the known rules of addition of $6j$-symbols. Thus, we find the rate of radiative transition to the given level of fine structure

$$w\left(J, L \rightarrow J', L'\right) = (2L+1)\left(2J'+1\right) \begin{Bmatrix} L' & 1 & L \\ J & S & J' \end{Bmatrix}^2 w\left(J, L \rightarrow L'\right) \quad (1.2.33)$$

One can average this expression over the total angular momentum J of the initial state. The probability of atom location in the initial state with a given value of the total atom momentum J is proportional to the statistical weight of this state of fine structure, i.e. to the ratio of the statistical weight of this state to the statistical weight of the electron term SL

$$w\left(J, L \rightarrow L'\right) = \frac{2J+1}{(2L+1)\,(2S+1)} w\left(L \rightarrow L'\right) \quad (1.2.34)$$

From formulas (1.2.33) and (1.2.34) we obtain

$$w\left(J, L \rightarrow J', L'\right) = \frac{(2J+1)\left(2J'+1\right)}{(2S+1)} \begin{Bmatrix} L' & 1 & L \\ J & S & J' \end{Bmatrix}^2 w\left(L \rightarrow L'\right) \quad (1.2.35)$$

We now sum this quantity over the initial total momentum J to obtain

$$w\left(L \rightarrow J', L'\right) = \sum_J w\left(J, L \rightarrow J', L'\right) = \frac{\left(2J'+1\right)}{(2L'+1)\,(2S+1)} w\left(L \rightarrow L'\right) \quad (1.2.36)$$

This formula is valid for the sum over intensities of all lines of a spectral multiplet with the same final level; it is proportional to the statistical weight of this final level.

In particular, let us consider the transition with $L' = L - 1$. Substituting the appropriate quantum numbers in the Wigner $6j$-symbol of formula (1.2.33) under the assumption $J \gg 1$, one can obtain the ratio

$$\frac{w\,(J,L \to J-1,L-1)}{w\,(J,L \to J,L-1)} = \frac{1}{2}\frac{(J+L)^2-S^2}{(J-L)^2-S^2};$$
$$\frac{w\,(J,L \to J,L-1)}{w\,(J,L \to J+1,L-1)} = 2\frac{(J+L)^2-S^2}{(J-L)^2-S^2}$$

Thus, among the lines of a multiplet, the most intense is the line with $\Delta J = \Delta L$, the so-called principal spectral line. The spectral line with $\Delta J = \Delta L + 1$ is approximately by J^2 times weaker, since at $J \gg 1$ we have $J \approx L$. Finally, the spectral line with $\Delta J = \Delta L + 2$ is approximately by J^4 times weaker than the principal line. These last two lines are called satellites. The greater the value of the total angular momentum J, the stronger the principal line in comparison to the satellite lines.

If the condition $J \gg 1$ is not fulfilled, the analysis is more complicated, except relatively simple expressions for alkali atoms. Let us consider an atom with a single valence electron located in the field of an atomic core with zero spin and zero total angular momentum. We then have $S = 1/2$. The initial term L splits into two sublevels with $J = L \pm 1/2$, and the final term with $L' = L - 1$ also splits into two sublevels, but with $J' = L - 1/2$ and $J' = L - 3/2$. The selection rules on J allow three transitions from the term L to the term $L - 1$. Their relative probabilities are found from the general expression (1.2.35)

$$\frac{w\left(J=L+1/2 \to J'=L-1/2\right)}{w\,(J=L-1/2 \to J'=L-1/2)} = (L+1)\,(2L-1)\,; \qquad (1.2.37)$$
$$\frac{w\left(J=L-1/2 \to J'=L-1/2\right)}{w\,(J=L-1/2 \to J'=L-3/2)} = \frac{1}{(L-1)\,(2L+1)}$$

The fourth transition with $J = L + 1/2 \to J' = L - 3/2$ is forbidden since it would require $\Delta L = 2$, in contradiction with the selection rule followed from formula (1.2.37). In the limit $L \gg 1$ we have for the first and third transitions $\Delta J = \Delta L = 1$. Thus, both these transitions correspond to principal spectral lines, and their intensities are comparable. In the case of the second spectral line, $\Delta J = \Delta L - 1 = 0$, so that this is a satellite line with an intensity approximately $2L^2$ times weaker than that the principal lines.

One can generalize this example as well as the previous one, where transitions from an initial state of a multiplet with a certain value of J were considered. It follows that accounting all possible transitions between states of two multiplets with $J \gg 1$, one can obtain the most intense lines with $\Delta J = \Delta L$, which are the principal spectral lines. The greater the difference $\Delta J - \Delta L$, the weaker is the intensity of the corresponding line. We can average over initial or final states of a multiplet, for example, for radiation in a gas with the large temperature compared to the fine structure splitting, but small compared to the energy difference for neighboring multiplets.

We now consider radiative transitions between components of hyperfine structure. Hyperfine structure of atomic levels occurs as a result of interactions of the atomic electrons with the nonzero nuclear spin. By analogy with the above consideration, where the total atomic momentum J is the sum the orbital momentum L and spin angular momentum S, and the interaction potential does not depend on the spin, one can present the total momentum F as a sum of the total electron momentum J and the nuclear spin I. Therefore, along with the above cited selection rule $J - J' = 0, \pm 1$ we have now the selection rule $F' = F, F \pm 1$.

Due to the similarity of these two problems, one can rewrite formula (1.2.35) by change $J \to F, L \to J$, and $S \to I$. This gives

$$w\left(F, J \to F', J'\right) = \frac{(2F+1)\left(2F'+1\right)}{(2I+1)} \begin{Bmatrix} J' & 1 & J \\ F & S & F' \end{Bmatrix}^2 w\left(J \to J'\right)$$

The total rate of transition from a given state of hyperfine structure follows from formula (1.2.34)

$$w\left(F, J \to J'\right) = \sum_{F'} w\left(F, J \to F', J'\right) = \frac{(2F+1)}{(2J+1)(2I+1)} w\left(J \to J'\right)$$

The radiative transition rate from all components of the hyperfine multiplet to a given hyperfine state of the other multiplet follows from formula (1.2.36)

$$w\left(J \to J', F'\right) = \sum_{F} w\left(F, J \to F', J'\right) = \frac{\left(2F'+1\right)}{(2J'+1)(2I+1)} w\left(J \to J'\right)$$

The statistical weights for these two expressions are the relative probabilities for filling of given states of the hyperfine structure.

Because of a weak interaction between electrons and nuclear spin, the selection rules with respect to electron total momentum J and electron parity remain valid for the above transitions. In particular, dipole transitions are forbidden between components of hyperfine structure for the same term because of the identical parity for these states. Thus, we have that radiative transitions of the type $J = 0 \to J' = 0$ are forbidden in the dipole approximation, as well as for the transition $F = 0 \to F' = 0$.

1.2.4 Polarization of Spontaneous Radiation of Atomic Particles

Polarization is one of the photon parameters at the interaction of a photon with atomic particles and atomic systems. Indeed, each emitted spontaneous photon is characterized by a direction of propagation that coincides with the wave vector **k** and by a polarization vector **s** that is perpendicular to the vector of propagation **k** of an electromagnetic wave. There are two polarizations, so that the their vectors are mutually perpendicular to each other and to the wave vector **k**, and in a general case the polarization vector is a superposition of these two vectors. For induced radiation, the ejected photon has the same polarization as the initial photon impacting on he atomic system. However, for spontaneous radiation, the problem of the polarization of the emitted photon remains open for investigation.

Let us obtain the spontaneous emission rate from an excited atom of a photon, expressed in terms of specific polarization and emission direction. The rate for the spontaneous transition $j \to i$ is given by formula (1.2.20), where we must put $n_\omega = 0$

$$w_{ji} = \frac{4\omega^3}{\hbar c^3} \left| \hat{\mathbf{D}}_{ji}\mathbf{s} \right|^2 g_i$$

The statistical weight of the final state is

$$g_i = \frac{1}{2}\frac{\mathrm{d}\Omega}{4\pi},$$

where $\mathrm{d}\Omega$ is the solid angle element for the emitted photon, and the factor 1/2 appears because of the two directions of the the photon polarization. We then obtain the spontaneous emission rate for a photon in a certain direction with a given polarization vector $\mathbf{s}$ as

$$\mathrm{d}w_{ij} = \frac{\omega^3}{2\pi\hbar c^3} \left| \hat{\mathbf{D}}_{ji}\mathbf{s} \right|^2 \mathrm{d}\Omega$$

Let us introduce spherical components s_q and D_q of vectors $\mathbf{s}$ and $\mathbf{D}$ that allows one to rewrite this expression in the form

$$\mathrm{d}w_{ij} = \frac{\omega^3}{2\pi\hbar c^3} \left| \sum_q s_{-q} \left\langle LM_L \left| D_q \right| L'M'_L \right\rangle \right|^2 \mathrm{d}\Omega, \tag{1.2.38}$$

where L, M_L are the orbital momentum and its projection onto the z axis respectively; the other quantum numbers for this state are omitted. The same quantities with the primes refer to the final state 0. We also introduce the quantity $\Delta M_L = M_L - M_{L'}$.

Only one term in the sum over q in formula (1.2.38) is nonzero for each of the three possible transitions with $\Delta M_L = 0, \pm 1$. If $\Delta M_L = 0$, then $q = 0$. It follows from formula (1.2.38) the rate for emission of a photon polarized in the plane defined by $\mathbf{k}$ and z axis

$$\mathrm{d}w\left(LM_L \to L'M_{L'}\right) = \frac{\omega^3 s_z^2}{2\pi\hbar c^3} \left| \left\langle LM_L \left| D_z \right| L'M_L \right\rangle \right|^2 \mathrm{d}\Omega \tag{1.2.39}$$

The quantity s_z is the projection of unit polarization vector on the z axis. We define θ to be the angle between the z axis and direction $\mathbf{n} = \mathbf{k}/k$ of the emitted photon. Then the direction of the vector $\mathbf{s}$ can be chosen to be either in the plane of $\mathbf{n}$ and the z axis or perpendicular to this plane. In the first case we have $s_z = \sin\theta$, since the vectors $\mathbf{s}$ and $\mathbf{n}$ are normal to each other. In the second case $s_z = 0$, since the vector $\mathbf{s}$ is perpendicular to the z axis. We thus obtain two spontaneous rates:

one corresponding to the emitted photon with the polarization in the plane of **n** and the z axis; and the second rate corresponding to the emission of a photon with a polarization perpendicular to this plane. These rates are

$$dw_1 = \frac{\omega^3}{2\pi\hbar c^3} \left|\left\langle LM_L \left|D_z\right| L'M_L\right\rangle\right|^2 \sin^2\theta d\Omega, \ dw_2 = 0 \tag{1.2.40}$$

Summarizing over two terms and integrating over angles, we get the total rate of photon emission for polarization in the plane of vectors **n** and the z axis

$$w\left(LM_L \to L'M_L'\right) = \frac{4\omega^3}{3\hbar c^3} \left|\left\langle LM_L \left|D_z\right| L'M_L\right\rangle\right|^2$$

If $\Delta M_L = -1$, then only one term is retained in the sum over q in formula (1.2.38) that gives

$$s_{-1}\left\langle LM_L \left|D_1\right| L'M_L + 1\right\rangle = \frac{1}{2}\left(s_x - is_y\right)\left\langle LM_L \left|D_x + iD_y\right| L'M_L + 1\right\rangle$$

This formula determines the photon emission in the xy plane with left circular polarization. Analogously, if $\Delta M_L = +1$, only one term is conserved in the sum over q in formula (1.2.38), which is

$$s_1\left\langle LM_L \left|D_{-1}\right| L'M_L - 1\right\rangle = \frac{1}{2}\left(s_x + is_y\right)\left\langle L, M_L \left|D_x - iD_y\right| L', M_L - 1\right\rangle$$

This determines photon emission in the xy plane with right circular polarization.

If there is no the reason for selection of the z axis, then an atom will be located in each of its M_L substates with equal probabilities. The transition rate $dw\left(L \to L'\right)$ is obtained by summarizing the expression for dw over $M_{L'}$ and averaging over M_L. There is a simpler alternative way to arrive at the same result. Averaging over a direction of the z axis in formula (1.2.38) with respect to the direction of the polarization vector **s**, one can obtain thereby

$$\begin{aligned} dw_{\mathbf{s}}\left(L \to L'\right) &= \frac{\omega^3}{2\pi\hbar c^3}\frac{1}{2L+1}\sum_{M_L, M_L'} \left|\left\langle LM_L \left|D_z\right| L'M_L'\right\rangle\right|^2 d\Omega = \\ &= \frac{\omega^3}{6\pi\hbar c^3}\frac{1}{2L+1}\sum_{M_L, M_L'} \left|\left\langle LM_L \left|\mathbf{D}\right| L'M_L'\right\rangle\right|^2 d\Omega = \\ &= \frac{e^2\omega^3}{6\pi\hbar c^3}\frac{1}{2L+1} \left|\left\langle L \left|r\right| L'\right\rangle\right|^2 d\Omega \end{aligned} \tag{1.2.41}$$

As is seen, emission is isotropic. This follows from the fact that all directions in space are equivalent. The quantity $\left\langle L \left|r\right| L'\right\rangle$ in formula (1.2.41) is the radial matrix element. Integration of formula (1.2.41) over angles is equivalent to multiplication by 4π because of isotropic emission. As a result, we have

$$w_{1,2}\left(L \to L'\right) = \frac{2e^2\omega^3}{3\hbar c^3}\frac{1}{2L+1}\left|\left\langle L\left|r\right|L'\right\rangle\right|^2$$

This expression determines the total rate of photon emission with a fixed polarization direction, but the result does not depend on this direction. Summarizing over two independent polarizations, one can obtain the rate w that determines the lifetime of the excited state and is equal

$$w\left(L \to L'\right) = \frac{1}{\tau} = \frac{4e^2\omega^3}{3\hbar c^3}\frac{1}{2L+1}\left|\left\langle L\left|r\right|L'\right\rangle\right|^2$$

The transitions considered above relate to an atom in the absence of an external field, and now we consider the case of a weak magnetic field. Let us evaluate the relative intensity of Zeeman spectral lines both along and perpendicular to the direction of the magnetic field. Let the z axis to be directed along the magnetic field and first the wave vector $\mathbf{k}$ of photon propagation be directed along the z axis also, so that the polarization vector is placed in the xy plane. According to formula (1.2.20), the photon emission rate is proportional to $\left|\left\langle LM_L\left|D_s\right|L'M_{L'}\right\rangle\right|^2$, where L, M_L and L', $M_{L'}$ are, respectively, the orbital angular momentum and its projection on the z axis for the initial and final states.

Let us choose two independent polarizations, $\mathbf{s}_1 = \mathbf{i}_x$, and $\mathbf{s}_2 = \mathbf{i}_y$, where $\mathbf{i}_x$ and $\mathbf{i}_y$ are unit vectors directed along the x and y axes. The rate $w_{\shortparallel}$ summed over photon polarizations has the form

$$w_{\shortparallel} \sim \left|\left\langle LM_L\left|D_x\right|L'M_L'\right\rangle\right|^2 + \left|\left\langle LM_L\left|D_y\right|L'M_L'\right\rangle\right|^2 = \sum_{q=\pm1}\left|\left\langle LM_L\left|D_q\right|L'M_L'\right\rangle\right|^2 \tag{1.2.42}$$

Let us consider right circularly ($\Delta M_L = 1$) and left circularly ($\Delta M_L = -1$) polarized light which propagates along the z axis. Then relative intensities (called σ components) correspond to terms in formula (1.2.42) with $q = -1$ and $q = +1$, respectively. These relative intensities are proportional to squares of the corresponding Clebsch-Gordan coefficients

$$\begin{aligned}\Delta M_L &= +1;\ w_{\shortparallel}^{+} \sim \left|\left\langle L, 1; M_L, -1\right|L', M_L - 1\right\rangle\right|^2; \\ \Delta M_L &= -1;\ w_{\shortparallel}^{-} \sim \left|\left\langle L, 1; M_L, +1\right|L', M_L + 1\right\rangle\right|^2\end{aligned} \tag{1.2.43}$$

We now turn our attention to observation of light in directions perpendicular to the magnetic field vector. One can select the direction $\mathbf{k}$ as the y axis. Then the polarization vector $\mathbf{s}$ of an emitted photon is placed in the xz plane. Taking $\mathbf{s}_1 = \mathbf{i}_z$, and then $\mathbf{s}_2 = \mathbf{i}_x$, one can find after summation, that the photon emission rate is proportional to the quantity

$$w_\perp \sim \left|\langle LM_L\,|D_z|\,L'M'_L\rangle\right|^2 + \left|\langle LM_L\,|D_x|\,L'M'_L\rangle\right|^2$$

By definition, $D_x = (D_1 + D_-1)/\sqrt{2}$, so that

$$\left|(D_x)_{ji}\right|^2 = \frac{1}{2}\left|\sum_{q=\pm1}(D_q)_{ji}\right|^2 = \frac{1}{2}\sum_{q=\pm1}\left|(D_q)_{ji}\right|^2 + \frac{1}{2}\left[(D_1)_{ji}(D_{-1})^*_{ji} + (D_1)^*_{ji}(D_{-1})_{ji}\right] \tag{1.2.44}$$

The last two terms of this expression must be zero, since for fixed values j and i both matrix elements $(D_1)_{ji}$ and $(D_{-1})_{ji}$ cannot be nonzero simultaneously. If we take $\Delta M_L = +1$, the matrix element $(D_{-1})_{ji}$ is zero, and if we take $\Delta M_L = -1$, the matrix element $(D_1)_{ji}$ is zero. We thus obtain

$$\left|(D_x)_{ji}\right|^2 = \frac{1}{2}\sum_{q=\pm1}\left|(D_q)_{ji}\right|^2$$

The rate $w_\perp$ after summation over polarizations of emitted photon can be written in the form

$$w_\perp \sim \left|\langle LM_L\,|D_z|\,L'M'_L\rangle\right|^2 + \frac{1}{2}\sum_{q=\pm1}\left|\langle LM_L\,|D_q|\,L'M'_L\rangle\right|^2$$

One can express these matrix elements through the Clebsch-Gordan coefficients

$$\Delta M_L = +1;\; w_\perp^+ \sim \frac{1}{2}\left|\langle L,1;M_L,-1|L',M_L-1\rangle\right|^2; \tag{1.2.45}$$
$$\Delta M_L = -1;\; w_\perp^- \sim \frac{1}{2}\left|\langle L,1;M_L,+1|L',M_L+1\rangle\right|^2$$

It follows from comparison of formulas (1.2.43) and (1.2.44) that the emission rate perpendicular to the magnetic field is two times less than that along the magnetic field.

The intensity of the component with $\Delta M_L = 0$ (the so-called π-component) is proportional to the quantity

$$w_\perp^0 \sim \left|\langle L,1;M_L,0|L',M_L\rangle\right|^2 \tag{1.2.46}$$

On the basis of properties of the Clebsch-Gordan coefficients, one can find the relative intensities of various components which are given in Table 1.4. As a supplement to this table, we note that

$$w_{\shortparallel}^0 = 0;\; w_{\shortparallel}^+ = 2w_\perp^+;\; w_{\shortparallel}^- = 2w_\perp^-$$

Table 1.4 Relative emission rates for components of Zeeman splitting

	$w_{\perp}^{+}$	$w_{\perp}^{0}$	$w_{\perp}^{-}$
$L' = L - 1$	$(L + M_L)(L + M_L - 1)$	$4\left(L^2 - M_L^2\right)$	$(L - M_L)(L - M_L - 1)$
$L' = L$	$(L + M_L)(L - M_L + 1)$	$4M_L^2$	$(L - M_L)(L + M_L + 1)$
$L' = L + 1$	$(L - M_L + 1)(L - M_L + 2)$	$4\left((L + 1)^2 - M_L^2\right)$	$(L + M_L + 1)(L + M_L + 2)$

Note also the absence of transitions $L' = L$ in the one-electron spectrum due to parity conservation.

Now we consider a relative intensity of spectral lines and their splitting that occurs when an atom is placed in an electric field (Stark effect). We find the relative intensity of Stark spectral lines when light is observed along the direction of the electric field vector compared to the direction perpendicular to the field. A constant electric field causes both shifting and splitting of atomic levels. Except the hydrogen atom case, shifts and splitting of electron levels are determined by the second-order perturbation theory. As above, the radiation polarization is determined by the observation direction. Let the z-axis be directed along the electric field. If radiation is observed along this direction, emitted photons are polarized in the xy-plane that corresponds to transitions $M_L \to M_L \pm 1$ which are called as σ-components. In the perpendicular direction with respect to the electric field vector, π-components are observed as well as σ-components. They correspond to transitions $M_L \to M_L$ and are polarized in xz-plane.

If $M_L = 0$, there is no difference with the previous case of the magnetic field, so that one can use the results of Table 1.4, taking $M_L = 0$. If $M_L \neq 0$, the energy levels are doubly degenerate with respect to the sign of M_L. Hence, intensities of the π-components are doubled compared to the magnetic field case for radiation polarization. The degeneracy arises from the fact that transitions with $M_L \to M_L$ and $-M_L \to -M_L$ are characterized by the same energies. In the case of σ-components, we find that the energies of transitions $M_L \to M_L - 1$ and $-M_L \to 1 - M_L$ are identical. The first transition is associated with emission of right-handed circularly polarized light and the second to emission of left-handed circular polarization of light. As it follows from data of Table 1.4, their intensities are identical, so that the sum doubles their intensities. Thus, the factor 2 is a general factor for Table 1.4 data and it may be omitted for relative values. Hence, the data of Table 1.4 may be used for the electric field case, if $w_{\perp}^{+}$ is replaced by $w_{\perp}(|M_L| \to |M_L - 1|)$ and $w_{\perp}^{-}$ is replaced by $w_{\perp}(|M_L| \to |M_L + 1|)$.

This analysis is simplified in the case of one-electron spectra of alkali metals. We consider the excitation of an alkali metal atom from the ground s-state with projections $\sigma = \pm 1/2$ on the reference axis to the excited ^2P_J state by a circularly polarized electromagnetic field. After emission of a spontaneous photon, the atom returns to the same s-state with the projections $\sigma' = \pm 1/2$ of its spin on the same reference axis. The number densities of atoms with spin projections $+1/2$ and $-1/2$

are labeled as N_+ and N_- respectively. The balance equations for populations of these atomic spin states in the initial s-state are

$$\frac{dN_+}{dt} = -w\,(1/2 \to -1/2)\,N_+ + w\,(-1/2 \to +1/2)\,N_-;$$
$$\frac{dN_-}{dt} = -w\,(-1/2 \to +1/2)\,N_- + w\,(1/2 \to -1/2)\,N_+$$

In the case of a steady process, one neglect the left hand side of these equations, and then the atomic average spin, which determines the atom polarization, is given by

$$\bar{S} = \frac{N_+ - N_-}{2\,(N_+ + N_-)} = \frac{1}{2}\frac{w\,(-1/2 \to +1/2) - w\,(1/2 \to -1/2)}{w\,(-1/2 \to +1/2) + w\,(1/2 \to -1/2)}$$

We now determine the relative probabilities of the radiative transitions. For definiteness, we assume the circular polarization of the excited electromagnetic field to be right, so that the projection of the orbital angular momentum increases by one as a result of excitation: $m \to m + 1$. The relative probability for atomic excitation from the state with spin 1/2 and spin projection σ into the state with total angular momentum J and its projection $M_J = \sigma + 1$ according to a general formula (1.2.20) for the transition rate is proportional to the squared Clebsch-Gordan coefficient $|\langle 1/2, 1; \sigma, 1|J, \sigma + 1\rangle|^2$. The relative probability for photon emission from the atom state JM-state for the transition back to an s-state with the spin projection σ' is proportional to the squared Clebsch-Gordan coefficient $\left|\langle J, 1; \sigma + 1, m|1/2, \sigma'\rangle\right|^2$. This leads to the following relative transition probabilities

$$w\,(-1/2 \to +1/2) = |\langle 1/2, 1; -1/2, 1|J, 1/2\rangle|^2\,|\langle J, 1; 1/2, 0|1/2, 1/2\rangle|^2\,;$$
$$w\,(+1/2 \to -1/2) = 0.$$

The conclusion is that the final average spin projection of a valence electron is $\sigma = 1/2$, so that an initially unpolarized atomic system becomes polarized through interaction with the circularly polarized light. From the physical standpoint this result may be explained as follows. Taking first $J = 1/2.$, we obtain that only atoms are excited with the initial spin projection $\sigma = -1/2$. As a result of their radiation, states with the spin projection $\sigma' = 1/2$ are formed, as well as with $\sigma' = -1/2$. However, because of the absence of transitions from the state $\sigma' = +1/2$, subsequently all the atoms transit in the state $\sigma' = -1/2$. Thus, the population of the $\sigma = -1/2$ state will diminish in time and all electrons will be finally in the $\sigma = 1/2$ state.

If the excited state has the total momentum $J = 1/2$, then excitation of both the $\sigma = 1/2$ and $\sigma = -1/2$ initial states take place. In the latter case, the excited state with $M = 3/2$ is produced. The spontaneous decay of this state according to the selection rule proceeds only to the state with $\sigma' = 1/2$. Thus, through a time the population of the state with $\sigma = -1/2$ decreases, whereas transitions from $\sigma = 1/2$ to the state with $\sigma' = -1/2$ are forbidden, and all the atoms transit in the $\sigma' = 1/2$

state. A typical time τ for establishment of the spin polarization is the inverse quantity with respect to the transition rate

$$\tau \sim \frac{1}{w\,(-1/2 \to 1/2)}$$

1.3 Oscillator Strength for Radiative Transition

1.3.1 Sum Rules

Let us introduce the oscillator strength f_{ij} for radiative transitions between i and j atomic states on the basis of the relation

$$f_{ij} = \frac{2m_e}{\hbar e^2}\omega_{ji}\left|(D_z)_{ji}\right|^2, \qquad (1.3.1)$$

where i and j are the initial and final states of the radiative transition, ω_{ji} is the radiation frequency for the transition between these states, and $(D_z)_{ji}$ is the matrix element of the projection of the dipole moment operator onto the polarization direction z which is taken between these states. The concept of the oscillator strength arises from a model of the electric properties of a matter in which we suppose that the atomic electrons are in equilibrium positions and react elastically to small perturbations. This means that a low-intensity electromagnetic field causes a weak harmonic motion of electrons around their equilibrium positions. In reality, electrons have not stable equilibrium positions in atoms, but there is the statistical distribution of such positions which is determined by the square modulus of the wave function. From the classical standpoint, there is a part of this square of the wave function modulus that characterizes the oscillator strength for a given frequency.

Let us introduce the real part of the oscillator strength f_{ij} on the basis of formula (1.3.1). Using formula (1.2.19), we have for the rate of a radiative transition

$$w_{ij} = \frac{2e^2\omega_{ji}^2}{m_e c^3} f_{ij} g_j \left(n_\omega + 1\right) \qquad (1.3.2)$$

Here we take the frame of reference in such a manner that the z axis is directed along the polarization vector **s**. Equation (1.3.2) shows that the oscillator strength is the dimensionless parameter that measures the rate of a given radiative transition.

The Hamiltonian $\hat{H}$ of the system of atomic electrons consists of the sum of the electron kinetic energies and the potential energies of their interactions with each other and with the atomic nucleus. The commutator between the projection of the momentum operator $\hat{p}_{jz}$ for an j-th electron along the field with the coordinate projection z_j in this direction is given by the expression

$$\left[\hat{p}_{jz}, z_j\right] = \hat{p}_{jz} z_j - z_j \hat{p}_{jz} = -i\hbar \tag{1.3.3}$$

From this it follows that the commutator of the Hamiltonian $\hat{H}$ and z_j arises from the kinetic energy part only and is equal

$$\left[\hat{H}, z_j\right] = \frac{1}{2m_e}\left[\hat{p}_{jz}^2, z_j\right] = -\frac{i\hbar}{m_e}\hat{p}_{jz} \tag{1.3.4}$$

For the commutator of the Hamiltonian $\hat{H}$ and the projection of the total dipole moment, we obtain expression

$$\left[\hat{H}, D_z\right] = -\frac{i\hbar}{m_e}\sum_j \hat{p}_{jz} \tag{1.3.5}$$

The matrix element in this relation taken between states i and j is equal

$$\omega_{ji}\left(D_z\right)_{ij} = \frac{ie}{m_e}\sum_j \left(\hat{p}_{jz}\right)_{ij} \tag{1.3.6}$$

Inserting this expression into formula (1.3.1) for the oscillator strength, one can obtain

$$f_{ij} = \frac{2i}{e\hbar}\sum_j \left(\hat{p}_{jz}\right)_{ij}\left(D_z\right)_{ji}, \tag{1.3.7}$$

which may be rewritten as

$$f_{ij} = -\frac{2i}{e\hbar}\sum_j \left(\hat{p}_{jz}\right)_{ji}\left(D_z\right)_{ij}, \tag{1.3.8}$$

with using the properties of operators $\hat{p}_{jz}$ and D_z at any conjugation of electrons.

We now express the oscillator strength as half the sum of two above expressions to give it the symmetrical form

$$f_{ij} = \frac{i}{\hbar}\sum_{j,j'}\left[\left(\hat{p}_{jz}\right)_{ij}\left(z_{j'}\right)_{ji} - \left(z_{j'}\right)_{ij}\left(\hat{p}_{jz}\right)_{ij}\right] \tag{1.3.9}$$

Summarizing this expression over all final states j, we obtain

$$\sum_j f_{ij} = \frac{i}{\hbar}\sum_{j,j'}\left[\hat{p}_{jz}, z_{j'}\right]_{ii} \tag{1.3.10}$$

Since the momentum operator of each electron commutes with the operator of space coordinates for of any other electron, then the double sum above reduces to the case $j = j'$ only. Furthermore, using the commutation relation (1.3.4), we find

$$\sum_j f_{ij} = n, \tag{1.3.11}$$

where n is the total number of electrons in the atomic system. If the sum is restricted only to such transitions in which valence electrons take part, then n is the number of valence electrons in the atom.

The sum (1.3.11) is taken over states of the discrete and continuum spectrum. As is seen, this sum is independent of quantum numbers of the initial atomic state i. If the state i is a highly excited state, then we encounter with the classical problem of the sum rule. Let us analyze this by analogy with the radiation intensity in the classical limit. Because of the symmetry of energy levels near a test one in the classical case, we obtain zero instead of n for the right-hand side of formula (1.3.11). This exhibits that the classical approach cannot not give the correct result for the oscillator strengths. This means that the main contribution to the sum (1.3.11) in the classical case follows from states which energies differs from the energy of the state i strongly, so that the symmetry in the level distribution is violated.

We now introduce the static polarizability of an atom α_i, which is found in the state i and is placed in a constant electric field of a strength E, in terms of the oscillator strengths as

$$\frac{\mathrm{d}\delta\varepsilon_i}{\mathrm{d}E} = -\alpha_i E, \tag{1.3.12}$$

where $\delta\varepsilon_i$ is the static Stark energy shift [26] for a state i under the action of an electric field of strength E. This energy shift may be determined as the second-order energy shift of the perturbation theory, that gives for a state i

$$\delta\varepsilon_i = \frac{E^2}{\hbar}\sum_j \frac{\left|(D_z)_{ij}\right|^2}{\omega_{ij}}. \tag{1.3.13}$$

This leads to the following quantum expression for the static polarizability of an atom

$$\alpha_i = -\frac{2}{\hbar}\sum_j \frac{\left|(D_z)_{ij}\right|^2}{\omega_{ij}} = \frac{e^2}{m_e}\sum_j \frac{f_{ij}}{\omega_{ij}^2}, \tag{1.3.14}$$

where we use the expression (1.3.1) for the oscillator strength.

Note that the static polarizability of a harmonic oscillator of a frequency ω in the classical limit is determined also from formula (1.3.12), where $\delta\varepsilon_i$ is the energy of the oscillator in the electric field. Then we have for the polarizability of a harmonic oscillator

$$\alpha = \frac{e^2}{m_e\omega^2} \tag{1.3.15}$$

From comparison of formulas (1.3.14) and (1.3.15) it follows these the atom polarizability is the sum of polarizabilities of atomic oscillators, and each of them appears in this sum with a weight f_{ij} that is the oscillator strength. This explains the origin of the term *oscillator strength.*

1.3.2 *Sum Rules for One-Electron Atom*

The matrix apparatus of quantum mechanics allows one to determine, along with the oscillator sum, the sums of similar quantities for a single-electron atom which are defined as

$$S_i^{(m)} = \frac{2m_e}{3e^2\hbar^2}\sum_k \left(\varepsilon_j - \varepsilon_i\right)^m \left|(\mathbf{D})_{ji}\right|^2, \quad m = 0, 1, 2, 3, 4 \tag{1.3.16}$$

for single-electron atomic states. Below we present these sum rules.

In the first case $m = 0$ we have for a single-electron atom

$$S_0^{(i)} = \frac{2m_e}{3\hbar^2}\sum_j \left|(\mathbf{r})_{ji}\right|^2 = \frac{2m_e}{3\hbar^2}\left(r^2\right)_{ii} \tag{1.3.17}$$

In the classical limit, the quantity $\left(r^2\right)_{ii}$ is the average value of the electron distance square $\overline{r^2(t)}$ from the center. For a hydrogen atom or hydrogenlike ion with a charge Z we have for the initial atomic state with the principal quantum number n and orbital momentum l

$$\left(r^2\right)_{00} = \frac{n^2\hbar^4}{2Z^2e^4m_e^2}\left(5n^2 - 3l(l+1) + 1\right) \tag{1.3.18}$$

From this we have

$$S_{nl}^{(0)} = \frac{n^2}{3Z^2Ry}\left(5n^2 - 3l(l+1) + 1\right), \quad Ry = \frac{m_e e^4}{\hbar^2} \tag{1.3.19}$$

For the case $m = 1$ in formula (1.3.16) the sum is given by the relation (1.3.11)

$$S_i^{(1)} = \sum_j f_{ij} = 1 \tag{1.3.20}$$

In the case $m = 2$ of formula (1.3.16) we use the condition of completeness of eigenfunctions

$$\sum_j \left|(\hat{\mathbf{p}})_{ji}\right|^2 = (p^2)_{ii} = 2m_e (E - V)_{ii} \tag{1.3.21}$$

and take into account that $(\hat{\mathbf{p}})_{ji} = i m_e \omega_{ji} \mathbf{r}_{ji}$. Thus, we get

$$S_i^{(2)} = \frac{2m_e}{3} \sum_k \omega_{ji}^2 \left|\mathbf{r}_{ji}\right|^2 = \frac{4}{3} (E - V)_{ii} \tag{1.3.22}$$

In the classical limit, this expression has the form

$$S_i^{(2)} = \frac{4}{3} \overline{T(t)}$$

where $\overline{T(t)}$ is the average value of the electron kinetic energy. For a hydrogenlike ion we have according to the virial theorem for an electron located in the Coulomb field $\overline{T(t)} = -E_n = Z^2 Ry/(2n^2)$ and

$$S_{nl}^{(2)} = \frac{2Z^2 Ry}{3n^2} \tag{1.3.23}$$

in the case $m = 3$ of formula (1.3.16), according to the Ehrenfest theorem [27] of quantum mechanics, we have

$$\frac{\mathrm{d}}{\mathrm{d}t} (\hat{\mathbf{p}})_{ji} = i\omega_{ji} (\hat{\mathbf{p}})_{ji} = -(\nabla V)_{ji} \tag{1.3.24}$$

Using these relations, we find

$$\sum_j \omega_{ji} \left|(\hat{\mathbf{p}})_{ji}\right|^2 = i \sum_j (\hat{\mathbf{p}})_{ij} (\nabla V)_{ji} = -i \sum_j (\nabla V)_{ij} (\hat{\mathbf{p}})_{ji} =$$
$$= \frac{i}{2} (\hat{\mathbf{p}} \cdot \nabla V - \nabla V \cdot \hat{\mathbf{p}})_{ii} = \frac{\hbar}{2} (\Delta V)_{ii}$$

Thus we have for the sum with $m = 3$ of formula (1.3.16)

$$S_i^{(3)} = \frac{2m_e \hbar}{3} \sum_j \omega_{ji} \left|\omega_{ji} (\mathbf{r})_{ji}\right|^2 = \frac{\hbar^2}{3m_e} (\Delta V)_{ii} \tag{1.3.25}$$

Since according to the Maxwell equation $\Delta V = 4\pi e\rho$, where ρ is the density of the positive charge producing the potential energy V, one can obtain

$$S_i^{(3)} = \frac{4\pi e\hbar^2}{3m_e}(\rho)_{ii} \tag{1.3.26}$$

In particular, taking $\rho = Ze\delta(\mathbf{r})$ for a hydrogenlike ion, we have from formula (1.3.25)

$$S_0^{(3)} = \frac{4\pi Ze^2\hbar^2}{3m_e}|\psi_0(0)|^2 \tag{1.3.27}$$

Since the limit $r \to 0$ we have $\psi_0(r) \sim r^l$, where l is the orbital angular momentum for the state 0, one can obtain that $S_0^{(3)} = 0$ for $l \neq 0$. If $l = 0$, one can obtain $\psi_0(0) = (\pi n^3 a_o^3)^{-1/2}$, where n is the principal quantum number and $a_o = \hbar^2/\left(Zm_e e^2\right)$. Thus we have

$$S_n^{(3)} = \frac{4Z^4 Ry^2}{3n^3} \tag{1.3.28}$$

Finally, we consider the case $m = 4$ of formula (1.3.16). Again using the Ehrenfest theorem, we have

$$\sum_j \omega_{ji}^2 \left|(\hat{\mathbf{p}})_{ji}\right|^2 = \sum_j |(\nabla V)_{ji}|^2 = \left[(\nabla V)^2\right]_{ii} \tag{1.3.29}$$

We then obtain

$$S_i^{(4)} = \frac{2m_e\hbar^2}{3}\sum_j \omega_{ji}^4 |(\mathbf{r})_{ji}|^2 = \frac{2\hbar^2}{3m_e}\left[(\nabla V)^2\right]_{ii} \tag{1.3.30}$$

For the hydrogenlike ion we have

$$(\nabla V)^2 = \frac{Z^2e^4}{r^4}, \tag{1.3.31}$$

that gives

$$S_0^{(4)} = \frac{2Z^2\hbar^2e^4}{3m_e}\left(\frac{1}{r^4}\right)_{00} \tag{1.3.32}$$

This expression diverges in the case $l = 0$, while it is finite for $l \neq 0$. Substituting the corresponding average value, one obtains for $l \geq 1$

$$S_{nl}^{(4)} = \frac{8Z^6Ry^3}{3n^5\left(4l^2-1\right)(2l+3)}\left[\frac{3n^3}{l(l+1)} - 1\right] \tag{1.3.33}$$

The sum rules may be used for estimation of matrix elements in the asymptotic limit. For example, from formula (1.3.16) it follows for $m = 3$ that, if $\varepsilon_j \to \infty$, part of this sum takes the form

$$\int \omega_{ji}^3 \left|(\mathbf{r})_{ji}\right|^2 \mathrm{d}\omega_{ji} \tag{1.3.34}$$

Because of convergence of this integral, one can conclude the following dependence for the matrix element $(\mathbf{r})_{ji}$

$$(\mathbf{r})_{ji} \sim \frac{1}{\varepsilon_j^p} \tag{1.3.35}$$

for $p > 2$.

1.3.3 Peculiarities of the Oscillator Strength

On the basis of the single-electron approximation one can determine the influence of the Pauli exclusion principle, which forbids location of two electrons in the same state, on the shell atom model. One can seen from the first glance that the Pauli principle should influence the sum rule (1.3.11), according to which the sum of the oscillator strengths is equal to the number of atomic electrons. In fact, in summation (1.3.11) over all possible states of electron in the self-consistent field of the atomic core, according to Pauli principle it is necessary to exclude filled electron states from this sum. However, we prove below that the Pauli principle does not violate the oscillator strength sum (1.3.11).

Indeed, the sum of oscillator strengths for a given electron, located in a state j, in the single-electron approximation taking into account the Pauli principle has the form

$$\sum_i f_{ji},$$

where index i refers to the empty electron states. Applying formula (1.3.11) for a given electron, we have

$$\sum_k f_{jk} = 1,$$

where k refers to all possible electron states. From this it follows

$$\sum_{j,k} f_{jk} = n,$$

where n is the number of electrons, and

$$\sum_{j,i} f_{ji} = \sum_{j,k} f_{jk} - \sum_{j,j'} f_{jj'},$$

where j and j' are filled states. According to definition (1.3.1) of the oscillator strength we have $f_{jj'} = -f_{j'j}$, that gives

$$\sum_{j,j'} f_{jj'} = 0$$

This leads to formula (1.3.11)

$$\sum_{j,i} f_{ji} = n$$

Thus, the Pauli principle does not affect the sum rule for the oscillator strengths. An analogous conclusion may be obtained for other sums (1.3.16) with odd values of m.

We below determine the dependence of the oscillator strength on the principal quantum number n of this state in the case of transitions for highly excited atom states $n \gg 1$. Indeed, expressing the oscillator strength between highly excited states (Rydberg states) $f_{nl,n'l'}$ through the matrix element of the electron dipole moment with using the radial wave functions, we have

$$f_{nl,n'l'} \sim \left| \int_0^\infty R_{nl}(r) R_{n'l'}(r) r^3 \mathrm{d}r \right|^2$$

Here R_{nl} and $R_{n'l'}$ are radial wave function in the single-electron approximation for the initial (nl) state and final $(n'l')$ highly excited state. The other factors in formula (1.3.1) for the oscillator strength $f_{nl,n'l'}$ including $\varepsilon_n - \varepsilon_{n'}$, do not depend on n' because of the condition $|\varepsilon_n| \gg |\varepsilon_{n'}|$. Thus, in order to find the dependence of the oscillator strength $f_{nl,n'l'}$ on the principal quantum number $n' \gg 1$ for the highly excited state, it is necessary to estimate the functions R_{nl} and $R_{n'l'}$. In addition, we must extract the most values of r which give the main contribution to the integral for the matrix element of the dipole moment.

For simplicity and for clarity improvement, we use the system of atomic units $m_\mathrm{e} = e = \hbar = 1$ in this operation. Since $n' \gg 1$, one can use the semiclassical expression [8, 9] for the radial wave function $R_{n'l'}$

$$R_{n'l'} = \frac{a}{r\sqrt{k_{n'l'}(r)}} \cos\left[\int_{r_1}^{r} k_{n'l'}\left(r'\right) \mathrm{d}r' - \frac{\pi}{4} \right],$$

where

$$k_{n'l'}(r) = \sqrt{2\left(E_{n'} - V_{l'}(r)\right)},$$

and $E_{n'} = -1/\left(2n'^2\right)$ is the energy of the term with the principal quantum number n'. This term is hydrogenlike one due to the condition $n' \gg 1$. The quantity

$$V_{l'}(r) = -\frac{1}{r} + \frac{l'(l'+1)}{2r^2}$$

is the effective potential for radial motion of an electron. The index l' is the orbital quantum number of the final state, and r_1 is the left turning point for the corresponding classical motion.

One can predict $n, l \sim 1$ for the initial atomic state. Then we find $r \sim a_o$ for the range r, where the wave function $R_{nl}(r)$ is relatively large, and hence this range of r gives the main contribution to the radial matrix element of the dipole moment. Furthermore, it follows from the dipole selection rule that at $l \sim 1$ also $l' \sim 1$. In the range $r \sim a_o$ we have $k_{n',l'} \sim 1$, and consequently, $R_{n',l'}(r) \sim b$. Under these conditions, we evaluate the normalization factor b of the radial wave function for a highly excited final state. The normalization condition has the form

$$\int_0^\infty R^2_{n'l'}(r) r^2 \mathrm{d}r = 1$$

In the semiclassical approximation this condition may be represented as

$$\frac{b^2}{2} \int_{r_1}^{r_2} \frac{\mathrm{d}r}{k_{n'l'}} = 1$$

From this condition it follows

$$\frac{1}{b^2} = \int_{r_1}^{r_2} \frac{\mathrm{d}r}{2\sqrt{-\frac{1}{(n')^2} + \frac{2}{r} - \frac{l'(l'+1)}{r^2}}},$$

where r_2 is the right turning point for the corresponding classical motion of an electron. Since for the important region in the integrand we have $r \sim a_o(n')^2 \gg a_o$, and $l' \sim 1$, one can neglect the contribution of the centrifugal potential in the normalization integral. We thus obtain

$$\frac{1}{b^2} = \int_{r_1}^{r_2} \frac{\mathrm{d}r}{2\sqrt{-\frac{1}{(n')^2} + \frac{2a_o}{r}}} = (n')^3 \int_{x_1}^{x_2} \frac{\mathrm{d}x}{2\sqrt{-1 + \frac{2}{x}}} \sim (n')^3$$

Our final result is $b \sim (n')^{-3/2}$.

In the range $r \sim a_o$ which gives the main contribution to the integrand $f_{nl,n'l'}$, we have $R_{nl} \sim 1$ and $R_{n'l'} \sim b$ that leads to the result

$$f_{nl,n'l'} \sim |R_{n'l'}|^2 \sim \frac{1}{(n')^3}$$

The coefficient in this dependence is of the order of one in atomic units. In particular, the accurate calculation for the ground state of the hydrogen atom $\left(n = 1, l = 0, l' = 1\right)$ yields

$$f_{10,n'1} = \frac{1.6}{(n')^3} \tag{1.3.36}$$

References

1. M.G. Veselov, L.N. Labtzovskij, *Theory of the Atom Structure of Electron Shells* (Nauka, Moscow, 1986) (in Russian)
2. H. Haken, A.C. Wolf, *The Physics of Atoms and Quanta* (Springer, Berlin, 2000)
3. J. Barrett, *Atomic Structure and Periodicity* (Wiley, Berlin, 2003)
4. B.H. Bransden, C.J. Joachain, *Physics of Atoms and Molecules* (Prentice Hall, Harlow, 2003)
5. H. Haken, H.C. Wolf, *The Physics of Atoms and Quanta* (Springer, Berlin, 2004)
6. C. Food, *Atomic Physics* (Oxford University Press, Oxford, 2004)
7. W. Demdroter, *Atoms, Molecules and Photons* (Springer, Berlin, 2006)
8. H.A. Bethe, *Quantenmechanik der Ein- und Zweielectronenprobleme. Handbuch der Physik*, vol. 24–1 (Springer, Berlin, 1933)
9. H.A. Bethe, E.E. Salpeter, *Quantum Mechanics of One and Two-Electron Atoms* (Springer, Berlin, 1957)
10. L.D. Landau, E.M. Lifshitz, *Quantum Mechanics* (Pergamon Press, Oxford, 1965)
11. W. Pauli, Zs. Phys. **31**, 765 (1925)
12. W. Pauli, *Exclusion Principle and Quantum Mechanics* (Editions du Griffon, Neuchatel, Switzerland, 1947)
13. I.G. Kaplan, *The Pauli Exclusion Principle* (Wiley, Chichester, 2017)
14. H.A. Bethe, *Intermediate Quantum Mechanics* (New York, Benjamin Inc., 1964)
15. B.M. Smirnov, *Physics of Atoms and Ions* (Springer NY, New York, 2003)
16. F. Hund, *Linienspectren and Periodisches System der Elemente* (Berlin, 1927)
17. S. Bashkin, J. Stoner, *Atomic Energy Levels and Grotrian Diagrams*, vol. 1–4 (Amsterdam, North Holland, 1975–1982)
18. A.A. Radzig, B.M. Smirnov, *Reference Data on Atoms, Molecules and Ions* (Springer, Berlin, 1985)
19. B.M. Smirnov, Phys. Usp. **44**, 221 (2001)
20. G. Racah, Phys. Rev. **61**, 186 (1942)
21. G. Racah, Phys. Rev. **62**, 438 (1942)
22. N. Bohr, *The Correspondence Principle* (Elsevier, Amsterdam, 1976)
23. A.R. Edmonds, *Angular Momentum in Quantum Mechanics* (Princeton University Press, Princeton, 1957)
24. D.A. Varshalovich, A.N. Moskalev, V.K. Khersonskii, *Quantum Theory of Angular Momentum* (World Scientific, Singapore, 1988)
25. O. Laporte, W. Meggers, J. Opt. Soc. Am. **11**, 459 (1925)
26. J. Stark, Ann. Phys. **43**, 965 (1914)
27. P. Ehrenfest, Zs. Phys. **45**, 455 (1927)

Chapter 2
Properties of Radiation Field

Abstract Mechanisms of broadening of spectral lines are presented, including that due to a finite lifetime of an emitting atom, the Doppler broadening, as well as the impact and quasistatic theories of broadening of spectral lines. Criteria for various mechanisms of line broadening are given and combined cases are considered. Properties of an equilibrium radiation field for a strong interaction with an atomic system are given, and laws of blackbody radiation are represented. This equilibrium causes the existence of stimulated and spontaneous emission. The parameters of interaction of photons with an individual atom and with an ensemble of gaseous atoms include the cross section for absorption and emission of photons by an atomic particle, the absorption coefficient of a gas for a given frequency, and the optical thickness of a gaseous layer. The outgoing partial radiative flux from a uniform and weakly nonuniform gas layer is analyzed and expressed through the radiative temperature at a given frequency. The character of emission of a gaseous layer is described.

2.1 Broadening of Spectral Lines

2.1.1 Broadening of Spectral Line for Isolated Atom

It is of importance that the width of spectral lines is small compared to a typical frequency difference between neighboring spectral lines. Therefore, the spectrum of each atom is specific, and the spectroscopy is a strong instrument for diagnostics of elements which allows one to determine the composition of materials. Therefore the spectroscopy gave a significant contribution for physics development. In particular, the spectroscopy of atoms became a basis for creation and development of quantum mechanics.

V. Krainov and B. M. Smirnov, *Atomic and Molecular Radiative Processes*,
Springer Series on Atomic, Optical, and Plasma Physics 108,
https://doi.org/10.1007/978-3-030-21955-0_2

Below we analyze broadening of spectral lines that determines the width and shape of spectral lines. These parameters determine the character of photon propagation through a matter and are included in the distribution function a_ω over frequencies ω, so that $a_\omega d\omega$ is the probability that the photon frequency is found between ω and $\omega + d\omega$. Correspondingly, this distribution function is normalized as

$$\int_0^\infty a_\omega d\omega = 1 \tag{2.1.1}$$

We below analyze the form of the photon distribution function a_ω under various physical conditions.

Here we consider the structure of the photon distribution function as a result of the elementary emission process. The shape of this function a_ω follows from the non-stationarity of the emission process, and the broadening of spectral lines in this case is called radiative broadening. Radiative broadening is a property of the emission process, so that the shape of the distribution function follows from a finite lifetime τ_j of the initial state j in the transition to the state i.

Let us consider a single-photon transition between the initial excited state j and the ground state i. If w is the transition rate which is determined by formula (1.2.20), according to which at times t such that $t \gg w^{-1}$, the atom will be in its ground state with a probability approaching unity. Due to the non-stationarity of the initial state j, photons can be emitted with various frequencies in the vicinity of the frequency ω_{ji}. We now determine the distribution function a_ω for emitted photon frequencies. Equation (1.2.10) for amplitudes c_m of atom location an atom in a state m have the form

$$i\hbar\frac{dc_j}{dt} = \sum_m V_{j,i\omega}c_{i\omega}\exp\left[i\left(\omega_{ji}-\omega\right)t\right]; \tag{2.1.2}$$
$$i\hbar\frac{dc_{i\omega}}{dt} = V_{i\omega,j}c_j\exp\left[-i\left(\omega_{ji}-\omega\right)t\right]$$

Here $c_{i\omega}$ is the probability amplitude for the atom location in a state i after emission of a photon with a frequency ω. The initial conditions for the equation set (2.1.2) are $c_j(0) = 1$ and $c_{i\omega}(0) = 0$. We express the solution of (2.1.2) in the form

$$c_j(t) = \exp\left(-\frac{wt}{2}\right)$$

Here w is the quantity to be determined. When we substitute this solution into the second equation of the set (2.1.2), one can obtain

$$c_{i\omega}(t) = V_{i\omega,j}\frac{\exp\left[-i\left(\omega_{ji}-\omega\right)t - wt/2\right]-1}{\hbar\left(\omega_{ji}-\omega-iw/2\right)} \tag{2.1.3}$$

The resonant character of this equation justifies the resonance approach we have employed in the interaction of an atomic electron with the electromagnetic field. Also, from formula (1.2.3), we have

$$V_{i\omega,j} = -\frac{1}{2}\mathbf{D}_{ij}\mathbf{E}$$

Substitution (2.1.3) into the first equation of the set (2.1.2) leads to the expression (1.2.20) for w with $n_\omega = 0$. Hence, $w = 1/\tau_j$, where τ_j is the lifetime of state j. Correspondingly, from (2.1.3), the probability for emission of a photon of a frequency ω is

$$a_\omega = |c_\omega(\infty)|^2 = \frac{1}{2\pi\tau_j}\frac{1}{\left(\omega_{ji}-\omega\right)^2 + \left[1/\left(2\tau_j\right)\right]^2} \qquad (2.1.4)$$

In derivation formula (2.1.4), we integrated over the solid angle of emitted photons and summed over their polarizations. This formula is valid under condition $\omega_{ji}\tau_j \gg 1$. Note that we neglect a small shift of the spectral line, so that formula (2.1.4) with the Lorenz shape of spectral lines describes the radiative mechanism of broadening.

In the case of two sources of the Lorentz line broadening with widths of $1/\tau_1$ and $1/\tau_2$, which relate to the upper and lower states, we find the shape and width of the compound line. In a general case, if we have two independent broadening mechanisms with the distribution functions $a_1(\omega)$ and $a_2(\omega)$, the distribution function for the compound line has the form

$$a_\omega = \int_{-\infty}^{\infty} a_1(\omega')a_2(\omega-\omega')\mathrm{d}\omega', \qquad (2.1.5)$$

Applying this expression several times, it is possible to take into account the arbitrary number of broadening mechanisms. Obviously, the shape of the final spectral line does not depend on the number of mechanisms under consideration.

Using formulas (2.1.4) and (2.1.5), we have

$$a_\omega = \frac{1}{4\pi^2\tau_1\tau_2}\int_{-\infty}^{\infty}\frac{\mathrm{d}\omega'}{\left[\left(\omega_{ji}-\omega'\right)^2 + [1/(2\tau_1)]^2\right]\left[\left(\omega_{ji}-\omega'\right)^2 + [1/(2\tau_2)]^2\right]} \qquad (2.1.6)$$

The integrand in this expression can be rewritten as

$$\frac{1}{\omega_{k0}-\omega-\frac{i}{2\tau_1}-\frac{i}{2\tau_2}}\left\{\frac{1}{\omega'-\omega_{k0}+\frac{i}{2\tau_1}} - \frac{1}{\omega'-\omega+\frac{i}{2\tau_2}}\right\}\times$$
$$\times\frac{1}{\omega_{k0}-\omega+\frac{i}{2\tau_1}+\frac{i}{2\tau_2}}\left\{\frac{1}{\omega'-\omega_{k0}-\frac{i}{2\tau_1}} - \frac{1}{\omega'-\omega-\frac{i}{2\tau_2}}\right\}.$$

The integral can then be accomplished in the complex plane by employing a path that encloses the two poles in the upper half plane to give the result

$$a_\omega = \frac{1}{2\pi\tau_{12}} \frac{1}{\left(\omega_{ji} - \omega\right)^2 + [1/(2\tau_{12})]^2}, \tag{2.1.7}$$

where the spectral width of the compound line is given by

$$\frac{1}{\tau_{12}} = \frac{1}{\tau_1} + \frac{1}{\tau_2} \tag{2.1.8}$$

Thus, the shape of the compound line is also the Lorentz one, and its width is equal to sum of widths of each lines. This result can be directly generalized on the arbitrary number of sources of the Lorentz broadening. The compound line has also the Lorenz profile, and its width is equal to the sum of widths of individual lines. In particular, one can generalize formula (2.1.4) of the above formula to the case where an excited state j may decay spontaneously not only into the state i, but also into other atomic states. Then the quantity w in the expression for $c_j(t) = \exp(-wt/2)$ is the total width of the state j. It is in accordance with first equation of the set (2.1.2), where we should add terms due to transitions into other states. As a result, we obtain again (2.1.2), where now the quantity τ_j takes into account transitions in all the states. A similar generalization can be made in the case if the final state i is not the ground one, but has some total width τ_i^{-1} of the spontaneous decay into other lower atomic states. According to formula (2.1.7) we obtain that the distribution function has the following form instead of (2.1.4)

$$a_\omega = \frac{1}{2\pi\tau_{ji}} \frac{1}{\left(\omega_{ji} - \omega\right)^2 + \left[1/\left(2\tau_{ji}\right)\right]^2} \tag{2.1.9}$$

Here the quantity

$$\frac{1}{\tau_{ji}} = \frac{1}{\tau_j} + \frac{1}{\tau_i} \tag{2.1.10}$$

defines the reduced width of the spectral line and the reduced lifetime τ_{ji}.

Another mechanism of broadening of spectral lines is determined by motion of a radiative particle. The Doppler broadening of spectral lines results from different velocities of the radiative source and receiver. We assume the radiation receiver to be motionless and the frequency of the emitted photon depends on the velocity of a radiating particle. Thus, the frequency distribution of the emitted photons is determined by the velocity distribution of radiating particles. We assume the emitted frequency is fixed strictly. Then the distribution function of received photons over frequencies is determined by the distribution of emitted particles over velocities. The basis of Doppler broadening of spectral lines is the Doppler law according to which a detected frequency ω is expressed through the frequency ω_0 of an emitted photon and the velocity of the radiation source v by the relation

$$\omega = \omega_0 \left(1 + v_x/c\right), \tag{2.1.11}$$

where c is the speed of light, and v_x is the projection of the relative velocity of the radiator and receiver onto the direction of the photon propagation. This formula corresponds to approach of the radiator and receiver.

In order to determine the frequency distribution of the emitted photons, we transfer to the laboratory frame of reference. The probability that an atom has a velocity in a range from v_x up to $v_x + \mathrm{d}v_x$ is $v_x \mathrm{d}v_x$, which is normalized to unity

$$\int\limits_{-\infty}^{\infty} f\left(v_x\right) \mathrm{d}v_x = 1.$$

On the basis of formula (2.1.11) connected the radiation frequency and the atomic velocity, one can obtain the distribution function over photon frequencies ω

$$a_\omega \mathrm{d}\omega = f(v_x)\mathrm{d}v_x = \frac{c}{\omega_0} f\left(\frac{\omega - \omega_0}{\omega_0}\right) \mathrm{d}\omega \tag{2.1.12}$$

Specifically, in the case of the Maxwell velocity distribution, the distribution function is

$$f\left(v_x\right) = \sqrt{\frac{M}{2\pi T}} \exp\left(-\frac{M v_x^2}{2T}\right),$$

where M is the mass of an emitting atom and T is the temperature expressed in energy units. Then formula (2.1.12) gives

$$a_\omega \mathrm{d}\omega = \frac{1}{\omega_0}\sqrt{\frac{Mc^2}{2\pi T}} \exp\left[-\frac{Mc^2\left(\omega - \omega_0\right)^2}{2\omega_0^2 T}\right] \tag{2.1.13}$$

From this expression, a typical Doppler width may be estimated as

$$\Delta\omega_D = \omega_0 \cdot \alpha, \;\; \alpha = \sqrt{\frac{T}{Mc^2}} \tag{2.1.14}$$

In order to understand the scale of values for the Doppler width$<$ we consider a certain case of sodium atoms located in a buffer gas at the temperature $T = 500\,\mathrm{K}$ and the radiative transition Na$((3s \to 3p)$ of the wavelength $\lambda = 589\,\mathrm{nm}$ (the frequency $\omega = 3.2 \cdot 10^{16}\,\mathrm{s}^{-1}$) which is characterized by the radiative lifetime of $\tau = 16\,\mathrm{ns}$ (see Fig. 1.4). In this case we have $\alpha = 1.4 \cdot 10^{-6}$, $\omega\tau = 5.1 \cdot 10^7$, that gives for the relation between the Doppler and radiative widths of spectral lines $\Delta\omega_D\tau = 72$.

As is seen, the Doppler width of the spectral line is larger than that due to the radiative one.

In considering the Doppler broadening of a spectral line, we assume an emitted atom to be moved as a free particle with a constant velocity. We now consider another case with the diffusion character of motion for a radiating particle, so that an excited atom experiences many collisions with other atoms during its radiative lifetime τ_{ji}. Also we assume that the line broadening is caused by the Doppler effect. In determination the shape of the spectral line, we fix the origin of coordinates $x_0 = 0$ on an atom at time $t = 0$ and introduce the probability $W(x_0, t)$ that this atom will be found at a distance x_0 at time t. We first study the one-dimensional atomic motion, so that the probability $W(x_0, t)$ satisfies the diffusion equation

$$\frac{\partial W}{\partial t} = D\frac{\partial^2 W}{\partial x^2}, \tag{2.1.15}$$

where D is the diffusion coefficient of an atom in the gas. The probability W is normalized by the condition

$$\int\limits_{-\infty}^{\infty} W(x_0, t)\,\mathrm{d}x_0 = 1$$

The electric field strength for the electromagnetic wave emitted spontaneously by an atom located at a coordinate x at time t, has the form

$$\mathbf{E} = \mathbf{E}_0 \exp\left[ik(x - x_0) - i\omega_{ji}t\right],$$

where the wave vector $k = \omega_{ji}/c$. Averaging this field strength over positions of the radiating atom, one can obtain

$$\overline{\mathbf{E}} = \mathbf{E}_0 \int\limits_{-\infty}^{\infty} W(x_0, t)\,\mathrm{d}x_0 = \mathbf{E}_0 J(t) \exp\left[ikx - i\omega_{ji}t\right],$$

where

$$J(t) = D\int\limits_{-\infty}^{\infty} \exp(-ikx_0)\, W(x_0, t)\,\mathrm{d}x_0$$

In order to find the function $J(t)$, let us multiply equation (2.1.15) by $\exp(-ikx_0)$ and integrate it over positions x_0. This gives

$$\frac{\mathrm{d}J}{\mathrm{d}t} = D\int\limits_{-\infty}^{\infty} \exp(-ikx_0)\frac{\partial^2 W(x_0, t)}{\mathrm{d}x_0^2}\mathrm{d}x_0$$

Integration by parts two times, with application of the boundary conditions $W(x_0 \to \pm\infty) = 0$, gives

$$\frac{\mathrm{d}J}{\mathrm{d}t} = -Dk^2 t$$

The solution of this equation with the initial condition $J(0) = 1$ (since $x_0(0) = 0$) is

$$J(t) = \exp\left(-Dk^2 t\right)$$

From this it follows that the average electric field strength of the emitted spontaneous photon is

$$\bar{\mathbf{E}} = E_0 \exp\left[ikx - i\omega_{ji}t - Dk_2 t\right]$$

In a classical treatment, this field is generated by a dipole moment that oscillates with a frequency ω_{k0} and depletes with time, that is, the dipole moment behaves as

$$\mathbf{d} \sim \exp\left[ikx - i\omega_{ji}t - Dk_2 t\right]$$

According to formula (1.2.20) for the rate of a radiative transition and, hence, the spectral distribution function a_ω, is determined by the square module of a Fourier component at a frequency ω for a dipole moment $\mathbf{d}$. The quantity a_ω may be expressed through the square of the second derivative for the Fourier component of the dipole moment (1.2.30) that (under the condition $\mathbf{d} = \mathbf{0}$ at $t < 0$) gives for the frequency distribution function which is normalized to unity

$$a_\omega = \frac{1}{\pi}\frac{k^2 D}{\left(\omega - \omega_{ji}\right)^2 + \left(k^2 D\right)^2}, \qquad (2.1.16)$$

As is seen, the distribution function has the Lorentz form, by analogy to the case of radiative broadening. The spectral width is equal to $2k^2 D$. Combining this effect with a finite lifetime τ_{ji} of the initial state due to the radiative transition $j \to i$, one can obtain the Lorentz spectral shape of the combined line with the width $2k^2 D + 1/2\tau_{ji}$.

Note that for the broadening mechanism under consideration only the atom motion in the direction of the photon propagation is significant, even in the case of the three-dimensional diffusion. There is an analogy for the Doppler broadening and that where the transverse motion of the radiating atom accompanies its displacement. Next, the criterion of validity of formula (2.1.16) is that a time τ_{ji} is large compared to the time of the mean free path of a radiating atom in a gas. In particular, this condition is fulfilled in masers. The opposite limit leads to the Doppler broadening considered above. It should be noted that the diffusion coefficient D in formula (2.1.16) is proportional to the mean thermal atomic velocity, i.e. to the square root of the gas temperature.

Let us find the spectral line shape resulting from simultaneous action of Doppler broadening and radiative broadening due to finite lifetime of the state, basing on formula (2.1.5). The central frequency $\omega_0 = \omega_{ji}$ is shifted according to the Doppler law because of the particle motion. Taking this shift into account and averaging over the velocity distribution of radiating particles, we obtain the general expression for the spectral line shape

$$a_\omega = \frac{1}{2\pi\tau_{ji}} \frac{c}{\omega_{ji}} \int_{-\infty}^{\infty} \frac{f\left[c\left(\omega' - \omega_{ji}\right)/\omega_{ji}\right]}{(\omega - \omega')^2 + \left(1/2\tau_{ji}\right)^2} d\omega' \tag{2.1.17}$$

Taking $f(v)$ the Maxwell velocity distribution function, we analyze limiting cases. We first consider the case where along with the broadening of the spectral line due to the radiative transition (2.1.4), its Doppler shift takes place, and the Doppler width is small compared to the natural (radiative) width

$$\Delta\omega_D = \omega_{ji}\sqrt{\frac{T}{Mc^2}} << \frac{1}{\tau_{ji}}.$$

Then on the basing of the replacement

$$\frac{c}{\omega_{ji}} f\left[\frac{c(\omega' - \omega_{ji}}{\omega_{ji}}\right] \rightarrow \delta(\omega' - \omega_{ji})$$

we arrive at formula (2.1.4). In the opposite limiting case the relation $\Delta\omega_D\tau_{ji} \gg 1$ the Doppler broadening (2.1.13) takes place for the central part of the spectral line. In this case it is convenient to separate the integral (2.1.17) in two parts, namely, $\omega' - \omega_{ji} \sim \Delta\omega_D$ and $\omega' - \omega_{ji} \sim 1/\tau_{ji}$. The contribution from the first range gives

$$\frac{1}{2\pi\tau_{ji}} \frac{1}{\left(\omega - \omega_{ji}\right)^2},$$

while the integral from the second range of formula (2.1.17) is equal to

$$\frac{1}{\sqrt{\pi}\Delta\omega_D} \exp\left[-\frac{\left(\omega - \omega_{ji}\right)^2}{(\Delta\omega_D)^2}\right]$$

Thus, the Doppler broadening of the spectral line takes place in a range not far from the central part, whereas if the following relation holds true

$$\exp\left[-\frac{\left(\omega - \omega_{ji}\right)^2}{(\Delta\omega_D)^2}\right] \ll \frac{1}{\tau_{ji}} \frac{\Delta\omega_D}{\left(\omega - \omega_{ji}\right)^2},$$

then the distribution function has the form

$$a_\omega = \frac{1}{2\pi\tau_{ji}} \frac{1}{\left(\omega - \omega_{ji}\right)^2}$$

Thus, in the case of competition between Lorentz and Doppler broadening of spectral lines, wings of the spectral lines are always determined by the Lorentz mechanism, whereas the central part of the spectral line is determined by the short-range mechanism of broadening.

If two sources of broadening lead of the Gaussian shape of spectral lines with widths Δ_1 and Δ_2, the frequency distribution function for the combined spectral line has the form

$$a_\omega = \frac{1}{\pi\Delta_1\Delta_2} \int\limits_{-\infty}^{\infty} \exp\left[-\frac{\left(\omega' - \omega_{ji}\right)^2}{\Delta_1^2} - \frac{\left(\omega' - \omega\right)^2}{\Delta_2^2}\right] \mathrm{d}\omega' = \frac{1}{\sqrt{\pi}\Delta} \exp\left[-\frac{\left(\omega - \omega_{ji}\right)^2}{\Delta^2}\right], \tag{2.1.18}$$

where $\Delta = \sqrt{\Delta_1^2 + \Delta_2^2}$. Thus the combine line has the Gaussian shape also, and its width is the root mean square of the partial widths.

2.1.2 Collision Broadening of Spectral Lines

The collision or impact broadening of spectral lines follows from a short-time interaction of a radiating atomic particle with surrounding gaseous particles in collisions. This is valid if the radiative lifetime of the upper state of the radiative transition $j \to i$ is large compared to a time between neighboring collisions involving a radiating atomic particles. In order to determine the width of a spectral line for this mechanism of broadening, let us expand the wave functions of transition states over stationary states as

$$\Psi_i = \sum_n c_{in}(t)\psi_n \exp\left(-i\varepsilon_n t\right)/\hbar; \ \Psi_j = \sum_m c_{jm}(t)\psi_m \exp\left(-i\varepsilon_m t\right)/\hbar, \tag{2.1.19}$$

where ψ_n, ψ_m are spatial parts of stationary wave functions of the the radiating atomic particle, and ε_n, ε_m are energies of these states, and the initial conditions for these amplitudes are $a_{in}(0) = \delta_{in}$, $a_{jm}(0) = \delta_{jm}$. The amplitudes $c_{in}(t)$, $c_{jm}(t)$ vary in time due to collisions with atoms or molecules of a surrounding gas.

Let us introduce the spectral function as

$$\varphi(t) \equiv f_{nm}(t) = c_{in}(t)c_{jm}^*(t) \exp\left[i\left(\varepsilon_j - \varepsilon_i\right)/\hbar\right], \tag{2.1.20}$$

and define the correlation function by the expression

$$\Phi\left(\tau\right)=\lim_{T\to\infty}\left[\frac{1}{T}\int\limits_{0}^{T}\varphi(t)^{*}(t)\varphi(t+\tau)\mathrm{d}t\right] \tag{2.1.21}$$

Let us express the photon frequency distribution function a_ω in terms of the correlation function $\Phi\left(\tau\right)$. The interaction (1.2.3) between an atomic electron and the radiation field has the dipole character, and the transition amplitude is proportional to the matrix element of the dipole moment operator between the initial and final states in the transition, that is,

$$\left\langle\Psi_{i}\left|\mathbf{D}\right|\Psi_{j}\right\rangle=\sum_{m,n}\mathbf{D}_{mn}c_{in}(t)c_{jm}^{*}(t)\exp\left[i\left(\varepsilon_{n}-\varepsilon_{m}\right)/\hbar\right], \tag{2.1.22}$$

where $\mathbf{D}_{mn}=\left\langle\psi_{n}\left|\mathbf{D}\right|\psi_{m}\right\rangle$. The Fourier component of formula (2.1.22) gives the amplitude for the emission of a photon of a given frequency. The probability for radiation of a photon of frequency ω is therefore proportional to

$$a_{\omega}\sim\left|\int\exp\left(i\omega t\right)\left\langle\Psi_{i}\left|\mathbf{D}\right|\Psi_{j}\right\rangle\mathrm{d}t\right|^{2} \tag{2.1.23}$$

Taking into account that the frequency range near ω_{ji} is of interest, one can restrict in formula (2.1.22) by terms with $m=j$ and $n=i$. Then on the basis of formula (2.1.19) and (2.1.20) we obtain

$$a_{\omega}\sim\left|\int\exp\left(i\omega t\right)\varphi\left(t\right)\mathrm{d}t\right|^{2} \tag{2.1.24}$$

This integral can be written in the form

$$a_{\omega}\sim\int\mathrm{d}t_{1}\int\mathrm{d}t_{2}\varphi\left(t_{1}\right)\varphi^{*}\left(t_{2}\right)\exp\left[i\omega\left(t_{1}-t_{2}\right)\right]$$

Introducing new variables $\tau=t_1-t_2$ and $t=t_2$, we have

$$a_{\omega}\sim\int\mathrm{d}\tau\exp\left(i\omega\tau\right)\int\mathrm{d}t\varphi\left(t\right)\varphi^{*}\left(t+\tau\right)$$

On the basis of the definition of the correlation function (2.1.21), we obtain

$$a_{\omega}=\frac{1}{2\pi}\int\limits_{-\infty}^{\infty}\Phi\left(\tau\right)\exp\left(i\omega\tau\right)d\tau, \tag{2.1.25}$$

where the normalization condition (2.1.1) for the distribution function a_ω is fulfilled. The relation (2.1.25) connects the frequency distribution function a_ω of emitted photons with the Fourier product of the correlation function for a given frequency of the radiating photon.

Our goal is to determine the shape of a spectral line if it is determined by collisions of an excited atom with atoms of a surrounding gas. This broadening mechanism for spectral lines allows us to express the photon distribution function a_ω through scattering phases in collisions of a radiating atom with gas atoms. Assuming atom motion to be classical, one can consider the wave function of a radiating atom in the initial excited state j as the solution of the Schrödinger equation

$$i\hbar\frac{\partial\Psi_j}{\partial t} = \varepsilon_j\Psi_j + V_{jj}(t)\Psi_j, \tag{2.1.26}$$

where $V_{jj}(t)$ is the diagonal matrix element of the interaction potential of the radiating atom with surrounding atomic particles of a gas. We neglect non-diagonal matrix elements which result in transitions from the state j to other atomic states. Then the Schrödinger equations for different states of a radiating atom are independently. On the basis of equation (2.1.26), we obtain for the amplitude $c_{jj}(t)$, defined by formula (2.1.19)

$$a_{jj}(t) = \exp\left[-\frac{i}{\hbar}\int\limits_{-\infty}^{t} V_{jj}\left(t'\right)\mathrm{d}t'\right] \tag{2.1.27}$$

Interaction between a radiating atom and perturbed gas atoms takes place during a short time and proceeds randomly. Let t_i be a time of i-th collision, $\mathbf{R}_i$ be a distance between a radiating atom and perturbed atomic particle in i-th collision. In considering quantum states of a radiating atom, we assume the motion colliding particles to be classical. As a result of a single collisions, the wave function of the initial state of a radiating atom acquires an additional phase

$$\chi_j^{(i)} = \frac{1}{\hbar}\int\limits_{-\infty}^{\infty} V_{jj}\left(\mathbf{R}_i\right)\mathrm{d}t' \tag{2.1.28}$$

This allows one to represent the amplitude (2.1.19) with accounting for collisions in the form

$$c_{jj}(t) = \exp\left[-i\sum_i \chi_j^{(i)}\eta\left(t-t_i\right)\right], \tag{2.1.29}$$

where the step Heaviside function $\eta(t)$ is defined as

$$\eta(t) = 0,\ t < 0;$$
$$\eta(t) = 1,\ t > 0.$$

This function accounts for a collision time smallness compared to a radiative lifetime. Formula (2.1.27) gives for the correlation function $\varphi(t)$

$$\varphi(t) = \exp\left[i\omega_{ji}t + i\sum_i \chi_i \eta(t - t_i)\right], \tag{2.1.30}$$

where

$$\chi_i = \chi_i(k) - \chi_i^{(0)} = \frac{1}{\hbar}\int_{-\infty}^{\infty}\left[V_{jj}(\mathbf{R}_i) - V_{ii}(\mathbf{R}_i)\right]\mathrm{d}t' \tag{2.1.31}$$

The quantity χ_i is the phase shift introduced by the difference between the interaction potentials for the upper and lower transition states with taking into account that collision times t_i are identical for states j and i.

The correlation function defined by formula (2.1.21) may be written as

$$\Phi(\tau) = \langle \varphi^*(t)\varphi(t+\tau)\rangle,$$

where a time average is carried out on time t. To find the correlation function $\Phi(\tau)$, we combine the combination of the correlation functions

$$\begin{aligned}\Delta\Phi(\tau) &= \Phi(\tau) - \exp\left(-i\omega_{ji}\Delta\tau\right)\Phi(\tau + \Delta\tau) = \\ &= \left\langle \varphi^*(t)\left[\varphi(t+\tau) - \exp\left(-i\omega_{ji}\Delta\tau\right)\varphi(t+\tau+\Delta\tau)\right]\right\rangle\end{aligned} \tag{2.1.32}$$

A time range $\Delta\tau$ is small compared to a typical time τ of collision broadening, which is given by the free path flight time of excited atom. However, $\Delta\tau$ is large compared to a collision time. Because of $\Delta\tau \ll \tau$, only one collision takes place during $\Delta\tau$ (or no collisions), and the probability of two collisions is negligibly small. Then expressing the quantity $\varphi(t+\tau+\Delta\tau)$ via $\varphi(t+\tau)$ on the basis (2.1.30), one can represent (2.1.32) in the form

$$\Delta\Phi(\tau) = \left\langle \varphi^*(t)\varphi(t+\tau)\left[1 - \exp\left(i\sum_i \chi_i\left[\eta(t+\tau+\Delta\tau - t_i) - \eta(t+\tau - t_i)\right]\right)\right]\right\rangle, \tag{2.1.33}$$

where the summation takes place over times t_i during a range between $t+\tau$ and $t+\tau+\Delta\tau$. Because of a random character of the collisions, one can average over various collisions independently. This means that in consideration of a given collision one can neglect previous collisions. Therefore the value (2.1.33) may be represented as a product of individual averages, that gives

$$\Delta\Phi(\tau) = \Phi(\tau)\langle 1 - \exp[i\chi(\rho)]\rangle;\ t+\tau < t_i < t+\tau+\Delta\tau, \tag{2.1.34}$$

where ρ is the impact parameter of collision which according to (2.1.31) determines the value χ. According to (2.1.34), an averaging over time t is equivalent to averaging over collision times t_i in a narrow time range. Since in the classical limit the impact parameter of collision is unambiguously connected with t_i, it is convenient to carry out the averaging in terms of ρ.

The volume related to one perturbed particle is equal $1/N_b$, where N_b is the number density of the gas atoms. The volume element in integration is $v\Delta\tau 2\pi\rho d\rho$, where v is the relative velocity of colliding particles. Thus, the averaging factor under conditions (2.1.34) may be rewritten in the form

$$\langle 1 - \exp[i\chi(\rho)]\rangle = v\Delta\tau N \int_0^\infty 2\pi\rho \mathrm{d}\rho \{1 - \exp[i\chi(\rho)]\} \quad (2.1.35)$$

In the limit $\Delta\tau \to 0$ in formula (2.1.32) one can obtain on the basis of formulas (2.1.34) and (2.1.35)

$$\begin{aligned} \Delta\Phi(\tau) &= \Phi(\tau) - \exp\left(-i\omega_{ji}\Delta\tau\right)\left[\Phi(\tau) + \frac{\mathrm{d}\Phi(\tau)}{\mathrm{d}\tau}\Delta\tau\right] = \\ &= -\Delta\tau\left[\frac{\mathrm{d}\Phi(\tau)}{\mathrm{d}\tau} - i\omega_{ji}\Phi(\tau)\right] = \\ &= \Phi(\tau)\,\Delta\tau N v \int_0^\infty 2\pi\rho \mathrm{d}\rho \{1 - \exp[i\chi(\rho)]\} \end{aligned} \quad (2.1.36)$$

This is equivalent to

$$\frac{\mathrm{d}\Phi(\tau)}{\mathrm{d}\tau} - i\omega_{ji}\Phi(\tau) = -\Phi(\tau)\,N_b v\left(\sigma' + i\sigma''\right) \quad (2.1.37)$$

with notations

$$\sigma' = \int_0^\infty 2\pi\rho \mathrm{d}\rho(1 - \cos\chi); \;\; \sigma'' = \int_0^\infty 2\pi\rho \mathrm{d}\rho \sin\chi \quad (2.1.38)$$

The solution of equation (2.1.37) with using these cross sections is

$$\Phi(\tau > 0) = \exp\left[i\omega_{ji}\tau - Nv\left(\sigma' + i\sigma''\right)\tau\right] \quad (2.1.39)$$

Substituting formula (2.1.39) into (2.1.25) and taking $\Phi(\tau < 0) = \Phi^*(\tau > 0)$, one can obtain for the photon frequency distribution function

$$a_\omega = \frac{\nu}{2\pi}\frac{1}{\left(\omega - \omega_{ji} + \Delta\nu\right)^2 + (\nu/2)^2}, \;\; \nu = Nv\sigma'; \;\; \Delta\nu = Nv\sigma'' \quad (2.1.40)$$

Formula (2.1.40) exhibits the Lorenz form of the spectral line. In contrast to the radiative broadening (2.1.21), a shift of spectral line $\Delta\nu = Nv\sigma$" takes place in this case, and its value is comparable with the width of the spectral line.

Let us consider the impact theory of line broadening for the dispersion interaction potential $U(R)$ radiating and perturbed atoms in the form

$$U(R) = -\frac{C_6}{R^6} \tag{2.1.41}$$

as it takes place at large distances R between neutral atomic particles. Then the total cross section of their collision is equal [1]

$$\sigma_t = 8.1\left(\frac{C_6}{\hbar v}\right)^{2/5} \tag{2.1.42}$$

Averaging over collision velocities with the Maxwell distribution function, one can obtain for the specific width of the spectral line in accordance with the Lindholm-Foley theory [2, 3]

$$\frac{\nu}{N_b} = 7.2\left(\frac{2T}{\mu}\right)^{3/10}\left(\frac{C_6}{\hbar}\right)^{2/5}, \tag{2.1.43}$$

where T is the gas temperature expressed in energetic units, and μ is the reduced mass of colliding molecules.

In order to estimate the scale of this broadening, let us return to the above example of sodium atoms located in a buffer gas at the temperature $T = 500\,\mathrm{K}$, and take neon as a buffer gas. Then the constant of dispersion interaction is $C_6 \approx 50e^2a_o^5$ (e is the electron charge, a_o is the Bohr radius), so that formula (2.1.43) gives at a temperature $T = 500\,\mathrm{K}$ for the specific width of the spectral line $\nu_b/N_b = 1.9\cdot10^{-9}\,\mathrm{cm}^3/\mathrm{s}$. Comparing with Doppler broadening which width for this case according to formula (2.1.14) is equal $\Delta\omega_D = 4.5\cdot10^9\,\mathrm{s}^{-1}$. The Doppler and collision widths are equalized in this case at the number density of neon atoms $N_b = 2.4\cdot10^{18}\,\mathrm{cm}^{-3}$, that corresponds to the neon pressure $p = 37$ Torr. Note that we assume the concentration of sodium atoms in neon to be small, that is, interaction between excited and nonexcited sodium atoms is ignored. In addition, we determine the average total cross section of scattering of a radiating sodium atom on a neon atom which is defined as

$$\sigma_t(T) = \frac{\nu}{\overline{(v)N_b}},$$

This gives $\sigma_t(T) = 2.8\cdot10^{-14}\,\mathrm{cm}^2$ for the case under consideration.

In the range of competition between Doppler and collision mechanisms of line broadening, so called the Voigt profile [4], in principle, is analogous to competition between the Doppler and radiative broadening mechanisms, where the frequency distribution function is given by formula (2.1.17), though the velocity dependence of the collision frequency change the classical Voigt profile [5–8]. The tables are composed [9] for the line profile in this range because this is of importance in reality. The general conclusion, as it was shown above, wings of the spectral line are determined by the collision mechanism of broadening.

2.1.3 Quasistatic Broadening of Spectral Lines

The quasi-static theory of broadening of spectral lines corresponds to physical conditions opposite to the collision mechanism of line broadening. In the quasistatic case, broadening is created during times that are small compared to times of atom motion. This allows us to consider the perturbed atoms to be motionless, and the spectral frequency shift resulted from interaction of a radiating atom with surrounding atoms in a gas is a sum of shifts due to pairwise interactions with each perturbed atom at the fixed spatial configuration of these atoms, and this interaction energy is small compared to a typical atomic energy. The perturbation theory is thus appropriate for treatment of this broadening. In addition, the pairwise interaction is averaged over states of gaseous particles.

A general goal of the quasi-static theory of spectral line broadening consists in determination the frequency distribution function of emitted photons a_ω; let us introduce the probability $w(R)$ for a perturbed atom to be located at a coordinate $\mathbf{R}$ per unit volume with respect to a radiating atom. In the first order perturbation theory, the frequency shift for an emitted photon due to interaction of the radiating atom particle with neighboring perturbing particles is

$$\omega_{ji} - \omega = -\frac{1}{\hbar}\sum_m U\left(\mathbf{R}_m\right) \quad (2.1.44)$$

Here, an index m described m-th perturbed atom, $\mathbf{R}_m$ is the coordinate of m-th perturbed atom with respect to a radiating atom, and $U\left(R_m\right)$ is the difference between the interaction potentials for the upper and the lower states of this radiative transition. Equation (2.1.44) is averaged over the quantum states of both particles, that is, it is the diagonal matrix element of the pairwise interaction between atoms.

Note that formula (2.1.44) implies the pairwise character for interaction of a radiating and gaseous atoms, where interaction with some perturbed atom does not influence on that with other one. This testifies about weak pair interactions which do not change the state of a radiating atom. This condition is also the criterion of the gaseousness of the system. From this one can estimate the width of the spectral line $\Delta\omega_S$ under the above conditions

$$\Delta\omega_S \sim U(N_b^{-1/3})\hbar \quad (2.1.45)$$

Equation (2.1.44) refers to a fixed coordinate $\mathbf{R}_m$ of a radiating atom. If this distance varies, the frequency shift varies also changes due to a change of the interaction potential $U(\mathbf{R}_m) = U_m$. Let us introduce the probability $\rho(U_m)\,dU_m$ that the pairwise interaction potential lies in an interval from U_m up to $U_m + dU_m$ for a perturbed particle described by an index m. According to definition of the distribution function of emitted photons a_ω, we have

$$a_\omega d\omega = \prod_m \rho(U_m)\,dU_m, \tag{2.1.46}$$

where it is assumed the pairwise character of a radiating atom with perturbed ones. According to definition, we have

$$\rho(U_m)\,dU_m = w(\mathbf{R}_m)\,d\mathbf{R}_m \tag{2.1.47}$$

Taking a normalization volume, where a radiating particle is located, to be one ($w = 1$), we have that a perturbed atom is located in spatial element $d\mathbf{R}_m$.

Let us evaluate first the Fourier component from the frequency distribution function a_ω that on the basis of formulas (2.1.44) and (2.1.46) gives

$$\mu(t) = e\int_{-\infty}^{\infty} \exp\left[i\left(\omega - \omega_{ji}\right)t\right] a_\omega d\omega = \prod_m \int \exp\left(\frac{iU_m t}{\hbar}\right)\rho(U_m)\,dU_m \tag{2.1.48}$$

Let us introduce the correlation function $F(t)$ as

$$F(t) = \int \exp\left(\frac{iUt}{\hbar}\right)\rho(U)\,dU$$

According to formula (2.1.47), this quantity may be represented in the form

$$F(t) = \int \exp\left(\frac{iU(R)\,t}{\hbar}\right) w(R)\,d\mathbf{R} \tag{2.1.49}$$

Combining formulas (2.1.48) and (2.1.49), we have

$$\mu(t) = \prod_m F(t)$$

If N_b is the number density of perturbed particles, the number of these particles in a normalization volume Ω is $N_b\Omega$

$$\mu(t) = F^{N_b\Omega}(t) = \lim\left\{1 + \frac{N_b}{N_b\Omega}\int\left[\exp\left(\frac{iU(R)\,t}{\hbar}\right) - 1\right] w(R)\,d\mathbf{R}\right\}^{N_b\Omega}$$
$$= \exp\left\{N_b\int\left[\exp\left(\frac{iU(R)\,t}{\hbar}\right) - 1\right] w(R)\,d\mathbf{R}\right\} \tag{2.1.50}$$

Returning to the frequency distribution function a_ω through the inverse Fourier transformation, we obtain

$$a_\omega = \frac{1}{2\pi}\int\limits_{-\infty}^{\infty} \exp\left[-i\left(\omega-\omega_{ji}\right)t\right]\mu\left(t\right)\mathrm{d}t$$
$$= \frac{1}{2\pi}\int\limits_{-\infty}^{\infty} \exp\left\{-i\left(\omega-\omega_{ji}\right)t + N_b\int\left[\exp\left(\frac{iU\left(R\right)t}{\hbar}\right)-1\right]w\left(R\right)d\mathbf{R}\right\}\mathrm{d}t \tag{2.1.51}$$

The same result may be obtained by the other way, by analogy with the method of determination of the correlation function $F(t)$ for the collision broadening mechanism. Namely, one can write the equation for variation of $\mu\left(t\right)$ for small time Δt, which is described by the differential equation for $\Delta t \to 0$. Solving of this equation leads to formulas (2.1.50) and (2.1.51). Note that the quantity $U(R)$ in expression (2.1.51) is identical to the corresponding quantity in the case of collision broadening.

As is seen, a general expression (2.1.51) for the distribution function a_ω has a complicated character and does not correspond to the Lorentz profile. Only if the exponent

$$\exp\left(\frac{iU\left(R\right)t}{\hbar}\right)$$

of formula (2.1.51) is expanded in the Taylor series, one can obtain the Lorentz line shape with zero width $\nu = 0$ and the Stark shift

$$\Delta\nu = -\frac{N_b}{\hbar}\int U(R)w\left(R\right)\,d\mathbf{R} \tag{2.1.52}$$

In particular, if we assume in formula (2.1.52) $w = 1$ and $R = vt$ which corresponds to a straight-line classical trajectory with $U \ll E$), formula (2.1.52) leads to formula (2.1.37), if $\sin\chi$ is replaced by χ. Thus, within the framework of the perturbation theory, the impact and quasi-static theories of the spectral line broadening lead to the same result. The second-order of the perturbation theory on the basis of (2.1.51) gives a nonzero addition to the broadening cross section, but it is not a Lorentz form. The additional term is proportional to t^2, whereas it must be proportional to t if it is to give rise to a Lorentzian line shape.

Let us give the criteria of the collision and quasistatic broadening mechanisms for free motion of interacted particles, that gives $\mathrm{d}t = dR/v$ where v is the relative velocity of colliding particles. Because of a weak interaction, a shift phase defined by formula (2.1.31), is $\chi \ll 1$, and then the broadening cross section in formula (2.1.37) is small compared to the line shift cross section. Because in this case according to formula (2.1.31) $\chi \sim RU(R)/\hbar v$, where R is a typical distance of approach of colliding particles, and smallness of the phase shift that is the criterion of collision broadening is

$$\frac{U(R)R}{\hbar v} \ll 1$$

In reality, the impact theory of broadening of spectral lines is valid, if a collision time ρ/v is small, where v is the relative velocity of colliding particles, and ρ is a typical impact parameter. The impact parameter can be estimated to be given by the Weisskopf radius $\sqrt{\sigma_t}$, where σ_t is the total elastic scattering cross section for interacting particles. The collision theory of broadening of spectral lines is valid if a collision time is small compared to the flight time of the mean free path for the colliding particles, that is $(N v \sigma_t)^{-1}$. A collision time should be also short compared to a detection time for the frequency shift the spectral line, which is given by the Heisenberg uncertainty principle as $\left|\omega - \omega_{ji}\right|^{-1}$. Thus the collision theory of spectral line broadening requires fulfilment of criteria

$$\frac{v}{\sqrt{\sigma_t}} \gg \max\left[\left|\omega - \omega_{ji}\right|, \; N v \sigma_t\right] \tag{2.1.53}$$

In particular, the criterion of the collision theory at wings of a spectral line has the form

$$\left|\omega - \omega_{ji}\right| \ll \frac{v}{\sqrt{\sigma_t}} \tag{2.1.54}$$

For the central part of the spectral line we have

$$\left|\omega - \omega_{ji}\right| \ll \nu \sim N_b v \sigma_t,$$

and the collision theory of line broadening is valid if

$$N_b \sigma_t^{3/2} \ll 1 \tag{2.1.55}$$

As is seen, the impact theory of spectral lines is not correct for far wings of a spectral line.

Let us apply this to the interaction potential $U(R) = -C_6/R^6$ in the above case, where a radiating sodium atom is located in neon as a buffer gas. Taking as above the neon temperature $T = 500\,\mathrm{K}$, that leads to the total cross section of scattering of a radiating sodium atom on a neon atom $\sigma_t(T) = 2.8 \cdot 10^{-14}\,\mathrm{cm}^2$, we obtain the criterion of validity of the impact theory of line broadening (2.1.55) $N_b \ll 2.1 \cdot 10^{20}\,\mathrm{cm}^{-3}$ or the neon pressure $p_b \ll 4$ atm. Thus, in this case we have a wide range of the neon number densities, where the collision line broadening dominates. Indeed, the width of spectral line increases in 90 times as a result of transition from the boundary, where the Doppler and collision widths are equal, up to that, where the collision width coincides with the quasistatic one.

We now turn to the validity criteria of the quasistatic theory of spectral line broadening. A typical time $\left|\omega - \omega_{ji}\right|^{-1}$, during which the frequency shift $\omega - \omega_{ji}$ is detected, is small compared to a typical time ρ/v of strong interaction with a perturbed atom during collisions. This gives the criterion of the quasistatic theory

$$\left|\omega - \omega_{ji}\right| \gg \frac{v}{\sqrt{\sigma_t}} \tag{2.1.56}$$

This is opposite to that (2.1.54) for the validity of the collision theory of broadening. There is an intermediate range

$$\left|\omega - \omega_{ji}\right| \sim \frac{v}{\sqrt{\sigma_t}},$$

where both collision theory and quasistatic theory are invalid. In addition, the quasistatic theory of line broadening becomes better at wings far from the central part of spectral line. One can obtain also for the central part of the quasistatic theory $\left|\omega - \omega_{ji}\right| \sim Nv\sigma_t$, that leads on the basis of (2.1.56) to the criterion of the validity of the quasistatic theory

$$N\sigma_t^{3/2} \gg 1, \tag{2.1.57}$$

that is also opposite to the criterion (2.1.55) of validity for the the impact theory of line broadening. Thus the impact and quasistatic theories of line broadening are opposite.

We also give formula for the photon distribution function at wings of the spectral line, where $\left|\omega - \omega_{ji}\right| \to \infty$, and typical times t which give the contribution to the integral (2.1.51), are of the order of $\left|\omega - \omega_{ji}\right|^{-1}$. Let us consider the integrand of formula (2.1.51) given by

$$\exp\left\{N_b \int \left[\exp\left(\frac{iU(R)\,t}{\hbar}\right) - 1\right] w(R)\, d\mathbf{R}\right\},$$

and estimate the exponent in this expression. This exponent can be represented as Nr_0^3, where r_0 is a typical size in the integrand which value is determined by the relation

$$\exp\left(\frac{iU(R)}{\hbar\left(\omega - \omega_{ji}\right)}\right) \sim 1$$

Taking $U(R)$ to be monotonous function of R, we have at large R, if $R > r_0$,

$$\exp\left(\frac{iU(R)}{\hbar\left(\omega - \omega_{ji}\right)}\right) - 1 \to 0,$$

so that the contribution to the integrand of large values R is small. Hence r_0 follows from the relation $U(r_0) \sim \hbar \left|\omega - \omega_{ji}\right|$. In the limit $|\omega - \omega_{ji}| \to \infty$ it follows $r_0 \to 0$, that gives $Nr_0^3 \ll 1$ for a far wing of the spectral line. In this limit on can expand the exponent (2.1.50) in a Taylor series, so that expression (2.1.51) takes the form

$$a_\omega = \frac{N_b}{2\pi} \int w(R) d\mathbf{R} \int_{-\infty}^{\infty} \exp\left[-i\left(\omega - \omega_{ji}\right)t\right] \left[\exp\left(\frac{iU(R)t}{\hbar}\right) - 1\right] \mathrm{d}t \tag{2.1.58}$$

Since $\omega - \omega_{ji} \neq 0$, the second term in this expression is zero. The first term gives a delta function, and so we obtain

$$a_\omega = N_b \int \delta\left(\omega - \omega_{ji} - \frac{U(R)}{\hbar}\right) w(R) d\mathbf{R} = 4\pi\hbar N_b R^2 \frac{w(R)}{dU/dR}|_{U(R)=\hbar(\omega-\omega_{ji})} \tag{2.1.59}$$

As is seen, the quasistatic theory gives the same dependence on the number density of perturbed atoms N_b as in the case of the collision broadening of spectral lines.

One can obtain the latter formula from a simple consideration using a small probability of location of perturbed atomic particles at low distances from a radiating. For the spherically symmetric interaction potential $U(R)$, the probability to find a perturbed atomic particle in a distance range between R and $R + dR$ is $N_b w(R) d\mathbf{R}$, and

$$a_\omega \mathrm{d}\omega = N_b w(R) d\mathbf{R}$$

Equation (2.1.44) gives $\mathrm{d}\omega = dU(R)/\hbar$, that leads to formula (2.1.59)

$$a_\omega = 4\pi\hbar N R^2 w(R) \left(dU/dR\right)^{-1}$$

Let us consider the case when the difference of interaction potentials for transition states is approximated as $U(R) = C/R^n$ and $w(R) = 1$. Then ignoring the shift of a spectral line, one obtain

$$a_\omega = \frac{4\pi\hbar N_b R^{n+3}}{Cn} = \frac{4\pi N_b}{n} \left(\frac{C}{n}\right)^{3/n} \frac{1}{\left|\omega - \omega_{ji}\right|^{1+3/n}} \tag{2.1.60}$$

One can see that that the Lorentzian shape of spectral lines at wings is possible only for $n = 3$. If $n > 3$, then a_ω decreases at wings slighter than that for a Lorentz profile.

Let us consider the case (2.1.41) for the interaction potential $U(R) = -C_6/R^6$ which relates to interaction of neutral atomic particles at large separations R. Then we have on the basis of formula (2.1.60) for the frequency distribution function at wings of the spectral line

$$a_\omega == \frac{2\pi N_b}{3\left|\omega - \omega_{ji}\right|^{3/2}} \left(\frac{C_6}{\hbar}\right)^{1/2n} \tag{2.1.61}$$

In the case of the collision broadening of the spectral line formulas (2.1.40) and (2.1.43)

$$a_\omega == \frac{7.2N_b}{2\pi\left|\omega-\omega_{ji}\right|^2}\left(\frac{2T}{\mu}\right)^{0.3}\left(\frac{C_6}{\hbar}\right)^{0.4} \qquad (2.1.62)$$

Equalizing these distribution functions, one can find the boundary shift $\Delta\omega_c$ of the spectral line above which the quasistatic theory of broadening holds true

$$\Delta\omega_c = 0.3\left(\frac{2T}{\mu}\right)^{0.6}\left(\frac{C_6}{\hbar}\right)^{-0.2} \qquad (2.1.63)$$

This shift corresponds to the line width (2.1.40) at the transiting number density of perturbed atoms $N_b = \sigma_t^{-3/2}$. In the case of a radiating sodium atoms in neon as a buffer gas this frequency shift is $\Delta\omega_c = 3\cdot 10^{11}\,\mathrm{s}^{-1}$.

Thus, if the impact theory of line broadening is valid for the central part of a spectral line in accordance with the criterion (2.1.55), and the transition takes place to the quasistatic theory, as we remove from the line center at $\Delta\omega_c$. According to formulas (2.1.54) and (2.1.56), transition from collision broadening to quasistatic one takes place at frequencies

$$\left|\omega-\omega_{ji}\right| \sim \frac{v}{\sqrt{\sigma_t}}$$

Since

$$\frac{v}{\sqrt{\sigma_t}} \gg N_b v\sigma_t,$$

then the transition from collision broadening to quasistatic one occurs at far wings of a spectral line. Let us analyze the transiting region of a spectral line. Since there

$$\left|\omega-\omega_{ji}\right| \sim \frac{v}{\sqrt{\sigma_t}},$$

we have

$$U(R) \sim \frac{\hbar v}{\sqrt{\sigma_t}},$$

in the transition region. We thus find that $R \sim \sqrt{\sigma_t}$. Hence, the frequency distribution function is of the order

$$a_\omega \sim \frac{\hbar N_b R^2}{U(R)} \sim \frac{N_b\sigma_t^{3/2}}{\left|\omega-\omega_{ji}\right|} \sim \frac{N_b\sigma_t^2}{v}$$

From another point of view, one can obtain within the framework of the collision broadening mechanism from formula (2.1.37) for the frequency distribution function

$$a_\omega \sim \frac{\nu}{\left|\omega - \omega_{ji}\right|^2} \sim \frac{N_b v \sigma_t}{v^2/\sigma_t} \sim \frac{N_b \sigma_t^2}{v}.$$

Thus, in the transition range where

$$\left|\omega - \omega_{ji}\right| \sim \frac{v}{\sqrt{\sigma_t}},$$

the results of the collision and quasistatic theories are of the same order of magnitude, as one can would expect. Outside this range, the dependence of a_ω on the frequency ω, and the order of magnitude of a_ω are different within the two theories of line broadening.

2.2 Equilibrium Radiation

2.2.1 *Laws of Blackbody Radiation*

The radiation field, i.e. a system of electromagnetic waves, is formed in processes of emission and absorption of photons as a result of interaction with atomic particles or atomic systems. We now consider an equilibrium radiation field, which follow from photon absorption and emission processes, using, for definiteness, photon interaction with atomic particles or with a solid surface. In particular, considering a system of photons as elementary particles of an electromagnetic field, one can present the equilibrium radiation as a system of photons which are found in an equilibrium due to interaction with a surface. In this case, we first consider this system of photons to be located in a cavity inside a vessel, so that photons are absorbed and emitted at vessel walls which temperature is T. This radiation inside the vessel is called blackbody radiation and may leave the cavity through a small hole. The latter process does not violate the equilibrium of the radiation field with walls, but allows one to obtain information about the radiation field.

We first determine the average number of photons $\overline{n_\omega}$ in a given state of a frequency ω taking into account that photons are subjected to the Bose-Einstein statistics. According to the Boltzmann formula, the probability for n photons of energy $\hbar\omega$ are found in a given state is equal to $\exp(-\hbar\omega n/T)$. From this it follows for the average number of photons $\overline{n_\omega}$ in this state

$$\overline{n_\omega} = \frac{\sum\limits_n n \exp(-\frac{\hbar\omega n}{T})}{\sum\limits_n \exp(-\frac{\hbar\omega n}{T})} = \frac{1}{\exp\left(\frac{\hbar\omega}{T}\right) - 1} \tag{2.2.1}$$

This is the Planck distribution which is the case of the Bose-Einstein distribution with zero chemical potential.

It is convenient to deal with the spectral radiation density U_ω which is the energy of radiation per unit time, per unit volume, and per unit frequency range. Below we obtain expressions for this quantity. The radiation field energy in a frequency range from ω to $\omega + d\omega$ according to the above definition is $VU_\omega d\omega$, where V is the cavity volume inside the vessel. On the other hand, the above specific energy of the radiation field is equal to $2\hbar\omega n_\omega V d\mathbf{k}/(2\pi)^3$, where $Vd\mathbf{k}/(2\pi)^3$ is the number of states in a given element of the phase space, **k** is the photon wave number, n_ω is the number of photons for one state, and the factor 2 accounts for the two polarizations of an electromagnetic wave, because it is the transverse wave. Using the dispersion relation $\omega = ck$ between the frequency ω and wave vector k of the photon (c is the velocity of light), one can obtain from the above relations

$$U_\omega = \frac{\hbar\omega^3}{\pi^2 c^3} n_\omega = \frac{\hbar\omega^3}{\pi^2 c^3[\exp(\hbar\omega/T) - 1]} \tag{2.2.2}$$

This is the Planck radiation formula which uses the Planck distribution (2.2.1).

Let us consider the limiting cases of the Planck formula. In the classical limiting case $\hbar\omega \ll T$ this formula is transformed into the Rayleigh-Jeans formula. This is a classical formula which does not contain the Planck constant $\hbar$

$$U_\omega = \frac{\omega^2 T}{\pi^2 c^3}, \quad \hbar\omega \ll T \tag{2.2.3}$$

The Wien formula describes another limiting case

$$U_\omega = \frac{\hbar\omega^3}{\pi^2 c^3} \exp\left(-\frac{\hbar\omega}{T}\right), \quad \hbar\omega \gg T \tag{2.2.4}$$

We now evaluate the radiative flux emitted by a blackbody surface on the basis of formula (2.2.3). A blackbody surface emits isotropically, and the flux per unit solid angle and per a frequency interval $d\omega$ is equal to $cU_\omega d\omega$. The resultant flux is directed perpendicular to the surface. Projecting the radiative flux onto this direction, we have for the total radiative flux from a blackbody surface

$$J = \int_{-1}^{1}\int_{0}^{\infty} cU_\omega d\omega \cos\theta \frac{d\cos\theta}{2} = \frac{c}{4}\int_{0}^{\infty} U_\omega d\omega = \sigma T^4, \tag{2.2.5}$$

where θ is the angle between a direction of an emitting photon motion and the perpendicular to the surface. Equation (2.2.5) is called the Stefan-Boltzmann law [10, 11]. The value σ is the Stefan-Boltzmann constant that equals

$$\sigma = \frac{1}{4\pi^2 c^2 \hbar^3} \int_0^\infty [\exp(x) - 1]^{-1} x^3 dx = \frac{\pi^2}{(60c^2\hbar^3)} = 5.67 \cdot 10^{-12} \frac{W}{cm^2 K^4} \tag{2.2.6}$$

Note that the dependence of the radiation flux (2.2.5) on the problem parameters may be obtained in the simplest way on the basis of dimensionality considerations. Indeed, the result—the energy flux J can depend on the radiative temperature T, the Planck constant $\hbar$, and the velocity of light c. From these parameters one can compose only one combination of the flux dimensionality which is given by formulas (2.2.5) and (2.2.6).

We above consider the flux of equilibrium radiation as a result of emission of a hard surface. It is clear that the same result relates to a gaseous system with a sharp boundary, such that radiation propagate outside this surface. Then the mean free path of photons of a given frequency inside the gas is small compared to a size of gaseous system. This means that radiation is absorbed and emitted inside a gas intensively. As a result, the energy flux I_ω at a given frequency ω which propagates inside the gaseous system, is equal

$$I_\omega = cU_\omega = \frac{\hbar\omega^3}{\pi^2 c^3 [\exp(\hbar\omega/T) - 1]} \tag{2.2.7}$$

The energy flux of photons J_ω at a given frequency which leave a gas volume through a flat surface, directs perpendicular to the separated surface and is equal

$$J_\omega = \frac{I_\omega}{4} = \frac{\hbar\omega^3}{4\pi^2 c^3 [\exp(\hbar\omega/T) - 1]} \tag{2.2.8}$$

From this one can find the isotropic flux i_ω of photons inside an equilibrium matter, and the photon flux j_ω which intersects the plane boundary of this matter

$$i_\omega = \frac{I_\omega}{\hbar\omega} = \frac{\omega^2}{\pi^2 c^3 \left[\exp(\hbar\omega/T) - 1\right]}, \quad j_\omega = \frac{J_\omega}{\hbar\omega} = \frac{\omega^2}{4\pi^2 c^3 \left[\exp(\hbar\omega/T) - 1\right]}, \tag{2.2.9}$$

2.2.2 Spontaneous and Stimulated Emission

We now analyze parameters described the rates of emission and absorption of an atomic gas which proceed according to the scheme

$$\hbar\omega + A \longleftrightarrow A^* \tag{2.2.10}$$

Denoting by n_ω the number of photons in one state of a given frequency ω, we have that the absorption rate W is proportional to this value

$$W(i, n_\omega \rightarrow j, n_{\omega-1}) = A_{ij} n_\omega, \tag{2.2.11}$$

where index i relates to the lower state of the transition (2.2.10), and index f refers to the upper state of this transition. As is seen, in the absence of photons ($n_\omega = 0$) the transition is absent. The quantity A does not depend on the electromagnetic field strength, i.e. on the number of photons in one state, and is determined only by the parameters of the atomic particle. In the same manner, we have for the rate of the elementary emission process (2.2.10)

$$W(j, n_\omega \rightarrow i, n_{\omega-1} + 1) = \frac{1}{\tau_{ji}} + B_{ji} n_\omega \tag{2.2.12}$$

Here $1/\tau_{ji}$ is the reciprocal lifetime of the upper state j with respect to the radiative transition in the lower state i or the rate of spontaneous emission of an excited atom which proceeds in the absence of an external field, and the quantity B refers to stimulated radiation by an external electromagnetic field. The quantities A_{ij} and B_{ji} are called the Einstein coefficients [12, 13]. Both Einstein coefficients depend only on properties of the atomic particle.

The connection between the parameters $1/\tau_{ji}$, A_{ij} and B_{ji} can be obtained from the analysis of equilibrium in atomic and photon systems. The relation between the number densities of atomic particles in the excited N_j and ground N_i states is given by the Boltzmann law

$$N_j = \frac{g_j}{g_i} N_i \exp\left(-\frac{\hbar\omega}{T}\right), \tag{2.2.13}$$

where g_i and g_j are the statistical weights of the lower and upper states, and the photon energy $\hbar\omega$ coincides with the energy difference $(E_j - E_i)$ between transition states (see Fig. 1.4). As a result of the equilibrium, we have from the equality of average emission and absorption rates

$$N_i\, W(i, \overline{n}_\omega \rightarrow j, \overline{n}_\omega - 1) = N_j\, W(j, \overline{n}_\omega - 1 \rightarrow i, \overline{n}_\omega) \tag{2.2.14}$$

On the basis of formulas (2.2.11) and (2.2.12) this relation is transformed to the form

$$N_i\, A\overline{n}_\omega = N_j\, (1/\tau_{ji} + B_{ji}\overline{n}_\omega) \tag{2.2.15}$$

Substituting the Planck distribution (2.2.10) and Boltzmann distribution (2.2.13) in formula (2.2.15), one can one can obtain the following expressions for the Einstein coefficients

$$A_{ij} = g_j/(g_i\tau_{ji}),\ B_{ji} = 1/\tau_{ji} \tag{2.2.16}$$

This leads to the following formulas for rates of the one-photon radiative processes

$$W(i, n_\omega \to j, n_\omega - 1) = A_{ij} n_\omega = \frac{g_j}{g_i \tau_{ji}} n_\omega, \; W(j, n_\omega \to i, n_\omega - 1) = \frac{1}{\tau_{ji}} + B_{ji} n_\omega = \frac{1}{\tau_{ji}} + \frac{n_\omega}{\tau_{ji}} \tag{2.2.17}$$

As it follows from this analysis, the stimulated radiation is of fundamental importance and is a part of equilibrium between the radiation field and an atomic gas.

2.2.3 Cross Section and Parameters of Radiative Processes

Our goal is to determine parameters of the elementary processes of interaction between radiation and atomic particles. The cross section of absorption or emission is the parameter that characterizes the transition between states of an atomic particle as a result of absorption or emission of one photon. At low intensities of incident radiation this parameter does not depend on the intensity. Below we determine the absorption cross section which corresponds to transition between two discrete states i and j of an atomic particle according to the following scheme in accordance with Fig. 1.4

$$A_i + n\hbar\omega \to A_j + (n-1)\,\hbar\omega$$

Here A is an atomic particle, the lower index shows its state; $\hbar\omega$ is the energy photon of a frequency ω. According to definition, the absorption cross section is the ratio of the photon absorption rate w_a to the flux of incident photons j_ω for a given frequency range, i.e.

$$\sigma_a = \frac{w_a}{j_\omega} \tag{2.2.18}$$

The process of induced radiation of photons proceeds according to the scheme

$$A_j + n\hbar\omega \to A_i + (n+1)\,\hbar\omega$$

The cross section of induced radiation may be determined by analogy with the absorption cross section as the ratio of the photon radiation rate w_r to the flow density of incident photons j_ω, i.e.

$$\sigma_r = \frac{w_r}{j_\omega} \tag{2.2.19}$$

The cross sections of absorption and induced radiation are characteristics of radiation propagation through a gas. Let I_ω be the intensity of radiation of a frequency ω propagated through a gas. The intensity of radiation I_ω varies due to absorption and induced radiation. The number of the emission and absorption events is proportional to the number of photons participated in these processes. Therefore variation of the intensity of the photon beam along the path is proportional to the intensity and is given by the Beer-Lamberte law [14, 15]

$$\frac{dI_\omega}{dx} = -k_\omega I_\omega, \tag{2.2.20}$$

where x characterizes the direction of propagation of the photon beam. The quantity k_ω is called the absorption coefficient.

We now use the formalism of radiative transitions between discrete states of atomic particles represented in the previous chapter. The rate of this radiative transition $i \to j$ is given by formula (1.2.19) and in scales of transition frequency ω_{ji} the rate can be represented in the form

$$w \sim \delta\left(\omega_{ji} - \omega\right),$$

where $\hbar\omega_{ji}$ is the energy difference between the states. The frequency distribution function a_ω for photon absorption or emission is of importance for description of these processes, where $a_\omega d\omega$ is the probability that the photon frequency is located in a range between ω and $\omega + d\omega$. In scales of the transition frequency it may be represented as

$$a_\omega d\omega = \delta\left(\omega_{ji} - \omega\right) d\omega \tag{2.2.21}$$

where the normalization condition is described by formula (2.1.1). Note that in scales of small frequencies compared with ω_{ji} formula (2.2.21) is violated, i.e. in these scales the structure of delta function is determined by a corresponding mechanism of line broadening.

Let us express the absorption and induced radiation cross sections through the frequency distribution function a_ω. The process of photon absorption proceeds in some range of frequencies in the vicinity of ω_{ji}, and the value $a_\omega d\omega$ is the probability that the frequency of an absorbed photon is found in a range between ω and $\omega + d\omega$. Replacing the discrete spectrum of absorption by a continuous one, we obtain the probability of photon absorption

$$dw_a = w_{ji} a_\omega d\omega,$$

where this probability is averaged over polarizations of incident photons.

One can introduce the radiative lifetime τ of an excited state j with respect to spontaneous decay into the state i in accordance with formula (1.2.20), where we take $n_\omega = 0$

$$\frac{1}{\tau_{ji}} = \frac{4\omega_{ji}^3}{3\hbar c^3}\left|\mathbf{D}_{ji}\right|^2 g_i,$$

where g_i is the statistical weight of the state i. Then one can rewrite formula for the absorption rate on the basis of formula (2.2.17)

$$\mathrm{d}w_a = A_{ij} n_\omega a_\omega \mathrm{d}\omega = \frac{1}{\tau_{ji}} \frac{g_j}{g_i} n_\omega a_\omega \mathrm{d}\omega \qquad (2.2.22)$$

Let an incident photon flux be interacted with an atomic particle. Taking the number of photons n_ω located in one state, we have for the number density of photons in this state per unit volume

$$\mathrm{d}N_\omega = \frac{n_\omega \cdot 2d\mathbf{k}}{(2\pi)^3} = \frac{n_\omega \omega^2 d!}{\pi^2 c^3},$$

where the factor 2 takes into account two photon polarizations. From this we have for the photon flux in a frequency range from ω up to $\omega + \mathrm{d}\omega$

$$\mathrm{d}j_\omega = c \cdot dN_\omega = \frac{n_\omega \omega^2 \mathrm{d}\omega}{\pi^2 c^2} \qquad (2.2.23)$$

The cross section of photon absorption is introduced by formula (2.2.19) as the ratio of the absorption rate to the photon flux. According to formulas (2.2.22) and (2.2.23) it is given by

$$\sigma_a = \frac{\mathrm{d}w_a}{\mathrm{d}j_\omega} = A_{ij} a_\omega \left(\frac{\pi c}{\omega}\right)^2 = \left(\frac{\pi c}{\omega}\right)^2 \frac{a_\omega}{\tau_{ji}} \frac{g_j}{g_i} \qquad (2.2.24)$$

for photon absorption with the atom transition from a state i to a state j. In the same manner one can determine the cross section of induced radiation as the ratio of the rate of induced radiation given by formula (1.2.20) to the photon flux (2.2.23). Expressing the rate (1.2.20) through the radiative lifetime τ of the upper state with respect to this transition, one can obtain for the cross section σ_r of induced radiation

$$\sigma_r = B_{ji} a_\omega \left(\frac{\pi c}{\omega}\right)^2 = \left(\frac{\pi c}{\omega}\right)^2 \frac{a_\omega}{\tau_{ji}} \qquad (2.2.25)$$

Let us express the absorption coefficient k_ω defined by formula (2.2.20) through the cross sections of radiative processes. We denote by N_i and N_j the number densities of atomic particles in the lower i and upper j states of the radiative transition. Because the product $N_i\sigma_a$ characterizes the lost of photons, and the value $N_j\sigma_r$ describes its increase due to stimulated emission, so that the absorption coefficient is determined by the expression

$$k_\omega = N_i \sigma_a - N_j \sigma_r \qquad (2.2.26)$$

Substituting formulas (2.2.24) and (2.2.25) for the absorption cross section and the induced radiation cross section into expression (2.2.26), one can represent the absorption coefficient in the form

$$k_\omega = N_j \left(\frac{\pi c}{\omega}\right)^2 \frac{a_\omega}{\tau_{ji}} \left(\frac{N_i}{N_j} \frac{g_j}{g_i} - 1\right) \qquad (2.2.27)$$

In the case of thermodynamic equilibrium between states i and j this formula gives

$$k_\omega = N_i \left(\frac{\pi c}{\omega}\right)^2 \frac{a_\omega}{\tau_{ji}} \frac{g_j}{g_i} \left[1 - \exp\left(-\frac{\hbar\omega}{T}\right)\right] = A_{ij} N_i \left(\frac{\pi c}{\omega}\right)^2 a_\omega \left[1 - \exp\left(-\frac{\hbar\omega}{T}\right)\right] \tag{2.2.28}$$

Let us consider integral relations for radiative cross sections. Integrating the cross section (2.2.24) over frequencies in the vicinity of the spectral line of transition and

$$\int \sigma_a (i \to j)\, \mathrm{d}\omega = \frac{4\pi^2 \omega_{ji}}{3\hbar c} \left|\mathbf{D}_{ji}\right|^2 , \tag{2.2.29}$$

where we put for simplicity $g_i = g_j = 1$. Note that integration over all possible frequencies of absorbed photons corresponds to summation over all possible states of the atomic particle in formula (2.2.28) (including states of the continuum spectrum). Using the sum rule and taking into account the relation $\left|\mathbf{D}_{ij}\right|^2 = 3\left|(\mathbf{D}_z)_{ij}\right|^2$, one can obtain

$$\int \sigma_r \mathrm{d}\omega = \frac{2\pi^2 e^2 n}{m_e c} \tag{2.2.30}$$

Here m_e is the electron mass, and n is the number of electrons in an atom which partake in the radiative process.

Note that (2.2.20) describes propagation of the radiative flux of a small intensity through a gas, and the gas is an absorber for the radiation. Let us consider this gas consisting of flat layers, and if the radiation propagates perpendicular to these layers, the measure of depleting of the radiation flux at a given frequency is the parameter

$$u_\omega = \int k_\omega \mathrm{d}x, \tag{2.2.31}$$

where the integral is taken along the path x in the transverse direction with respect to the layer. The quantity u_ω is the optical thickness of this layer at a given frequency. The optical thickness is the characteristic of the gas layer as an absorber at a given frequency. It is convenient to express the equilibrium flux outside the gaseous absorber on the basis of its optical thickness. Indeed, let an absorbed gas be separated from a vacuum or a transparent gas by a flat boundary, and in accordance with Fig. 2.1 the photon flux j_ω, i.e. the number of photons crossed the surface per unit area and unit time, is given by

$$j_\omega = \int_0^1 d\cos\theta \int_0^{u_o} du_\omega \exp\left(-\frac{u_\omega}{\cos\theta}\right) i_\omega, \quad u_\omega(z) = \int_0^z k_\omega \mathrm{d}z, \tag{2.2.32}$$

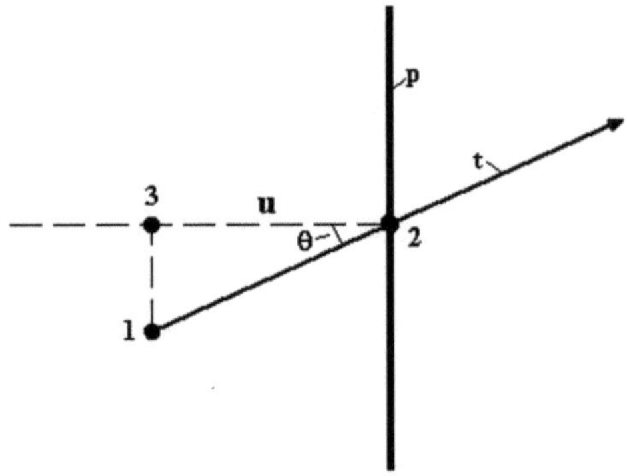

Fig. 2.1 Geometry for an outgoing photon in the course of photon emission from a plane gas layer where the gas temperature depends on a distance from the boundary only. Here t is trajectory of a propagated photon, p is the plane boundary of an emitted gas, 1 is the origin, 2 is a point of its intersection with a line of photon motion, 3 is the projection of this point onto the boundary plane direction

where i_ω is the equilibrium photon flux at a given frequency, u_o is the total optical thickness of the gas, and the flux in formula (2.2.32) results from collection of photons at the gas boundary. In the equilibrium case, where a gas is uniform with the temperature T and infinite optical thickness $u_o = \infty$ at a given frequency, we have for the partial photon flux inside the gas $i_\omega = I_\omega/(\hbar\omega)$, where i_ω is given by formula (2.2.26). Then the total photon j_ω flux is equal

$$j_\omega = \frac{i_\omega}{4} = \frac{\omega^2}{4\pi^2 c^3 \left[\exp\left(\hbar\omega/T\right) - 1\right]} \tag{2.2.33}$$

As is seen, the partial photon flux is equal $j_\omega = J_\omega/(\hbar\omega)$, where the partial energy flux due to photons J_ω is given by formula (2.2.9).

We now generalize the expression (2.2.33) for the partial photon flux on the case of the local thermodynamic equilibrium. This means that the gas temperature varies in a space, but the temperature gradient is relatively small, i.e.

$$\lambda \frac{d \ln T}{dx} \ll 1, \tag{2.2.34}$$

where λ is the mean free path for atomic particles of the gas, and under this condition the stationary state of the gas is supported. We also assume that the radiation flux at each point is characterized by the gas temperature there. Therefore the flux of photons in this case is given by formula (2.2.9)

$$j_\omega = \frac{\omega^2}{4\pi^2 c^3 \left\{\exp\left[\hbar\omega/T(\mathbf{r})\right] - 1\right\}}, \tag{2.2.35}$$

where **r** is the coordinate of a point from which photon is emitted. Let us rewrite this formula for the case of a flat gas layer, where gas parameters depend only on a distance z from its boundary, as

$$j_\omega = \frac{\omega^2}{2\pi^2 c^2}\int_0^1 d\cos\theta \int_0^{u_o} du_\omega \exp\left(-\frac{u_\omega}{\cos\theta}\right) F(u_\omega),\ u_\omega(z) = \int_0^z k_\omega dz,\ F(u_\omega) = \left\{\exp\left[\frac{\hbar\omega}{T(z)}\right]-1\right\}^{-1}, \tag{2.2.36}$$

where u_*, the total optical thickness of the layer at a given frequency, is large $u_* \gg 1$. If the gas temperature is constant over this layer, this is transformed into formula (2.2.9) for the equilibrium radiation.

In the case under consideration, where the gas temperature varies at removal from the boundary, but the local thermodynamic equilibrium is conserved, i.e. the criterion (2.2.34) holds true, one can expand the integrand of formula (2.2.36) over a small parameter (2.2.34). This is equivalent to expansion of the function $F(u_\omega)$, namely,

$$F(u_\omega) = F(u_o) + (u_\omega - u_o)F'(u_o) + \frac{1}{2}(u_\omega - u_o)^2 F''(u_o)$$

We choose the parameter u_o such, that the second term of the expansion vanishes after integration. Finally we obtain for the energy flux of radiation [16, 17]

$$J_\omega = I_\omega(u_o)(1-\alpha);\ u_o = \frac{2}{3};\ \alpha = \frac{5F''(u_o)}{18F(u_o)}\ I_\omega = \frac{\omega^3}{4\pi^2 c^3\left\{\exp\left[\hbar\omega/T_\omega\right]-1\right\}}, \tag{2.2.37}$$

where I_ω is the energy flux transported by photons from the surface of a blackbody in accordance with formula (2.2.20). This means that the flux of outgoing radiation at a given frequency is the equilibrium flux, which temperature T_ω is equal to the temperature of a layer with the optical thickness of $2/3$ from the boundary. In other words, we have [16, 17]

$$T_\omega = T(z_\omega),\ u(z_\omega) = \int_0^{z_\omega} k_\omega dz = \frac{2}{3} \tag{2.2.38}$$

The above results relate to a large optical thickness u_o of the total gas layer. We below generalize them to the case of a finite optical thickness u_o at a given frequency. According to the geometry of photon propagation toward the boundary, given in Fig. 2.1, the total radiative flux of the layer with the same temperature T is equal

$$j_\omega = j_\omega g(u_\omega),\ g(u_\omega) = 2\int_0^1 d\cos\theta \int_0^u dx \exp\left(-\frac{x}{\cos\theta}\right), \tag{2.2.39}$$

where u is the total optical thickness for a given frequency, and $g(u)$ represented in Fig. 2.2 tends to one in the limit $u \to \infty$. It is convenient to use the following approximation for $g(x)$

$$g(x) = 1 - \exp(-1.6x), \tag{2.2.40}$$

and both functions are represented in Fig. 2.2. Thus, the function $g(u)$ characterizes the probability to emit a photon of a given frequency, where this probability is one for an optically thick layer. In addition, $g(u)$ is the probability for a photon to reach the opposite boundary of the layer if it emits at another boundary. Hence the probability to survive for the photon in the course of propagation through the layer is equal $1 - g(u)$.

For a slightly nonuniform gas layer, the radiative temperature is given by formula (2.2.38), if this layer is optically thick. By analogy with derivation of this formula, in the general case we introduce the quantity $u_{\rm ef}$ as the optical depth of a layer; its temperature coincides with the radiative temperature of radiation at a given frequency. Repeating operations at derivation this formula, we have

$$u_{\rm ef} = 2\int_0^1 d\cos\theta \int_0^u x\mathrm{d}x \exp\left(-\frac{x}{\cos\theta}\right) \tag{2.2.41}$$

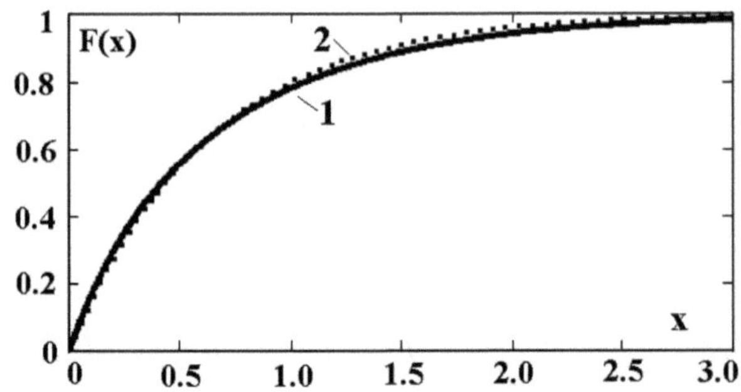

Fig. 2.2 Function $g(u)$ according to formulas (2.2.39) and (2.2.40) [18]

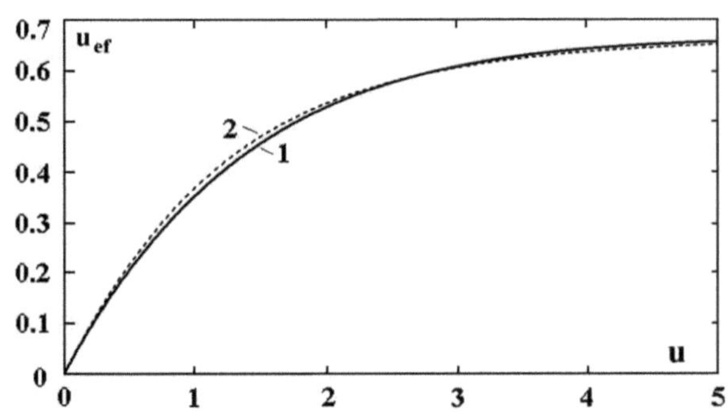

Fig. 2.3 Effective optical thickness $u_{\rm ef}(u)$ in accordance with formulas (2.2.41) and (2.2.42) [18]

In the limit of a large optical thickness u at a given frequency this formula is transformed into (2.2.38). The dependence $u_{ef}(u)$ is presented in Fig. 2.3 together with the following approximation of this dependence

$$u_{ef}(u) = \frac{u}{2\exp(-u) + 1.5u} \quad (2.2.42)$$

It is clear that if we operate with radiative temperature as a parameter of the radiation, a certain equilibrium takes place in a gas through which photons propagate. Let us consider the character of this equilibrium. Evidently, in the case of the local thermodynamic equilibrium of a gaseous system, an equilibrium is established at each spatial point of a gas under action of external field and as a result of collisions of gaseous atoms or molecules. Then the radiation field does not influence on this equilibrium because of large times of radiative transitions. On contrary, since the number density of atomic particles in the upper state of the radiative transition is determined by the equilibrium, the radiative temperature of the photon flux is associated with this equilibrium. The case when radiative transitions influence on the distribution of excited atomic particles requires a special consideration. In particular, this case is realized if resonance radiation propagates in rarefied gases.

References

1. L.D. Landau, E.M. Lifshitz, *Quantum Mechanics* (Pergamon Press, Oxford, 1965)
2. E. Lindholm, Ark. Mat. Astron. Fys. **32A**, 1 (1946)
3. H.M. Foley, Phys. Rev. **69**, 616 (1946)
4. W. Voigt, S. B. Bauer Acad. Wiss., 603–620 (1912)
5. J.I. Gersten, H.M. Foley, J. Opt. Soc. Am. **58**, 933 (1968)
6. H.M. Foley, C.C. Sung, J. Opt. Soc. Am. **65**, 949 (1975)
7. C.D. Rodgers, Appl. Opt. **15**, 714 (1976)
8. P.L. Varghese, R.K. Hanson, Appl. Opt. **23**, 2376 (1984)
9. D.W. Posener, Austr. J. Phys. **12**, 184 (1959)
10. J. Stefan, Sitz. math-natur. Classe Kaiz. Acad. Wiss. Vienna. Math. Natur. Classe. **79**, 391 (1879)
11. L. Boltzmann, Ann. Phys. Chem. **258**, 291 (1884)
12. A. Einstein, Ver. Deutsch Phys. Ges. **18**, 318 (1916)
13. A. Einstein, Zeit. Phys. **18**, 121 (1917)
14. A. Beer, Annalen der Physik und Chemie. **86**, 78 (1852)
15. J.H. Lambert, *Photometry, or, on the Measure and Gradations of Light, Colors, and Shade* (Augsburg, Eberhardt Klett, 1760)
16. B.M. Smirnov, *Physics of Weakly Ionized Gas* (Moscow, Nauka, 1979). (in Russian)
17. B.M. Smirnov, *Physics of Weakly Ionized Gases* (Moscow, Mir, 1980)
18. B.M. Smirnov, JETP **126**, 446 (2018)

Chapter 3
Resonant Radiation in Atomic Gases

Abstract Emission of resonance radiation proceeds as a result of dipole radiation from the resonantly excited states of atoms to their ground state. Because of a strong interaction between atoms in the ground and resonantly excited states, this interaction determine the broadening of spectral lines for resonance radiation even at a low concentration of radiating atoms. Propagation of resonance radiation in a weakly ionized gas is analyzed for the regime where quenching of excited atoms in collisions with electrons or atoms is negligible. Transport of resonance radiation results in reabsorption of resonant photons at wings of the spectral line. The phenomenon of self-reversal of spectral lines results due to a high optical thickness of a gas for resonance radiation and also due to low temperature of electrons near the plasma boundary.

3.1 Radiation Involving Resonantly Excited Atoms

3.1.1 Broadening of Resonant Spectral Lines

One can extract the resonantly excited atom states as lower excited atom states, so that the dipole radiative transition is possible from these states in the ground atom states. Atoms in these states are of importance for properties of an excited gas [1]. Correspondingly, the resonant radiation is emitted and is absorbed as a result of radiative transition between those and ground atom states. The operator of interaction between atoms in the ground and resonantly excited states at large separations R is given by

$$V(R) = \frac{\mathbf{D}_1\mathbf{D}_2 - 3(\mathbf{D}_1\mathbf{n})(\mathbf{D}_2\mathbf{n})}{R^3}, \qquad (3.1.1)$$

where $\mathbf{D}_1$, $\mathbf{D}_2$ are the operators of the dipole moment for the first and second atoms correspondingly, $\mathbf{n}$ is the unit vector along the direction connected neighboring atoms. The matrix element from the interaction operator (3.1.1) is nonzero for a system consisting of two atoms of the same sort in the ground and resonantly excited states, so that the interaction potential depends on a distance R between atoms as $\sim R^{-3}$

V. Krainov and B. M. Smirnov, *Atomic and Molecular Radiative Processes*,
Springer Series on Atomic, Optical, and Plasma Physics 108,
https://doi.org/10.1007/978-3-030-21955-0_3

and depends on the momentum projection onto the axis for a molecule consisting of these atoms, as well as on the symmetry of this state.

The total cross section σ_t of scattering of atoms in the ground and resonantly excited states includes processes of elastic collision of atoms, excitation transfer and turn of the atom momentum. This cross section depends also on the momenta of colliding atoms. Being guided by alkali metal atoms, we assume the momenta of colliding atoms to be $L = 0$ and $L = 1$. In this case the total collision cross section of atoms σ_t in the ground and resonantly excited states, which is expressed in terms of the matrix element of the dipole moment operator $\mathbf{D}$ between the ground and excitation states of the atom, is equal [2, 3]

$$\sigma_t = \frac{4.8\pi d^2}{\hbar v}, \tag{3.1.2}$$

where v is the relative collision velocity of the atoms, and d^2, the square of the matrix element from the operator dipole moment between the ground state of the sodium atom with a moment 0 and its projection 0, and the excited sodium atom with a moment 1 and its projection $M = 0, \ \pm 1$, is given by formula

$$d^2 = \frac{1}{3}\sum_M |\langle 00|\mathbf{D}|1M\rangle|^2 \tag{3.1.3}$$

and in the case of a sodium atom this quantity is $d^2 = 6.28e^2a_o^2$, where e is the electron charge, a_o is the Bohr radius. Accordingly, the width of the absorption or radiation spectral line for the impact mechanism of broadening is [3, 4]

$$\nu = 4.8\pi d^2 N \tag{3.1.4}$$

where N is the number density of these atoms formed a vapor. Table 3.1 contains the parameters of radiative transitions involving resonance states of alkali atoms and broadening of corresponding spectral lines. Namely, in this Table λ is the photon wavelength for a given transition, τ is the spontaneous lifetime for resonantly excited atoms, $\Delta\omega_D$ is the Doppler width, $\Delta\omega_L$ is the collision width, N_{DL} is the atomic number density, at which the Doppler width is equal to the collision one, N is the atomic number density, N_* is the atomic number density, at which the quasistatic width is equal to the collision width. These data correspond to the temperature of 500 K.

In the above analysis we will be guided by the case of sodium located in neon as a buffer gas. In the case of a sodium vapor the specific width of the resonant spectral line is $\nu/N = 5.8 \cdot 10^{-7}\,\text{cm}^3/\text{s}$, where N is the number density of sodium atoms. If a sodium vapor is located in a buffer gas–neon, an additional broadening of a spectral line results from interaction of a radiating sodium atom with neon atoms. As it follows from formula (2.1.43), at the temperature $T = 500\,\text{K}$ for the specific width of the spectral line is $\nu_b/N_b = 1.9 \cdot 10^{-9}\,\text{cm}^3/\text{s}$. Comparing it with the broadening

Table 3.1 The broadening parameters for spectral lines of alkali atoms

Element	Li	Na	Na	K	K	Rb	Rb	Cs	Cs
Transition	2^2s–3^2p	$3^2S_{1/2}$–$3^2P_{1/2}$	$3^2S_{1/2}$–$3^2P_{3/2}$	$4^2S_{1/2}$–$4^2P_{1/2}$	$4^2S_{1/2}$–$4^2P_{3/2}$	$5^2S_{1/2}$–$5^2P_{1/2}$	$5^2S_{1/2}$–$5^2P_{3/2}$	$6^2S_{1/2}$–$6^2P_{1/2}$	$6^2S_{1/2}$–$6^2P_{3/2}$
λ, nm	670.8	589.59	589.0	769.0	766.49	794.76	780.03	894.35	852.11
τ, ns	27	16	16	27	27	28	26	31	27
$\Delta\omega_D$, $10^9\,s^{-1}$	8.2	4.5	4.5	2.7	2.7	1.7	1.8	1.2	1.3
$\Delta\omega_L/N$, $10^{-7}\,cm^3/s$	2.6	1.6	2.4	2.0	3.2	2.0	3.1	2.6	4.0
N_{DL}, $10^{16}\,cm^{-3}$	16	15	9.4	6.5	4.2	4.5	2.9	2.43	1.6
N_*, $10^{18}\,cm^{-3}$	3.2	2.5	1.4	1.2	0.6	0.7	0.4	0.3	0.2

of resonant spectral lines due to interaction of a radiating sodium atom with other sodium atom one can find that the contributions to the broadening due to the above broadening channels are identical at the concentration of sodium atoms in neon of 0.3%.

It is of importance the character of equilibrium of resonantly excited atoms in an excited gas. This equilibrium is determined by collisions involving atoms in the ground and resonantly excited states and usually takes place in a weakly ionized gas, so that it results from collisions involving electrons. Figure 3.1 [5] represents the radiative and collision processes which influence on this equilibrium. In particular, the thermodynamic equilibrium involving excited atoms and electrons of this plasma, takes place at high number density N_e of electrons

$$N_e \gg N_q = \frac{1}{k_q \tau}, \tag{3.1.5}$$

where k_q is the rate constant of quenching of the resonantly excited state in collisions with electrons, τ is the lifetime of this excited state with respect to its location in the plasma volume. It is of importance that at small electron energies the rate

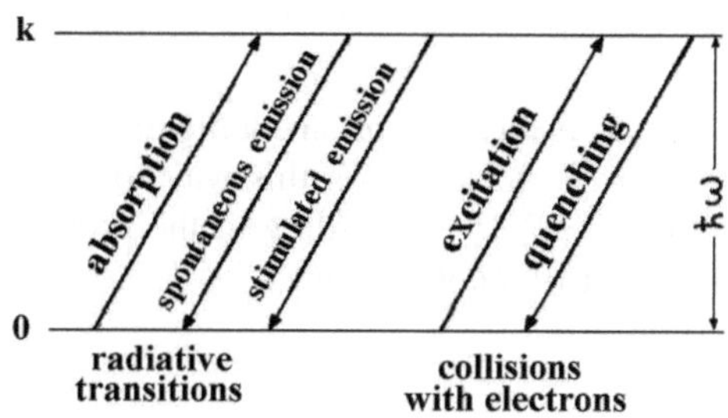

Fig. 3.1 Radiative and collision processes which connect the ground and resonantly excited states of an atom

constant of quenching is independent on this energy. In particular, for resonantly excited sodium atoms in states $Na(3^2P)$ we have $k_q = (2.0 \pm 0.3) \cdot 10^{-7}\,cm^3/s$ [6]. If resonant radiation is not closed in the plasma volume, the lifetime of an excited atom is the radiative lifetime, and then the boundary number density of electrons is $N_q \approx 3 \cdot 10^{15}\,cm^{-3}$. In the case, if resonant radiation is closed in the plasma volume, the limited electron number density N_q is lower. Nevertheless, thermodynamic equilibrium may be fulfilled only in arc, but in glow discharge it is violated. This means that a resonant excitation formed in a plasma is conserved in the course of reabsorption of resonant radiation in a plasma. We below consider namely this case of equilibrium in a weakly ionized gas.

3.1.2 Propagation of Resonant Radiation in Excited Gas

We now consider propagation of resonance radiation in an excited gas in the opposite regime to (5.4.5) where the lifetime of resonant excitation in this gas is determined by outgoing of the resonant radiation outside the gas. In this case we should consider the behavior of elementary excitations in a plasma [7–9], rather than resonant photons. It is used that transitions between resonant photons and excited atoms lead to conservation of a number of excitations, because quenching of excited atoms in collisions with electrons is a weak process. Let us introduce the probability $P(r)$ that a resonant photon survives at a distance r from a point of its formation, and this distance exceeds significantly the mean free path of the photon for the center of a spectral line $1/k_o$ (k_o is the absorption coefficient in the center of the resonant line). Assuming the statistical character for an emitting frequency in the limit of the small line width, we obtain that this probability is the product of the probability $a_\omega\, d\omega$ of photon emission with a given frequency and the probability $\exp(-k_\omega r)$ of photon surviving during its propagation between these points, that is

$$P(r) = \int a_\omega d\omega \exp(-k_\omega r) \qquad (3.1.6)$$

Let us use this formula for the Lorentz (2.1.4) and Doppler (2.1.13) line shape. In the case of the Lorentz shape of spectral lines we have, introducing a new variable $s = (\omega - \omega_o)/\nu$,

$$a_\omega d\omega = \frac{ds}{\pi(1+s^2)}\,, \quad k_\omega = \frac{k_o}{1+s^2}\,, \qquad (3.1.7)$$

where k_o is the absorption coefficient for the spectral line center. Let us use $u = k_o r$ the optical thickness on this way for the center line frequency that is defined by formula (2.2.31); according to the problem condition, $u \gg 1$. Substituting this in formula (3.1.7), we obtain

$$P(r) = \frac{1}{\pi} \int_{-\infty}^{\infty} \frac{ds}{(1+s^2)} \exp\left(-\frac{u}{1+s^2}\right)$$

For a dense plasma the main contribution in this integral follows from large $s \gg 1$, and for the probability $P(r)$ for the photon to pass a given distance r we have in this limit

$$P(r) = \frac{1}{\sqrt{\pi u}}, \quad u \gg 1 \tag{3.1.8}$$

For the Doppler shape of the spectral line we introduce a new variable

$$t = u \exp\left[-\frac{mc^2}{2T}\left(\frac{\omega - \omega_o}{\omega_o}\right)^2\right]$$

that gives for the probability to pass a distance r for the resonant photon

$$P(r) = \frac{1}{\sqrt{\pi u}} \int_0^u e^{-t} dt \left(\ln \frac{u}{t}\right)^{-1} = \frac{1}{\sqrt{\pi u}\sqrt{\ln u + C}}, \tag{3.1.9}$$

where $C = 0.577$ is the Euler constant. Since the frequency distribution function drops at wings of the Doppler spectral line sharper than that in the case of the Lorentz spectral line, the probability to propagate on large distances for the Lorentz spectral line is larger than that for the Doppler one at the same optical thickness of the layer.

Reabsorption of a resonant photon results in its formation and subsequently in its absorption in other spatial point. We will characterize this process by the probability $G(\mathbf{r}', \mathbf{r})$, so that $G(\mathbf{r}', \mathbf{r})d\mathbf{r}'$ is the probability for a photon formed at point $\mathbf{r}'$ to be absorbed in a volume $d\mathbf{r}'$ near the point $\mathbf{r}'$. This probability satisfies to the normalization condition

$$\int G(\mathbf{r}', \mathbf{r})d\mathbf{r}' = 1,$$

where the integral is taken over all the space. We obtain the expression for this probability by analogy with that in formula (3.1.7) which has the form

$$G(\mathbf{r}', \mathbf{r}) = \int \frac{a_\omega k_\omega d\omega}{4\pi |\mathbf{r} - \mathbf{r}'|^2} \exp\left(-\int k_\omega dx\right) \tag{3.1.10}$$

One can use the reabsorption process in the balance equation for the number density N_* of excited atoms located in a plasma where the reabsorption process is included along with the processes of atom excitation and quenching by electron impact and spontaneous remission of an excited atom. This equation has the form [7, 10, 11]

$$\frac{\partial N_*}{\partial t} = N_e N_o k_{ex} - N_e N_* k_q - \frac{N_*}{\tau} + \frac{1}{\tau}\int N_*(\mathbf{r}')d\mathbf{r}' \int \frac{a_\omega k_\omega d\omega}{4\pi |\mathbf{r}-\mathbf{r}'|^2} \exp\left(-\int k_\omega dx\right) \tag{3.1.11}$$

and is named the Biberman–Holstein equation.

In particular, for the Lorentz profile of the spectral line (2.1.4) and for a uniform plasma, this probability is

$$G(R)=\frac{k_o}{4\pi R^2}\int \frac{ds}{(1+s)^2}\exp\left(-\frac{k_o R}{1+s^2}\right),$$

where we use the variable $s = 2(\omega - \omega_o)/\nu$ and k_o is the absorption coefficient at the line center. In the limiting case of an optically thick plasma $k_o R \gg 1$ this formula gives

$$G(R) = \frac{1}{(4\pi k_o)^{1/2} R^{7/2}} \tag{3.1.12}$$

Correspondingly, in the case of the Doppler profile of the spectral line (2.1.13), using a new variable $s = [(\omega - \omega_o)/\omega_o](mc^2/T)^{1/2}$, we have $a_\omega d\omega = \pi^{-1/2}\exp(-s^2)ds$ and $k_\omega = k_o \exp(-s^2)$. Substituting this in formula (3.1.10) and introducing the variable $t = k_o R \exp(-s^2)$, one can obtain

$$G(R) = \frac{k_o}{4(\pi)^{3/2}R^2}\int_{-\infty}^{\infty} ds \exp(-s^2 - t) = \frac{1}{4(\pi)^{3/2}k_o R^4}\int_0^{k_o R}\frac{te^{-t}dt}{\sqrt{\ln(k_o R/t)}}$$

Replacing the upper limit of integration by infinity in the limiting case of an optically thick plasma $k_o R \gg 1$ and accounting for $\ln(k_o R) \gg 1$, one can obtain

$$G(R) = \frac{1}{4(\pi)^{3/2}k_o R^4\sqrt{\ln(k_o R/t_o)}},$$

where t_o is the solution of the equation

$$\int_0^{t_o} te^{-t}dt\sqrt{\ln(t/t_o)} = 0,$$

i.e. $t_o = 1.52$. From this we have

$$G(R) = \frac{1}{4(\pi)^{3/2}k_o R^4\sqrt{\ln(k_o R) - 0.42}} \tag{3.1.13}$$

We obtain that propagation of resonance radiation in a gas or plasma has a specific character. In contrast to diffusion transfer of atomic particles in a gaseous matter, radiation transfer proceeds at wings of the spectral line for the radiative transition, and the larger is the optical gas thickness, the more remote wings of the spectral line are responsible for radiation transfer. As an example of the radiative transfer of thid type, we consider emission of a cylinder plasma region for the Lorentz profile (2.1.4) of the spectral line. In the limit of an optically thin plasma ($k_o d \ll 1$, where d is the tube diameter), if all the formed photons leave the plasma region, the flux j of photons is [12]

$$J = \frac{N_*}{\tau_r}\frac{V}{S} = \frac{N_* d}{4\tau_r},$$

where $V = L \cdot \pi d^2/4$ and $S = L \cdot \pi d$ are the volume and surface area for a plasma region of a length L, respectively. In the limit of an optically thick plasma $k_o d \gg 1$ the photon flux is

$$j = \frac{N_*}{\tau_r}\int \frac{P(r)\cos\theta d\mathbf{r}}{4\pi r^2},$$

where r is a distance of the point of photon formation from the surface, $P(r)$ is the probability of photon surviving at a distance r from its formation, and θ is the angle between the photon direction and perpendicular to the plasma surface (Fig.2.1). Evaluating this integral, we obtain for the photon flux [12]

$$j = 0.39\frac{N_* R^{1/2}}{k_o^{1/2}\tau_r} \tag{3.1.14}$$

As is seen, the photon flux from an optically thick plasma is lower in $(k_o d)^{1/2}$ than that from an optically thin plasma.

Let us introduce the effective radiative time τ_{ef} of the resonantly excited state as

$$\tau_{\text{ef}} = \tau_r\frac{J}{j},$$

and τ_r is the radiative lifetime for an isolated atom; the flux J corresponds to an optically thin gas, i.e.,

$$J = \frac{N_* V}{\tau_r S},$$

where S is the area of the plasma surface, V is its volume, and this plasma assumes to be uniform. The effective radiative time depends both on the geometry of the volume occupied by the plasma and on the character of broadening of the spectral line. In particular, if an optically thick uniform plasma is located inside a cylinder tube, and the Lorentz character (2.1.4) of spectral line broadening occurs, on the basis of the above result we have [6]

$$\tau_{\text{ef}} = 1.3\tau_r(k_o R)^{1/2} \tag{3.1.15}$$

The above analysis shows that the system under consideration consisting of a buffer gas (neon) and the radiating impurity (sodium) exists at the buffer gas pressure of the order of 1 Torr, and presence of a buffer gas does not affect the resonant radiation of the impurity. Let us analyze the character of radiation output from this system. We have the following expression for the absorption cross section σ_ω at the given frequency [13, 14]

$$\sigma_\omega = \frac{\pi^2 c^2}{\omega^2}\frac{a_\omega}{\tau}\left[1 - \exp\left(-\frac{\hbar\omega}{T}\right)\right], \tag{3.1.16}$$

where a_ω is the distribution function of photons over frequencies, τ is the radiative time, and the last factor takes into account the induced radiation. Restricted to the impact mechanism of broadening of spectral lines, we have for the absorption cross section σ_o and for the absorption coefficient $k_o = N\sigma_o$ at the spectral line center, ignoring the induced radiation

$$\sigma_o = \frac{\pi^2 c^2}{\omega^2}\frac{2}{\pi\nu\tau}, \; k_o = \frac{\lambda^2}{2\pi\tau\nu/N}, \tag{3.1.17}$$

where $\lambda = 2\pi c/\omega$ is the radiation wavelength. According to this formula, the absorption coefficient k_o at the center of a spectral line does not depend both on the number density of irradiating atoms, and on their mass. Next, since according to formula (2.1.40), the specific width of the spectral line ν/N for a resonant transition is proportional to the square of the matrix element of the dipole moment d^2 between states of the radiative transition, as well as the rate of the radiative transition $1/\tau$, the absorption coefficient k_o does not depend on this value. Let us determine this parameter for a sodium vapor, where $\lambda = 589\,\mathrm{nm}$, $\tau = 1.6 \cdot 10^{-8}\,\mathrm{s}$, $\nu/N = 5.8 \cdot 10^{-7}\,\mathrm{cm}^3/\mathrm{s}$. This gives $k_o = 5.8 \cdot 10^4\,\mathrm{cm}^{-1}$. Correspondingly, the absorption coefficient of photons k_ω at frequencies ω at the spectral line wing, in accordance with the formula (2.1.40), is

$$k_\omega = k_o \frac{\nu^2}{4(\omega - \omega_o)^2} \tag{3.1.18}$$

Let us consider the spectral line profile of outgoing resonance radiation from a nonuniform plasma under real conditions where the temperature at plasma boundaries is less than that far from the boundary. Radiation emitted by a plasma at frequencies near the center of the spectral line is formed in a plasma region near its boundaries. If we are based on formula (2.2.33) for the radiative flux j_ω from the equilibrium plasma, it is necessary to use there the temperature T of the layer that gives the main contribution to the radiation flux. This layer lies at the distance x_ω from the boundary such that the optical thickness of this region is of the order of one

$$u(x_\omega) = \int_0^{x_\omega} k_\omega \mathrm{d}x \sim k_\omega x_\omega \sim 1$$

This formula describes self-reversal of spectral lines with a dip at the line center as it is shown in Fig. 3.2 for the case when the plasma temperature drops sharply near

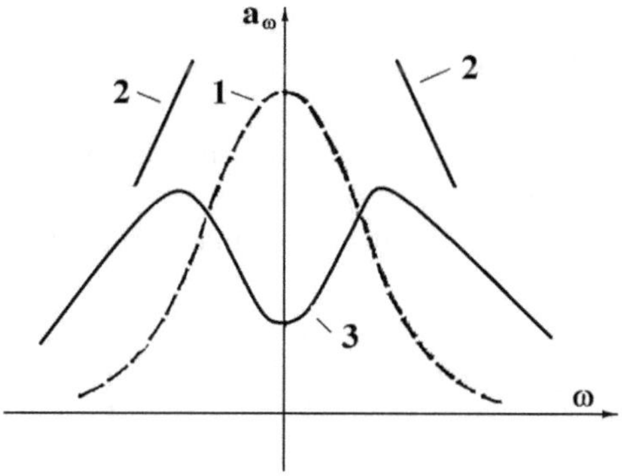

Fig. 3.2 Character of self-reversal for the spectral line, where the radiative temperature near the boundary of an excited gas is low compared to that far from the boundary. 1—the shape of the spectral line for an isolated atom, 2—the shape for the constant temperature, 3—the shape of the spectral line for a nonuniform excited gas

the plasma boundary. Evidently, the condition for a strong minimum of the radiation flux for the spectral line center has the following form

$$\frac{1}{k_o}\frac{\hbar\omega}{T}\frac{\mathrm{d}T}{\mathrm{d}x} \gg 1, \tag{3.1.19}$$

in the case $\hbar\omega > T$.

Thus, propagation of resonance radiation in an optically thick gas results from radiative transitions involving resonantly excited atoms and atoms in the ground state, and according to the above analysis the nature of this process is not diffusive. Therefore transport of photons on large distances compared with the mean free path of photons for the spectral line center is determined by photon frequencies at spectral line wings. This is explained by many events of photon absorption at the line center and a large mean free path for photons at far wings. Let us consider the behavior inside a gas for photons with a small mean free path compared the gas size, where the following criterion holds true

$$k_\omega L \gg 1 \tag{3.1.20}$$

Here the absorption coefficient k_ω is equal $k_\omega = N_o\sigma_a(\omega) - N_*\sigma_r(\omega)$ according to formula (2.2.26). Because of the effective absorption and emission of photons with this frequency inside the gas, the equilibrium is established for these photons, i.e. rates of the absorption and emission events are equal. Let i_ω be the flux of photons of frequency ω inside the gas, so that the number of photons absorbed per unit volume per unit time in a frequency range from ω to $\omega + \mathrm{d}\omega$ is equal $i_\omega k_\omega \mathrm{d}\omega$. The reduced number of absorbing photons is equal to the corresponding number of emitting photons, which is given by $N_j a_\omega \mathrm{d}\omega/\tau_r$. From this we obtain for the isotropic flux of photons at a given frequency

$$i_\omega = \frac{a_\omega N_*}{k_\omega \tau_r} = \frac{\omega^2}{\pi^2 c^2}\left(\frac{N_o}{N_*}\frac{g_*}{g_o} - 1\right)^{-1} \tag{3.1.21}$$

If atoms in the ground and excited states are in thermodynamic equilibrium, this formula coincides with formula (2.2.9).

As for the outgoing photon flux through a flat surface at a given frequency, it is equal

$$j_\omega = \int_0^1 i_\omega \cos\theta d\cos\theta \left(\int_-^1 d(\cos\theta)\right)^{-1} = \frac{i_\omega}{4}, \tag{3.1.22}$$

This coincides with formula (2.2.9). Here θ is the angle between the normal to the gas surface and the direction of photon propagation; and we have taken into account that the total photon flux outside the system is normal to the gas surface.

Formula (3.1.9) for the partial outgoing flux can be presented in the form

$$j_\omega = \frac{\omega^2}{2\pi^2 c^2}\int_0^1 d(\cos\theta)\int_0^{u_\omega} \mathrm{d}x\mathrm{d}u \exp\left(-\frac{u}{\cos\theta}\right)\left(\frac{N_o}{N_*}\frac{g_*}{g_o} - 1\right)^{-1}, \tag{3.1.23}$$

Here the current optical thickness of a layer is defined as $u_\omega = \int_0^x k_\omega \mathrm{d}x'$ and the total optical thickness of the layer at a given frequency is $u_\omega = \int_0^L k_\omega \mathrm{d}x$, i.e. $\mathrm{d}u_\omega = k_\omega \mathrm{d}x$. This formula allows us to estimate the width of the spectral line of radiation that leaves the plasma. The boundaries of the spectral line can be estimated from the relation

$$u_\omega = \int_0^L k_\omega \mathrm{d}x \sim k_\omega L \sim 1 \tag{3.1.24}$$

This formula determines the width of the spectral line $\Delta\omega$ of the total radiation.

Indeed, in the case of Lorentz broadening of the spectral line we have from formula (2.1.4) for the absorption coefficient at the line wing

$$k_\omega = k_o \frac{\nu^2}{(\omega - \omega_o)^2},$$

and the width of the spectral line for the total radiation flux is estimated as

$$\Delta\omega \sim \nu\sqrt{k_o L}, \quad k_o L \gg 1 \tag{3.1.25}$$

Correspondingly, for the Doppler profile (2.1.13) of the atom spectral line we obtain the width of the spectral line of total radiation

$$\Delta\omega \sim \Delta\omega_D\sqrt{\ln(k_o L)}, \quad k_o L \gg 1, \tag{3.1.26}$$

where $\Delta\omega_D$ is the Doppler width of the spectral line in the case of a small optical thickness of the gas. Thus, the yield of the resonance radiation from the optically thick gas plasma is characterized by a broad spectral line compared with the spectral line from an individual atom, because the main contribution to the outgoing radiation flux arises from wings of the spectral line for individual atoms.

3.2 Applying Aspects of Resonant Photons

3.2.1 Optical Pumping

Radiative transitions from the ground atomic state to the resonantly excited state results from the strongest interaction between the radiation states and free atoms; hence they are realized in effective processes of gas excitation [1, 15]. Therefore excitation of atoms by resonance radiation is a simple and strong process. Excitation of the resonantly excited state is the first stage of optical pumping [16–18], so that first resonantly excited atoms are formed, and these atoms are used subsequently for various goals. We below consider briefly principles of various applications of excited atoms [19].

It is clear that if the excited state includes several sublevels and excitation leads to population of one of these sublevels, this process allows one to create a laser or a maser. In particular, as an example of a device on the basis of optical pumping we consider below the rubidium maser which uses the isotope ^{87}Rb [20–23]. The orbital momentum is zero for the ground state of the rubidium atom, its electron spin equals 1/2, and the nuclear angular momentum is 3/2. Therefore, two hyperfine states exist for the rubidium atom in the ground state with the total momentum $F = 2$ for the upper state and $F = 1$ for the lower state. If rubidium atoms are excited by circularly polarized resonant radiation, subsequent emission of excited atoms leads to formation of atoms in the ground electron state and mostly in the superfine state with $F = 2$. This inverse population of levels is used in the rubidium maser which is an element of the optical frequency standard (see, in particular, [24, 25]). We neglect relaxation processes due to atomic collisions which mix states of the superfine structure. This example demonstrates also the possibility to separate certain nuclear states in spite of a small energy difference for corresponding levels.

In addition, this method of formation of rubidium atoms in a given superfine state may be used for measurement of low magnetic fields (for example, [26–30], if the splitting and shift of magnetic field is compensated by transitions under the action of a high-frequency electromagnetic field. We below consider another example of optical alignment is a magnetometer created on the isotope ^{4}He (its prevalence is nearly 100%) [31–33]. Gas discharge is initiated in the cell with the concentration of helium atoms in the ground state of the order of $10^{16}\,\mathrm{cm}^{-3}$. As a result, metastable atoms He(2^3S) are produced with a concentration of the order $10^{10}\,\mathrm{cm}^{-3}$. Resonance polarized radiation excites atoms into state He(2^3P). Inverse process is the spontaneous transition $He(2^3P) \to He(2^3S)$. Thus, atoms in the state $He(2^3S)$ acquire maximum projection of angular momentum to the direction of radiation. In an external magnetic field this state is splitting in three sublevels. Resonance transitions between these magnetic sublevels are produced by a radio-frequency electromagnetic field. On the basis of measured resonant radio-frequencies one can determine weak magnetic fields of the order of 10^{-7} G.

Other approach for measurements of magnetic fields as a result of optical pumping is so called the double radio-optical resonance method. It was first realized using resonance excitation of Hg vapors [17, 34]. In this case mercury atoms are excited firstly from the ground state $Hg(6^3S_0)$ by polarized resonance radiation with the wavelength 253.7 nm into the state $Hg(6^3P_1)$ with the projection of orbital momentum $M = 1$. The polarization of radiation emitted by atoms in the direction which differs from the direction of the initial radiation is measured. If the cell is found in a magnetic field, then the atomic level is splitting into three sublevels. Measurement the frequency of an applied radio-frequency field which produces the most strong variation of emitted radiation, it is possible to determine the value of the magnetic field strength.

We consider above only some examples of optical pumping by the polarized radiation. In this case absorption of radiation changes the atomic angular momentum. Oppositely, spontaneous emission of an atom does not change its average angular momentum. As a result, after several processes of radiation absorption and spontaneous radiation an atom returns to the initial atomic state; there arises so called optical alignment of atoms. The above concepts are used for various atomic objects [35–37].

3.2.2 Cooling of Atoms in Laser Field

Though spectroscopy methods are used during several centuries and played an important role in development of physics, in particular, in creation of quantum mechanics. Subsequent development of laser physics related to laser spectroscopy [19, 37–40] led to sources of monochromatic radiation with very narrow width of the spectral line, and this allows one to solve principally new problems. Lasers as a sources of monochromatic radiation open additional possibilities in gas excitation, though basic methods of optical pumping and subsequent methods for the action on atomic systems were developed just before creation of lasers. Nevertheless, lasers allow one to use new methods of gas excitation. Below we demonstrate this on the example of atom cooling as a result of optical pumping.

The method of optical pumping, where atoms are transferred into a resonantly excited state by means of a resonant excitation, and then return to the ground state due to spontaneous emission, is the basic of atomic cooling and creation of optical traps for atoms. Radiation of a tunable laser is tuned for this goal to the tail of the velocity distribution function of atoms. Then absorption of these photons and subsequent atomic transition to the ground state due to spontaneous emission diminishes the atom velocity and, thus, decreases the atomic temperature. Deceleration of atoms results from interaction between atoms and an electromagnetic field. Since atoms are moving in various directions, in the best methods six laser jets are used, where laser beams propagate towards each other in three mutual perpendicular directions. In the course of motion in a laser field, atoms breaks [41], so that the electromagnetic field is similar to viscous medium in this interaction. The motion of atoms in a viscous fluid of photons or optical molasses [42] is analogous to diffusion in classical Brownian motion. It should be noted that this configuration of electromagnetic fields does not produce a trap for atoms, but leads to their braking.

Developed methods of atomic cooling based on braking of atoms in the optical molasses allow one to cool a gas up to ultra-low temperatures of the order of 10^{-7} K. We below consider application of this method to sodium atoms. Transition of these atoms from the ground $3^2S_{1/2}$ state to the excited $3^2P_{1/2}$ state results from resonant radiation with the wavelength of $\lambda = 589\,\text{nm}$. The optimal atomic confinement can to be realized under low densities, if the width of the spectral radiation line Γ is the natural width $\Gamma = 1/\tau$, where τ is the radiative lifetime of the excited atom state. Then in the sodium case the temperature of $T_D = 240\,\mu\text{K}$ may be attained. Of course, this is an enough costly method, so that the number of photon emission and absorption events per atom can excess 10^4.

In consideration of interaction between atoms and electromagnet field in the course of atom cooling, in the first approximation one can neglect the influence of radiation on the spatial distribution of atoms. Then the lowest atomic temperature is determined by interaction of atoms with individual photons. As a result of an elementary act of absorption or spontaneous emission an atom acquires the momentum

$$\Delta p = \hbar k = \frac{\hbar\omega}{c},$$

where k is the photon wave number, c is the light speed of and $\omega = kc$ is the radiation frequency. Equalized this quantity to variation of the atomic momentum as a result of the radiative process, one can obtain for variation of the atomic velocity

$$\Delta v_R = \frac{\hbar k}{M} = \frac{\hbar\omega}{Mc},$$

where M is atomic mass. From this one can obtain the energy acquired by an initial motionless atom or the recoil energy

$$E_R = \frac{M(\Delta v_R)^2}{2} = \frac{(\hbar k)^2}{2M}$$

The temperature $T_R = 2E_R$ is called the limiting recoil temperature. In the case of sodium atoms we have $\Delta v_R = 3\,\text{cm/s}$, $E_R = 2.4\,\mu\text{K}$ and the limiting recoil temperature is $T_R = 2E_R = 5\,\mu\text{K}$. A thermal velocity of the sodium atom at room temperature is $5.0 \cdot 10^4\,\text{cm/s}$. However, if the temperature is equal to $T_D = \Gamma/2$ and the distribution function for atomic velocities is determined by the natural width of the spectral line, the thermal atomic velocity is $\Delta v_D = 30\,\text{cm/s}$ and the recoil velocity is $\Delta v_R = 3\,\text{cm/s}$.

Let us consider the character of interaction between laser radiation and sodium atoms. The maximum absorption cross section takes place at the natural width of spectral line

$$\sigma_{\max} = \frac{3\lambda^2}{2\pi} = 1.6 \cdot 10^{-9}\,\text{cm}^2,$$

where $\lambda = 589\,\text{nm}$ is the wavelength of resonant radiation; the coefficient 3 is the ratio of statistical weights for upper and lower transition states. The mean free path of resonant photons is of the order of 100 cm at the typical atomic concentration of $10^7\,\text{cm}^{-3}$ that is large compared to the size of a cell that is of the order 1 cm. Further, a small intensity of laser radiation $1\,\text{W/cm}^2$ corresponds to the photon flux of $3 \cdot 10^{18}\,\text{cm}^{-2}\text{s}^{-1}$ for sodium atoms. Hence, the rate of radiative transitions is estimated as $5 \cdot 10^4\,\text{s}^{-1}$. The thermal velocity of atoms is $5 \cdot 10^4\,\text{cm/s}$ at room temperature, and the mean free path for atoms which interact with resonant radiation is of the order of $0.1\,\mu\text{m}$. Note that this estimation corresponds to the low limit of the atom mean free path, and from this it follows that many events of absorption and emission including an individual atom occur under typical conditions where atoms move in a viscous medium of the optical molasses.

In order to create a trap for atoms based on the optical molasses it is convenient to use the combination of a polarized radiation and magnetic field. In the case of the quadrupole magnetic trap with zero magnetic field strength at the cell center, atoms tend to the cell center at a certain detuning of laser radiation from the line center, if polarized laser jets directed in opposite sides each to other. This type of magneto-optical trap permits to capture about of $4 \cdot 10^{10}$ sodium atoms [43].

Another method of atom capture into a trap on the basis of the optical molasses and polarized jets of laser radiation is the modulation of laser radiation. The spatial field distribution of each from two plane polarized waves is shown in Fig. 3.3. A laser beam of the first polarization excites atoms with the spin projection $+1/2$ on the beam direction, and the second beam excites atoms with the spin projection $-1/2$ with respect to the direction of the second beam. The electromagnetic field is taken as a stationary wave and it is tuned to the center of spectral line. Hence only atoms having zero kinetic energy during the excitation process interact with the field. After photon absorption, an atom emits a photon with other wavelength and therefore it loses the kinetic energy. Thus, atoms are captured by the stationary electromagnetic wave and oscillate in the potential well of this wave which lose their energy. Of course, atoms with the kinetic energies higher than the top of the potential well move free. However, these atoms move in the optical molasses and therefore they are decelerated. Finally, atoms are captured by such a trap, and their energy is of the order of the recoil energy.

The recoil energy is not the principal limit for atomic energies. First, it is possible to use two-photon absorption for oppositely propagating photons. Then recoils at the absorption of two photons with opposite momenta compensate each other. Second, a

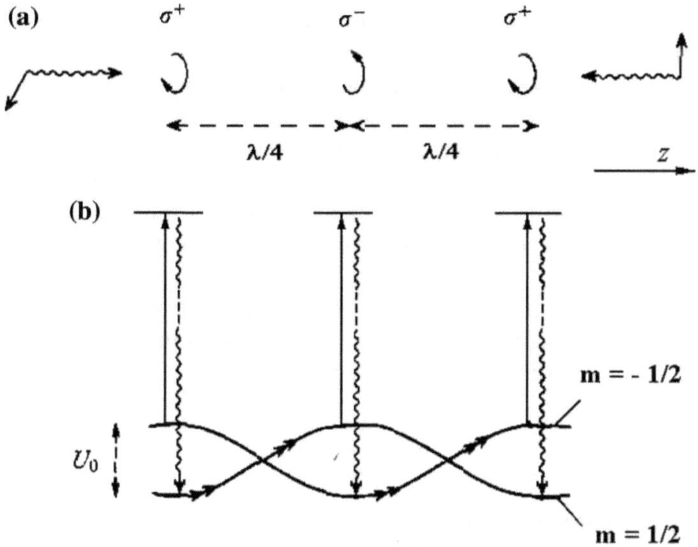

Fig. 3.3 Sisyphus effect in interaction between atoms and two oppositely directed laser jets with different polarizations. λ is the period of the field intensity variation in a space (**a**). Detuning of laser radiation from the spectral line center leads to absorption in such a way that an atom "climbs" to the potential hill and thus gives part of its kinetic energy to the wave (**b**)

recoil occurs in atom-photon interaction in the limit of low intensities of laser radiation. At high intensities of laser radiation and a specific direction and polarization of beams the optical lattice [44, 45] is formed, and atoms may be captured by potential wells of this lattice. The process of photon scattering on atoms which are located in a potential well created by the radiation field is analogous to the scattering of X-ray on a crystal lattice. We have considered above the most prospective methods to obtain ultra-low temperatures. They are based on processes of interaction between atoms and laser radiation.

3.2.3 *Light Induced Drift*

Optical pumping is a strong method to input an energy into a certain degree of freedom of an atomic system, and then this system develops along a certain channel that leads to a nonequilibrium state of this atomic system. We below consider the method of light induced drift. The idea of this method [46] is that if monochromatic radiation is tuned to a wing of the resonance spectral line, than at not high gas pressure where the Doppler broadening of spectral lines is realized, absorption at a wing of the spectral line relates to atoms of a certain velocity. Because an excited atom emits in a random direction, the action of radiation creates a force which acts on atoms of a certain kind.

This method was checked experimentally and developed (for example, [47–50]) and may be used practically twofold. The first application is separation of isotopes. This process may be realized in a gasdynamic beam of atoms where laser radiation is directed perpendicular to a beam. Then the space distribution of an exciting isotope becomes nonuniform in the transverse direction; if the beam after irradiation is divided in two beams, the concentration of a given isotope in two beams will be different.

This method may be used practically. Indeed, the relative excitation energy is of the order of m_e/M, where m_e is the electron mass, M is the nucleus mass, and the relative difference of excitation energies δ is

$$\delta \sim \frac{\Delta M}{M}\frac{m_e}{M}, \tag{3.2.1}$$

where ΔM is the difference of the masses for isotopes under consideration. In particular, in the case of lithium isotopes ^{6}Li and ^{7}Li we have $\delta \sim 10^{-4}$, while the Doppler width of a spectral line (2.1.12) the ratio of the width to the excitation energy is of the order of 10^{-6}. From this it follows the possibility to separate isotopes in this method. This method of isotope separation was realized for various elements (for example, [51–54]). But in analogy with other methods based on optical pumping, this method of isotope separation requires a relative large energy, and hence its application in reality is problematic.

Another application of the method of light induced drift is separation of elements. The character of this process is similar to electrophoresis [55–58], where space separation of elements results from the action of the electric field. Then in a mixture of some components the electric field acts of the ions formed from atoms or molecules of a certain component. The similar situation takes place in the method of light induced drift, where the radiation creates a force which acts on one component only. As a result, a space separation of components proceeds. But in contrast to electrophoresis, which may be used for gases and liquids, the method of light induced drift is used for separation of elements which are located in the gaseous matter.

3.2.4 Photoresonant Plasma

Because resonant spectral lines are narrow, it gives an effective instrument for excitation of gases and their diagnostics. Lasers as a sources of monochromatic radiation open additional possibilities in gas excitation, though basic methods of optical pumping and subsequent methods for the action on atomic systems were developed just before creation of lasers. Nevertheless, it is convenient to use tunable lasers for excitation resonantly excited atom states.

If the gas or vapor pressure is enough high, excited atoms are destroyed as a result of collisions, and the radiative energy absorbed by this gas remains there. Under these conditions the specific input energy may be high. As a result of collisions of resonantly excited atoms with each other and also with electrons, so called photoresonant plasma is formed in a small region. Hence, at a certain level of the gas pressure and radiative flux the exciting gas is heated and is transformed in a plasma. Therefore this plasma is compact and has a high specific internal energy that results in various applications [59–61]. Hence, this plasma may be used as a source of negative and multicharged ions. In addition, since a fast heating takes place in a small space region, it is a convenient acoustic source.

There are various regimes of realization of the photoresonant plasma depending on the intensity of the absorbed photon flux. We first consider the regime of low photon intensities [62, 63] where an absorption energy is compensated by spontaneous emission of resonantly excited atoms, i.e. processes in the photoresonant plasma proceed according to the scheme

$$\hbar\omega + A \leftrightarrow A^*, \tag{3.2.2}$$

and the appropriate balance equation for the number density of resonantly excited atoms N_* has the form

$$\frac{dN_*}{dt} = j_\omega \sigma_{\rm abs} N_0 - j_\omega \sigma_{\rm em} N_* - \frac{N_*}{\tau}, \tag{3.2.3}$$

where $\hbar\omega$ is the photon energy, $j_\omega = J_\omega/(\hbar\omega)$ is the photon flux, so that J_ω is the energy radiative flux, $\sigma_{\rm abs}$ is the absorption cross section, $\sigma_{\rm em} = \sigma_{\rm abs} g_0/g_*$ is the cross section of stimulated emission, so that $g_0,\ g_*$ are the atom statistical weights in the ground and excited states, τ is the radiative lifetime of excited atoms in a plasma that in the absence of reabsorption processes is equal to the radiative lifetime of an individual atom. One can introduce the temperature T_* of excitation as

$$\frac{N_*}{N_0} = \frac{g_*}{g_0} \exp\left(-\frac{\hbar\omega}{T_*}\right) \tag{3.2.4}$$

On the basis of this temperature, the stationary regime of the balance equation (3.2.2) takes the form

$$j_\omega k_o \left[1 - \exp\left(-\frac{\hbar\omega}{T_*}\right)\right] = \frac{N_*}{\tau}, \tag{3.2.5}$$

where $k_o = \sigma_{\rm abs} N_0$ is the absorption coefficient which does not depend on the number density of atoms N_0 for the resonance interaction. One can present this equation as

$$j_\omega = \frac{j_o}{\exp\left(\frac{\hbar\omega}{T_*}\right) - 1}; \quad j_o = \frac{N_0 g_*}{g_0 k_o \tau} \tag{3.2.6}$$

Table 3.2 Parameters of interaction of resonant radiation with vapors of alkali metals [62]

	k_o, $10^5\,\text{cm}^{-3}$	T_*, eV	j_o/N_0, 100 cm/s	I_o/N_0, 10^{-17} W · cm
$Li(2^2P)$	1.6	2.61	2.3	6.8
$Na(3^2P_{1/2})$	1.1	3.04	5.6	19
$Na(3^2P_{3/2})$	1.4	3.04	4.4	15
$K(4^2P_{1/2})$	0.85	2.32	4.5	15
$K(4^2P_{3/2})$	1.1	2.33	3.6	9.4
$Rb(5^2P_{1/2})$	0.91	2.25	3.9	10
$Rb(5^2P_{1/2})$	1.2	2.29	3.2	8.0
$Cs(6^2P_{1/2})$	0.77	2.00	4.3	9.6
$Cs(6^2P_{3/2})$	1.0	2.10	3.7	8.6

This relation connects the flux of resonance photons in the spectral line center and the temperature of excitation. This relation can be represented in the form

$$T_* = \frac{\hbar\omega}{\ln\frac{1+\eta}{\eta}};\ \eta = \frac{j_\omega}{j_o} = j_\omega \sigma_{\text{abs}} \tau \frac{g_0}{g_*} \tag{3.2.7}$$

In particular, if $j_\omega = j_o$, we have $T_* = 1.44\hbar\omega$. Table 3.2 contains values of this excitation temperature T_*, of the specific photon flux j_o/N_0 for $g_*/g_0 = 1$ and the specific intensity of incident radiation $I_o/N_0 = \hbar\omega \cdot j_o/N_0$ for alkali metal vapors at the Lorenz shape of a spectral line. These data allow one to estimate the range of parameters for this regime of the photoresonant plasma. Indeed, taking the pressure of alkali metal vapor $p \sim 1$ Torr, that corresponds to the number density of atoms in the ground state $N_0 \sim 10^{16}\,\text{cm}^{-3}$, we find a typical radiative flux $J_o \sim 1\,\text{W/cm}^2$.

This regime of the photoresonant plasma corresponds to the scheme (3.2.2); it is realized at low radiative fluxes j_ω. An increase of the number density of resonantly excited atoms opens new channels of the energy loss. As a result, free electrons are formed, at the beginning as a result of associative ionization

$$2A^* \to A_2^+ + e \tag{3.2.8}$$

When the number density of electron becomes sufficiently high, the electron subsystem is formed, where the equilibrium in the electron subsystem is established faster then that in the atomic subsystem. Then one can introduce the electron temperature T_e, and the energy distribution function of electrons has the Maxwellian form. Under these conditions, another regime of the photoresonant plasma is realized in accordance with the following processes

$$e + A^* \leftrightarrow e + A,\ e + A^*,\ e + A \leftrightarrow 2e + A^+,\ e + A_* \to 2e + A^+ \tag{3.2.9}$$

where A, A_* mean an atom in the ground and resonantly excited states. According to the first equation, the temperature of excitation and the electron temperature are identical $T_* = T_e$. The second and third equations mean an increase of the electron number density. Let us restrict by the impact mechanism of broadening of spectral lines, so the absorption coefficient at the center of the spectral line is equal (ignoring the induced radiation)

$$k_o = \frac{\lambda^2}{2\pi\tau\nu/N}, \tag{3.2.10}$$

where $\lambda = 2\pi c/\omega$ is the radiation wavelength. According to this formula, the absorption coefficient k_o at the center of a spectral line does not depend both on the number density of irradiating atoms, and on their mass. Next, the specific width of the spectral line ν/N is determined by collision of atoms in the ground and resonantly excited states, so that it is proportional to the square of the matrix element of the dipole moment d^2 between states of the radiative transition, similar to the rate of the radiative transition $1/\tau$. Therefore, the absorption coefficient k_o does not depend on this value. Correspondingly, the absorption coefficient of photons k_ω at spectral line wings according to formula (3.1.18) is given by

$$k_\omega = k_o \frac{\nu^2}{4(\omega - \omega_o)^2} \tag{3.2.11}$$

Considering transport of resonant radiation in a gas, where the photon mean free path at the line center $\lambda_o = 1/k_o$ is small compared to a system size, we have that the character of this process differs from that for transport of particles in a dense gas which has the diffusion character. In the case of diffusion particle transport of a particle displacement in one transfer act takes place for a distance of the order of the particle mean free path in a gas, which is small compared to a system size. Propagation of resonant radiation in a gas occurs in other manner and is described by the Biberman-Holstein equation [7, 10, 11]. As it follows from this equation, the displacement of a resonant photon for long distances in the gas does not occur as a result of many reabsorptions near the spectral line center, but it proceeds due to radiation at a spectral line wing, and the probability of such an event is small. It is convenient to use the Veklenko concept [8, 9] for the analysis of resonant radiation propagation in a gas, where the subject of consideration is an individual excitation, rather than a resonant photon. Correspondingly, the lifetime of the excited state $\tau_{\rm ef}$, i.e., the time of the location of excitation inside the volume gas, is estimated as $\tau_{\rm ef} \sim \tau/P_*$, where τ is the radiative lifetime of an isolated atom, P_* is the probability of emission of a photon at a frequency ω^*, for which the mean free path of a photon is comparable to a size of the system L, that is, $k(\omega_*)L \sim 1$. Since $k_oL \gg 1$, we have $P_* \sim \nu/|\omega_* - \omega_o|$, and the effective lifetime of excitation residence inside a gas volume is estimated as

$$\tau_{\rm ef} \sim \tau\sqrt{k_oL}, \; k_oL \gg 1 \tag{3.2.12}$$

In particular, if this gas is located inside a cylindrical discharge tube of a radius R, then at the uniform gas distribution the residence time of an excitation inside the tube volume is equal to [6]

$$\tau_{\text{ef}} = 2.6\tau\sqrt{k_o R} \tag{3.2.13}$$

Summarizing the above analysis, one can find the following character of creation of the photoresonant plasma at high intensities of resonant radiation [63, 64]. On the first stage of this process, atoms are excited in a narrow space region of micron sizes $\sim 1/k_o$, i.e. radiation acts on a small region that causes large specific powers injected in the gas. When the number density of atoms in the ground and resonantly excited states become comparable, the gas is cleared with respect to absorption of resonant radiation, so that radiation penetrates in more deep regions. As the excitation wave propagates inside the gas, the number density of electrons increases. Thus creation of photoresonant plasma proceeds in the form of an ionization wave, and in the case of a pulse resonant radiation, a photoresonant plasma is formed in a restricted space region of the gas.

References

1. A.C.G. Mitchell, M.W. Zemansky, *Resonance Radiation and Excited Atoms* (Cambridge University Press, Cambridge, 1934)
2. T. Watanabe, Phys. Rev. **139A,** 1375 (1965)
3. Y.A. Vdovin, V.M. Galitskii, ZhETF **52**, 1345 (1967)
4. A.A. Vlasov, V.S. Fursenko, ZhETF **6**, 750 (1936)
5. B.M. Smirnov, *Physics of Ionized Gases* (Wiley, New York, 2001)
6. B.M. Smirnov, *Theory of Gas Discharge Plasma* (Springer, Heidelberg, 2014)
7. L.M. Biberman, Dokl. AN SSSR **49**, 659 (1948)
8. B.A. Veklenko, ZhETF **33**, 629 (1957)
9. B.A. Veklenko, ZhETF **33**, 817 (1957)
10. L.M. Biberman, J. Exper. Teor. Fiz. **17**, 416 (1947) (in Russian)
11. T. Holstein, Phys. Rev. **72**, 1212 (1947)
12. B.M. Smirnov, *Physics of Weakly Ionized Gas* (Nauka, Moscow, 1972) (in Russian)
13. L.I. Sobelman, *Atomic Spectra and Radiative Transitions* (Springer, Heidelberg, 1979)
14. V.P. Krainov, H.R. Reiss, B.M. Smirnov, *Radiative Processes in Atomic Physics* (Wiley, New York, 1997)
15. A.C.G. Mitchell, M.W. Zemansky, *Resonance Radiation and Excited Atoms* (Cambridge University Press, Cambridge, 1961)
16. A. Kastler, J. Phys. Radium **11**, 255 (1950)
17. J. Brossel, F. Bitter, Phys. Rev. **86**, 308 (1952)
18. R. Bernheim, *Optical Pumping: An Introduction* (Benjamin, New York, 1965)
19. W. Demdroter, *Laser Spectroscopy. Basic Principles* (Springer, Heidelberg, 2008)
20. N. Knabe, Bul. Am. Phys. Soc. **6**, 68 (1961)
21. P. Davidovits, N. Knabe, J. Appl. Phys. **35**, 3042 (1964)
22. P. Davidovits, Appl. Phys. Lett. **6**, 20 (1965)
23. J. Vanier, Phys. Rev. **168**, 129 (1968)
24. J. Vanier, C. Audoin, *The Quantum Physics of Atomic Frequency Standards* (CRC Press, Boca Raton, 1989)

25. F.G. Major, *The Quantum Beat: The Physical Principles of Atomic Clocks* (Springer NY, New York, 1998)
26. W. Bell, A. Bloom, Optical detection of magnetic resonance in alkali metal vapor. Phys. Rev. **107**, 1559–1565 (1957)
27. W. Bell, A. Bloom, Phys. Rev. Lett. **6**, 280 (1961)
28. A. Bloom, Appl. Opt. **1**, 61 (1962)
29. J. Dupont-Roc, S. Haroche, C. Cohen-Tannoudji, Phys. Lett. A **28**, 638 (1969)
30. E.B. Aleksandrov, M.V. Balabas, A.K. Vershovskii, A.S. Pazgalev, Tech. Phys. **49**, 779 (2004)
31. A.R. Keiser, J.A. Rice, L.D. Schearer, J. Geophys. Res. **66**, 4163 (1961)
32. L.D. Schearer, F.D. Colegrove, G.K. Walters, Rev. Sci. Instrum. **34**, 1363 (1963)
33. D.D. McGregor, Rev. Sci. Instrum. **58**, 1067 (1987)
34. J. Brossel, A. Kastler, Compt. Rend. Acad. Sci. **229**, 1213 (1949)
35. E.B. Alexandrov, V.A. Bonch-Bruevich, Opt. Eng. **31**, 711 (1992)
36. S. Groeger, G. Bison, J.L. Schenker, R. Wynands, A. Weis, Eur. Phys. J. D **38**, 239 (2006)
37. C. Cohen-Tannoudji, D. Guery-Odelin, *Advances in Atomic Physics: An Overview* (World Scientific Publishing, New Jersey, 2011)
38. V.S. Letokhov, *Laserspektroskopie* (Vieweg, Braunschweig, 1977)
39. S. Stenholm, *Foundation of Laser Spectroscopy* (Dover, New York, 2005)
40. H. Abramczyk, *Introduction to Laser Spectroscopy* (Elsevier, Amsterdam, 2005)
41. T.W. Hansch, A.L. Schawlow, Opt. Commun. **13**, 68 (1975)
42. S. Chu, L. Hollberg, J. Bjorkholm, A. Cable, A. Ashkin, Phys. Rev. Lett. **55**, 48 (1985)
43. K. Gibble, S. Kasapi, S. Chu, Opt. Lett. **17**, 526 (1992)
44. D.S. Weiss, E. Riis, Y. Shery, P.J. Ungar, St. Chu, J. Opt. Soc. Am. **B6**, 2072 (1989)
45. C.I. Westbrook et al., Phys. Rev. Lett. **65**, 33 (1990)
46. F.K. Gel'mukhanov, A.M. Shalagin, JETP Lett. **29**, 711 (1979)
47. V.D. Antsygin, S.N. Atutov, F.K. Gel'mukhanov, JETP Lett. **30**, 243 (1979)
48. H.G.C. Werij, J.E.M. Haverkort, P.C.M. Planken et al., Phys. Rev. Lett. **58**, 2660 (1987)
49. S.N. Atutov, I.M. Ermolaev, A.M. Shalagin, Sov. Phys. JETP **65**, 679 (1987)
50. H.G.C. Werij, J.P. Woerdman, Phys. Rep. **169**, 145 (1988)
51. V.Yu. Baranov, E.P. Velikhov, A.M. Dykhne et al., JETP Lett. **31**, 445 (1980)
52. W.A. Hamel, A.D. Streater, J.P. Woerdman, Opt. Commun. **63**, 32 (1987)
53. A.D. Streater, J. Mooibroek, J.P. Woerdman, Opt. Commun. **64**, 137 (1987)
54. T.F. Gallagher, *Rydberg Atoms* (Cambridge University Press, Cambridge, 1994)
55. A. Tiselius, Trans. Faraday Soc. **33**, 524 (1937)
56. K. Hannig, Electrophoresis **3**, 235 (1982)
57. A.S. Cohen, B.L. Karger, J. Chromatogr. **397**, 409 (1987)
58. R. Westermeier, *Electrophoresis in Practice* (Wiley VCH, Weinheim, 2005)
59. I.M. Beterov, A.V. Eletskii, B.M. Smirnov, Sov. Phys. Usp. **31**, 535 (1988)
60. O.A. Gorshkov, R.N. Rizakhanov, Rev. Sci. Instr. **63**, 2466 (1992)
61. B.A. Knyazev, Nucl. Instr. Meth. Phys. **415**, 525 (1998)
62. B.M. Smirnov, *Plasma Processes and Plasma Kinetics* (Wiley, Weinheim, 2007)
63. B.M. Smirnov, *Fundamental of Ionized Gases. Basic Topics of Ionized Gases* (Wiley, Weinheim, 2012)
64. A.V. Eletskii, Y.N. Zaitsev, S.V. Fomichev, Zh. Exp. Teor. Fiz. **94**, 98 (1988)

Chapter 4
Radiative Processes in Molecular Gases

Abstract Selection rules are analyzed for vibration-rotation radiative transitions. Within the framework of the harmonic model for atomic oscillations the selection rule corresponds to a change of the vibrational number by one in radiative transitions, whereas rotational number may be conserved or be changed by one in vibrational-rotational transitions. The expression is presented for the absorption coefficient as a result of radiative vibrational-rotational transitions in diatomic molecules, as well as the expression for the spectral line intensity. The absorption coefficient is obtained for radiative vibrational-rotational transitions in atmospheres of the Earth and Venus due to atmospheric carbon dioxide molecules.

4.1 Selection Rules for Radiation of Molecular Gases

4.1.1 Selection Rules for One-Photon Transitions Between Vibrational States in Molecules

Molecules consisting of bound atoms are the simplest atomic systems; their properties are determined by interaction between atoms. Large difference in masses of nuclei and electrons gives the possibility to split the problem of calculation of molecular energy levels in two parts. First we determine a surface of potential energy, i.e., energy levels at fixed positions of nuclei and obtain molecular energy with infinitely heavy nuclei which depend on the nuclear configuration. In the case of diatomic molecules the potential energy surface is the electron term (potential curve), where the molecular electron energy depends on a distance between nuclei. Analogously to infinite numbers of atomic electron levels, the infinite number of electron terms relates to each molecule. As an example of electron terms for low excited states of the NH molecule are shown in Fig. 4.1.

The other part of the molecular energy levels results from nuclear motion within a given electron term. We consider atomic nuclei as particles which interact with

V. Krainov and B. M. Smirnov, *Atomic and Molecular Radiative Processes*,
Springer Series on Atomic, Optical, and Plasma Physics 108,
https://doi.org/10.1007/978-3-030-21955-0_4

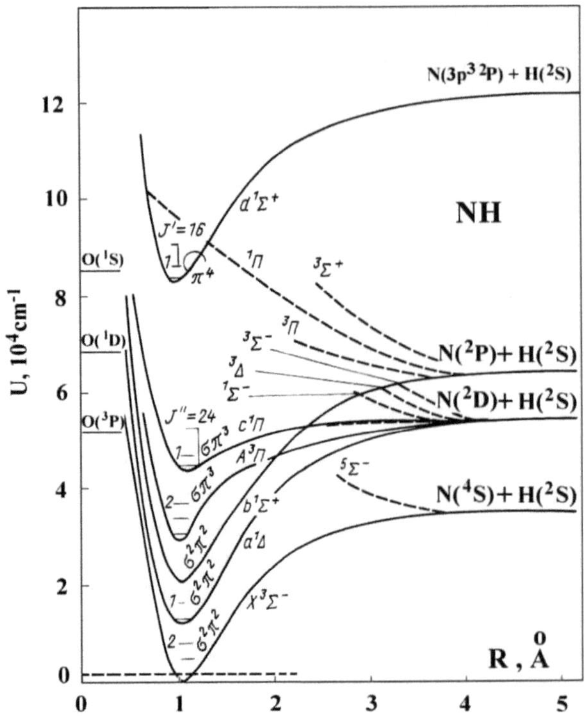

Fig. 4.1 Electron terms of the molecule NH resulted from interaction of hydrogen and nitrogen atoms. At large separation these electron terms correspond to isolated nitrogen atom which is found in various electron states, and the hydrogen atom in the ground state [1]

each other by means of some interaction potential $U(R)$, where R describes the sum of nuclear coordinates. Below we consider only the motion of nuclei with respect to their center of inertia which can be connected with radiative transitions. This motion within the framework of each potential well of a diatomic molecule consists of the vibrational and rotational degrees of freedom. Emission and absorption of molecules in the infrared spectrum range are determined by large number of vibrational-rotational transitions, and the molecule spectrum consists of a large number of broadened spectral lines due to these transitions. We below are restricted by diatomic and triatomic molecules where three atoms are located in the same line. Then under thermodynamic equilibrium the number density of molecules $N_{v,J}$ in a given vibrational-rotational state is equal

$$N_{v,J} = N_o \frac{B}{T} \exp\left(-\frac{\hbar v \omega_o}{T}\right) \exp\left[-\frac{BJ(J+1)}{T}\right], \qquad (4.1.1)$$

this thermodynamic equilibrium in a gas is supported by collisions involving these molecules. Here N_o is the molecular number density in the ground vibrational and rotational states, $\hbar\omega_o$ is the excitation energy of the vibrational level, v is the quantum number of this level, J is the rotational quantum number, $B = \hbar^2/2\mu \cdot r_o^2$ is the rotational constant (μ is the reduced mass of nuclei, r_o is the distance between nuclei), T is the gas temperature expressed in the energy units. Usually $B \ll T$, and below we assume this relation to be fulfilled.

Let us consider the character of interaction in diatomic molecules and its influence on radiative transitions in molecules. Evidently, the interaction potential $U(r)$ in the molecule depends on a distance between nuclei r. Excluding the angle dependence of the wave function of nuclei, one can reduce the problem of the nuclear motion to their one-dimensional motion in the standard method by addition the centrifugal potential and introduce in this manner the effective interaction potential as

$$\hat{U}_{\text{eff}}(r) = U(r) + \frac{\hbar^2}{2\mu r^2}\hat{K}^2$$

Here μ is the reduced mass of the molecule, $\hat{K} = \hat{J} - \hat{L}$ is the nuclear rotational momentum, so that $\hat{J}$ is the total molecular angular momentum, $\hat{L}$ is the electron orbital momentum. One can average the effective interaction potential $\hat{U}_{\text{eff}}(r)$ in the adiabatic approximation over the electron state at a fixed value of r. Thus, the effective interaction potential takes the form

$$U_{\text{eff}}(r) = U(r) + \frac{\hbar^2}{2\mu r^2}K(K+1)$$

The nuclear motion within one electron term may be described as small oscillations with respect to the equilibrium nuclear position. In the lowest order of expansion we have

$$U_{\text{eff}}(r) = U(r) + \frac{\hbar^2}{2I}K(K+1) + \frac{\mu\omega_o^2(r - r_0)^2}{2}$$

Here r_o is the equilibrium distance between nuclei, $I = \mu r_0^2$ is the molecular moment of inertia, ω_o is the frequency of the classical oscillator, i.e.

$$\omega_o = \sqrt{\frac{U''(r_o)}{\mu}}$$

The last term in the expression for $U_{\text{eff}}(r)$ presents the potential of one-dimensional harmonic oscillator. Therefore molecular energy levels are of the form

$$E_{K,v} = E_0 + \hbar\omega_0\left(v + \frac{1}{2}\right) + \frac{\hbar^2}{2I}K(K+1); \;\; v = 0, 1, 2, \ldots; \; K = 0, 1, 2, \ldots \tag{4.1.2}$$

This gives the following estimation for the vibrational frequency

$$\omega_0 \sim \frac{1}{\sqrt{M}}.$$

Since $\mu \gg m_e$, where m_e is the electron mass, typical vibrational energies are small compared to a difference between the neighboring electron terms which are of the order of an atomic unit.

Let us consider electron terms at fixed positions of nuclei and ignoring the molecular rotation. Along with the projection Λ of the electron orbital momentum onto the molecular axis and quantum number v of vibrational motion, it is necessary to include into consideration the projection Σ of the total electron spin S onto the molecular axis which values are $\Sigma = -S, -S+1, \ldots, +S$. The projection of the total rotational momentum of electrons on the molecular axis is $\Omega = \Lambda + \Sigma$ and its values are $\Omega = \Lambda + S, \Lambda + S - 1, \ldots, \Lambda - S$. Thus, an electron level with the quantum number Λ splits into $2S+1$ sublevels of the fine structure with various values of Ω. Since the spin-orbital interaction potential is proportional to $\hat{L}\hat{S}$, the corresponding spin-orbital splitting can be presented as $A(r)\Sigma$, since the vector $\left\langle \hat{\mathbf{L}} \right\rangle$ is directed along the molecular axis. At a certain value Λ the electron energy is equal to $U(r) + A(r)\Sigma$. Hence, the energy difference of neighboring levels is the same. It should be noted that the above consideration is valid for molecules of light elements where relativistic effects are negligibly small.

We now compare the rates of radiative transitions between vibrational and electron molecular states. The difference of these rates is, on the one hand, due to difference in transition energies. On the other hand, the matrix elements of the molecular dipole moment are different. We have seen above that the ratio of photon energies for the vibrational transition $\hbar\omega_o$ and the electron transition $\hbar\omega_e$ is equal to

$$\frac{\hbar\omega_o}{\hbar\omega_e} \sim \sqrt{\frac{m_e}{\mu}}$$

The matrix element of the molecular dipole moment for the transition between two neighboring vibrational states is estimated as

$$\langle v\,|D|\,v-1\rangle = e\sqrt{\frac{\hbar v}{2\mu\omega_o}}$$

Hence, its ratio to the matrix element of the dipole moment D_e for the electron transition is of the order

$$\frac{\langle v\,|D|\,v-1\rangle}{D_e} = \sqrt{v\frac{m_e}{M}\frac{\omega_e}{\omega_o}}$$

It is seen that the rates of vibrational transitions are much less than the rates of electron transitions.

One can prove that the matrix element of the dipole moment for radiation transitions between vibrational states of a diatomic molecule consisting of two identical atoms is equal to zero. We first determine the dipole moment of the molecule, considering the nuclei to be in fixed positions, and viewing these nuclei as sources of a potential field. The molecule is symmetrical for reflection with respect to the plane which is perpendicular to the molecule axis and bisects it. In addition, the electron density also has axial symmetry with respect to the molecular axis. Therefore the electron density ρ is invariant with respect to inversion of all electrons. The dipole moment is thus

$$\mathbf{D} = \int \sum_i e\mathbf{r}_i \rho d\mathbf{r}_1 d\mathbf{r}_2 ... d\mathbf{r}_n = - \int \sum_i e\mathbf{r}_i \rho d\mathbf{r}_1 d\mathbf{r}_2 ... d\mathbf{r}_n = 0$$

for a fixed internuclear distance. Therefore, the matrix element of this operator between vibrational states is zero, and radiative vibrational transitions are absent in this case.

This conclusion holds true also, if nuclei are of different isotopes because the symmetry is determined by fields which results from interaction of these nuclei with electrons. However, this statement is violated, if the molecular rotation influences the electron state. Then the inversion symmetry of the electron wave function is lost because two nuclei have different masses. But, even for identical nuclei of a diatomic molecule this violation is not strict, because a very weak interaction of the nuclear spins with the electrons disrupts the above symmetry of the electron wave function, since two nuclear spins may have a different direction and then their influence on the electron density is different. In this case interaction of electrons with the total nuclear spin will lead to a weak mixing of electron states of opposite parity, so that dipole transitions become possible.

Let us consider the selection rules for radiative vibrational transitions in diatomic molecules which are determined by properties of the matrix element $\langle v | \bar{\mathbf{D}} | v' \rangle$, where v and v' are vibrational quantum numbers, and $\bar{\mathbf{D}}$ is the dipole moment averaged over that part of the electron configuration, that does not change as a result of the dipole transition; the value $\bar{\mathbf{D}}$ is taken at a certain distance between nuclei and then it is averaged. This corresponds to the adiabatic approximation, where the motion of the nuclei proceeds slower than the electron motion. We use that the amplitude of vibrations of nuclei is small compared to a distance between them, that allows us to employ the expansion

$$\bar{\mathbf{D}} = \bar{D}_0 + \sum_i \left(\frac{\partial \bar{\mathbf{D}}}{\partial Q_i} \right)_0 Q_i + \frac{1}{2} \sum_{i,k} \left(\frac{\partial^2 \bar{\mathbf{D}}}{\partial Q_i \partial Q_k} \right)_0 Q_i Q_k + \cdots \tag{4.1.3}$$

Here the Q_i are normal coordinates of nuclei, and an index i enumerates the type of vibrations. The quantity $\bar{\mathbf{D}}_0$ describes the dipole moment of the molecule at the equilibrium configuration of the nuclei. Derivatives of the normal coordinates are also evaluated for the equilibrium configuration.

Within the framework of the harmonic oscillator model for nuclear vibrations, we have that the matrix element of the normal coordinate Q_i for the second term of the right-hand side of relation (4.1.15) is nonzero only for transitions with a change of the vibrational quantum number v by one. This matrix element is of the form

$$\langle v\,|Q_i|\,v-1\rangle = \sqrt{\frac{\hbar v}{2M_i\omega_i}},$$

Here M_i is the reduced mass of the molecule for a given type of vibrations, and ω_i is the frequency of this vibration for the mode i. As is seen from formula (4.1.15) for the mean dipole moment, transitions with a change of vibrational quantum number by two are possible owing to the third term on the right-hand side of relation (4.1.15).

We now compare the expressions for the radiative transition rates with a change of vibrational quantum number by two to those with a change by one. The ratio of the rates of these radiative transitions is given by

$$\frac{w\,(v\rightarrow v-2)}{w\,(v\rightarrow v-1)} \sim \left|\frac{\langle v\,|\bar{\mathbf{D}}|\,v-2\rangle}{\langle v\,|\bar{\mathbf{D}}|\,v-1\rangle}\right|^2 \sim$$

$$\sim \left|\frac{\sum\limits_{i,k}\left(\frac{\partial^2\bar{\mathbf{D}}}{\partial Q_i\partial Q_k}\right)_0\langle v\,|Q_iQ_k|\,v-2\rangle}{\sum\limits_i\left(\frac{\partial\bar{\mathbf{D}}}{\partial Q_i}\right)_0\langle v\,|Q_i|\,v-1\rangle}\right|^2 \sim \left|\frac{\sum\limits_{i,k}\left(\frac{\partial^2\bar{\mathbf{D}}}{\partial Q_i\partial Q_k}\right)_0\sqrt{\frac{\hbar v(v-1)}{M_iM_k\omega_i\omega_k}}}{\sum\limits_i\left(\frac{\partial\bar{\mathbf{D}}}{\partial Q_i}\right)_0\sqrt{\frac{\hbar v}{M_i\omega_i}}}\right|^2,$$

where we have used the rule of matrix multiplication

$$\langle v\,|Q_iQ_k|\,v-2\rangle = \langle v\,|Q_i|\,v-1\rangle\,\langle v-1\,|Q_k|\,v-2\rangle$$

The oscillator frequency is given in atomic units, it is estimated as

$$\omega_i \sim \frac{1}{\sqrt{M_i}}$$

The derivation of the dipole moment is of the order of the atomic value. Hence, we obtain the following estimation for the ratio of probabilities

$$\frac{w\,(v\rightarrow v-2)}{w\,(v\rightarrow v-1)} \sim v\sqrt{\frac{m_e}{\mu}} \tag{4.1.4}$$

Numerically, this ratio is of the order of 10^{-2}–10^{-3}. Thus, the most probable radiative transitions between vibrational states take place with change of the vibrational quantum number by one. Transitions with change of the vibrational quantum number by two are relatively weak if the vibrational quantum number v is not too large. The ratio of the rates (4.1.16) becomes of the order of one if the quantum number v is of the order of

$$v \sim \sqrt{\frac{\mu}{m_e}} \gg 1$$

If this criterion holds true, the vibrational energy is of the order of the electron energy, $\hbar v\omega_i \sim \varepsilon_e$ which would imply that the harmonic oscillator approximation is not valid. Consequently, the above analysis is not suitable for this case.

Transitions with $v \to v-2$ also take place in the first-order term in the expansion of the dipole moment $\bar{\mathbf{D}}$, if we take into account the anharmonicity of the nuclear oscillations. To estimate this effect, we introduce the anharmonic term αQ^3 into the Hamiltonian, describing vibrations. The value of α is of the order of an atomic value. For simplicity, we shall consider only one type of vibrations with the frequency ω_o, so we below omit the indexes i and k. In the first-order perturbation theory, the correction to the harmonic wave function $\psi_{v-2}^{(0)}$ of the state with the quantum number $v-2$ is of the form

$$\psi_{v-2}^{(1)} = \alpha \sum_{v'} \frac{\langle v' \,|Q^3|\, \nu - 2\rangle}{\varepsilon_{v-2}^{(0)} - \varepsilon_{v'}^{(0)}}.$$

We now extract the term with $v' = v-1$ (other terms with $v' = v+1,\ v-3,\ v-5$ have the same estimate). Then we find that the matrix element $\langle v' \,|Q^3|\, v-2\rangle$ is of the order of

$$\langle v' \,|Q^3|\, v-2\rangle \sim [\langle v-1 \,|Q|\, v-2\rangle]^3 \sim \left(\frac{\hbar v}{M\omega_o}\right)^{3/2}$$

and

$$\left\langle \psi_v^{(0)} \,|Q|\, \psi_{v-2}^{(1)} \right\rangle \sim \alpha \left\langle \psi_v^{(0)} \,|Q|\, \psi_{v-1}^{(0)} \right\rangle \frac{\langle v-1 \,|Q^3|\, v-2\rangle}{\varepsilon_{v-2}^{(0)} - \varepsilon_{v-1}^{(0)}} \sim \frac{\alpha}{\hbar\omega_o} \left(\frac{\hbar v}{M\omega_o}\right)^2$$

Let us estimate the ratio of the radiative transition rates

$$\frac{w\,(v \to v-2)}{w\,(v \to v-1)} \sim \left| \frac{\langle v \,|Q|\, v-2\rangle}{\langle v \,|Q|\, v-1\rangle} \right|^2 \sim \frac{\alpha^2}{(\hbar\omega_o)^2} \left(\frac{\hbar v}{\mu\omega_o}\right)^4 \frac{\mu\omega_0}{\hbar v} \sim \frac{v^3}{\mu^3\omega_o^5} \sim v^3 \sqrt{\frac{m_e}{\mu}}$$

This ratio becomes of the order of one, if the vibrational quantum numbers $v \sim (\mu/m_e)^{1/6}$. Then the vibrational energy is of the order of

$$v\hbar\omega_o \sim \left(\frac{m_e}{\mu}\right)^{1/3} \varepsilon_e \ll \varepsilon_e$$

It is small compared to a typical electron energy ε_e. As is seen, the correction to the rate of the transition $v \to v-2$ due to anharmonicity of nuclear vibrations is larger than that due to the dependence of the mean dipole moment on a distance between nuclei at high vibrational quantum numbers. However, both corrections are of the same order of magnitude for small vibrational numbers v.

It is seen, that the rates for transitions with $v \to v-2$ and $v \to v-1$ become of the same order of magnitude if the correction to the wave function

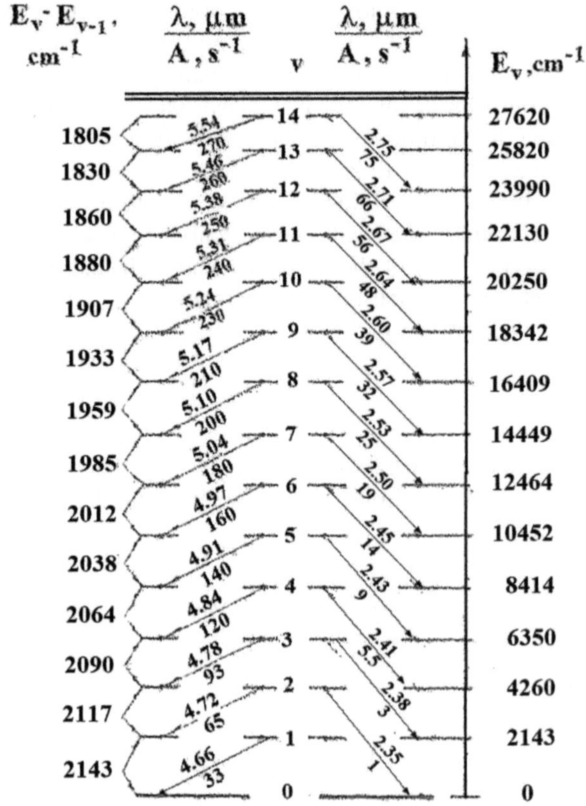

Fig. 4.2 Parameters of radiative vibrational transitions for the CO molecule [2]. Here E_v is the excitation energy of the vibration level with the vibration number v, A is the Einstein coefficient for transition between indicated levels, λ is the wavelength for this transition

$$\psi_{v-2}^{(1)} \sim v^{3/2}\left(\frac{m_e}{\mu}\right)^{1/4}\psi_{v-1}^{(0)}; \quad v \sim \left(\frac{\mu}{m_e}\right)^{1/6}$$

due to anharmonicity is comparable to the unperturbed harmonic wave function $\psi_{v-1}^{(0)}$. The harmonic approximation is inapplicable under such circumstances, and these quantum numbers v are indeed absent even though the corresponding energies are still small compared to the typical electron energies ε_e. From the above analysis one can conclude that the vibrational number v is a "good" quantum number, the most effective radiative transitions take place with the change of v by one. But with growth of the vibrational quantum number transitions $v \to v \pm 2$ become remarkable. This is shown in Fig. 4.2, where the rates of radiative transitions (Einstein coefficients) are given for radiative transitions of the CO molecule with the change of the vibrational quantum number both by one and by two. The results of Fig. 4.2 confirm the above conclusion that the two-photon radiative transitions between vibrational states of a diatomic molecule are weaker than the single-photon ones; however, their role rises as the vibrational quantum number increases.

4.1.2 *Selection Rules for Transitions Between Rotational States of Diatomic Molecules*

We now analyze radiative transitions between rotational states of a diatomic molecule and determine the selection rules. For the sake of simplicity, we consider first only the electron terms for which the total molecular spin is zero. We denote by **J** the total angular momentum of the molecule in the initial state. It is composed of the orbital electron momentum and the rotational angular momentum of the nuclei. The projection of the total angular momentum on a fixed axis is denoted by M. The orbital momentum Λ conserves its projection onto the molecular axis at the transition due to the axial symmetry of the molecule. Since the rotational angular momentum is perpendicular to the molecular axis, then the quantity Λ also presents the projection of the total angular momentum of the molecule onto its axis. Analogous quantities for the final state of the molecule are marked by a prime.

Let us consider the transition $JM \to J'M'$ between rotational states of the molecule for a given electron state, that is, for fixed quantum number Λ. The problem is reduced to calculation of the matrix element for the dipole moment operator. The matrix element of the component D_q (where q is a spherical component) of the dipole moment vector in the rest system can be expressed via the analogous matrix element in the rotating coordinate system in which the z axis is along the direction of the molecular axis

$$\left\langle J'M'\Lambda \left| D_q \right| JM\Lambda \right\rangle = \sqrt{\frac{2J'+1}{2J+1}} \left\langle J'1, M'q | JM \right\rangle \left\langle J'1, \Lambda 0 | J\Lambda \right\rangle \langle \Lambda \left| D_z \right| \Lambda \rangle \tag{4.1.5}$$

The index q takes the values $0, \pm 1$. Is is clear that the matrix element of the dipole moment operator does not depend on the rotational quantum numbers in the frame of reference associated with the molecular axis; it is determined only by the electronic state of the molecule. Thus, this matrix element is diagonal with respect to rotational transition; it is equal to the mean dipole moment of the molecule, $\bar{D} = \langle \Lambda \left| D_z \right| \Lambda \rangle$. The selection rules for dipole rotational transitions follow from the properties of the Clebsch-Gordan coefficients contained in the (4.1.5) as

$$J - J' = \pm 1; \;\; M - M' = q = 0, \pm 1 \tag{4.1.6}$$

Since the energies ε_J of the rotational states are determined by formula $\varepsilon_J = BJ(J+1)$, then the spontaneous transition from the state with angular momentum J to the lower state is possible only with $J' = J - 1$. On the basis of formula (1.2.18) for the rate of radiative processes, we obtain

$$w\left(J \to J' = J - 1\right) = \frac{4\omega_{JJ'}^3 \bar{D}^2}{3\hbar c^3} \langle J-1, 1; \Lambda 0 | J, \Lambda \rangle^2 \sum_{M',q} \left\langle J-1, 1; M', q | J, M \right\rangle^2 \tag{4.1.7}$$

This expression is averaged over polarizations of emitted photons and integrated over the angle of emission. When we carry out the sum in (4.1.7) and use explicit values of the Clebsch-Gordan coefficients, we find the spontaneous transition rate

$$w\left(J \to J-1\right) = \frac{32 B^3 \bar{D}^2}{3\hbar c^3} \frac{J^2\left(J^2 - \Lambda^2\right)}{2J+1}; \quad B = \frac{\hbar^2}{2\mu r_o^2} \tag{4.1.8}$$

This result relates to the transition at which the vibrational and electronic states of the molecule do not change.

If the spin of the molecule is nonzero, the results must be modified somewhat. One can consider two limiting cases. If the spin interaction with the molecular axis due to spin of electrons of both atoms is large compared to the difference of neighboring rotational levels, the molecule rotation does not destroy spin coupling, and the projection of the total spin onto the molecular axis Σ is conserved. Then the projection of the total angular momentum onto the molecular axis is $\Omega = \Lambda + \Sigma$. In this case it is necessary to replace Λ by Σ in formula (4.1.8). In the opposite limiting case, where the spin interaction with the molecular axis is small compared to the difference of rotational energies for neighboring states, rotation destroys the spin-axis coupling. Then one can introduce the conserved orbital and rotational quantum numbers. The total angular momentum $\mathbf{J} = \mathbf{K} + \mathbf{S}$ is also conserved, where $\mathbf{S}$ is the spin vector of the molecule, and $\mathbf{K}$ is the rotational momentum. Each rotational level splits into a multiplet with $2S + 1$ components, which have angular momenta ranging from $J = K - S$ to $J = K + S$. If we do not specify the component of the multiplet, then the total rotational transition rate is obtained from (4.1.8) by replacing J with K. However, for relative probabilities involving individual lines of the multiplet, then we obtain expressions by analogy with those in the case of radiative transitions between components of fine structure. Namely, in this case the angular momenta $\mathbf{K}$ and $\mathbf{S}$ are coupled by analogy to the summation of momenta $\mathbf{L}$ and $\mathbf{S}$ in the fine structure of light atoms. In this case there is no coupling of the angular momenta $\mathbf{K}$ and $\mathbf{S}$ with the molecular axis.

Let us compare the rates of radiative transitions between rotational and electronic molecular states. It follows from formula (4.1.8) for $J \gg 1$ that

$$\frac{w_{\rm rot}}{w_e} \sim J^3 \left(\frac{m_{\rm e}}{M}\right)^3 \ll 1 \tag{4.1.9}$$

From this one can conclude that the rate of rotational transitions is small compared to that of electron ones. One can compare this with the above ratio of the rates of vibrational and electronic transitions that is given by

$$\frac{w_{\rm vib}}{w_e} \sim v \left(\frac{m_{\rm e}}{\mu}\right)^2 \ll 1 \tag{4.1.10}$$

Here v is the vibrational quantum number. It follows from comparison of formulas (4.1.9) and (4.1.10) that the rate of rotational transitions is less also than that of the vibrational transitions.

Let us discuss the selection rules for rotational quantum numbers in vibrational-rotational transitions with a change of the vibrational quantum number. This problem is a generalization of the previous one, where the rotational transition rate takes place without a change of the vibrational state. In that case, the matrix element of the projection of the dipole moment onto the molecular axis was equal to the mean dipole moment of the molecule at the distance $r = r_o$ between the nuclei, where r_o is the equilibrium distance. This matrix element is zero for transitions associated with a change in vibrational state as a consequence of the orthogonality of vibrational wave functions. Therefore now we present the next term of expansion of the dipole moment averaged over an electron state over a small differences $Q = r - r_0$. The term proportional to Q leads to matrix elements arising from the linear harmonic oscillator coordinate that is nonzero for transitions between neighboring vibrational states only. Thus, the selection rule $v' - v = \pm 1$ is valid for the vibrational quantum number v.

Considering the radiative vibrational-rotational transitions, we now concentrate on the change of the rotational state. For a diatomic molecule, the dipole moment operator of the molecule is directed along its axis with the unit vector $\mathbf{n}$ in this direction. The matrix element of the dipole moment is proportional to the quantity $\langle JM \left| n_q \right| J'M' \rangle$, where the component n_q is connected with the vector $\mathbf{n}$ in the same manner, as D_q relates to $\mathbf{D}$. Correspondingly, the rate of this spontaneous transition with a given change of quantum numbers is proportional to the square of this matrix element, i.e.,

$$w\left(v, J, M \to v', J', M'\right) \sim \left|\langle JM \left| n_q \right| J'M' \rangle\right|^2 ,$$

where v, J, M are the vibrational quantum number, the total angular momentum of the linear molecule, and its projection onto a fixed axis in a space respectively. The primed quantities v', J', M' are the same quantum numbers for the final molecular state.

The total transition rate into all rotational states is an inverse lifetime of this state, or

$$\frac{1}{\tau} = \sum_{J',M',q} w\left(v, J, M \to v', J', M'\right)$$

On the basis of the normalization condition

$$\sum_{J',M',q} \left|\langle JM \left| n_q \right| J'M' \rangle\right|^2 = \sum_q \langle JM \left| n_q^2 \right| JM \rangle = 1,$$

one can obtain from the last two expressions

$$w\left(v, J, M \to v', J', M'\right) = \frac{1}{\tau} \left|\langle JM \left| n_q \right| J'M' \rangle\right|^2 ; \; q = M' - M$$

Table 4.1 The probability $\Phi_{JM}\left(J', M'\right)$ for molecular vibrational-rotational transitions

	$J' = J - 1$	$J' = J + 1$
$M' = M - 1$	$\frac{(J+M)(J+M-1)}{2(2J-1)(2J+1)}$	$\frac{(J-M+1)(J-M+2)}{2(2J+1)(2J+3)}$
$M' = M$	$\frac{J^2-M^2}{(2J-1)(2J+1)}$	$\frac{(J+1)^2-M^2}{(2J+1)(2J+3)}$
$M' = M + 1$	$\frac{(J-M)(J-M-1)}{2(2J-1)(2J+1)}$	$\frac{(J+M+1)(J+M+2)}{2(2J+1)(2J+3)}$

We now calculate the matrix element from the projection of the unit vector n_q. Using the Clebsch-Gordan coefficients, one can obtain

$$\Phi_{JM}\left(J'M'\right) = \left|\left\langle JM\left|n_q\right|J'M'\right\rangle\right|^2 = \frac{2J'+1}{2J+1}\left\langle J'1, M'q|JM\right\rangle^2\left\langle J'1, 00|J0\right\rangle^2, \tag{4.1.11}$$

where q is equal to 0, -1, $+1$. The second factor in expression (4.1.11) is nonzero only if $J' = J \pm 1$. Therefore the only transition is possible with a change of the rotational quantum number by one. Values of the function $\Phi_{JM}\left(J'M'\right)$ for various values of J' and M' are given in Table 4.1.

The function $\Phi_{JM}\left(J'M'\right)$ is satisfied the following sum rules

$$\sum_{M'}\Phi_{JM}\left(J-1, M'\right) = \frac{J}{2J+1}; \quad \sum_{M'}\Phi_{JM}\left(J+1, M'\right) = \frac{J+1}{2J+1}$$

From this we also have

$$\sum_{J'M'}\Phi_{JM}\left(J', M'\right) = 1$$

One can obtain from formula (4.1.11) the average rates of the radiative spontaneous transitions $v \to v-1$ over projections M' of the molecular angular momentum. These rates for spontaneous emission for transitions with an increase and decrease by one of the rotational quantum number J are equal

$$w\left(v, J \to v-1, J+1\right) = \frac{J+1}{2J+1}\frac{1}{\tau}, \quad w\left(v, J \to v-1, J-1\right) = \frac{J}{2J+1}\frac{1}{\tau}, \tag{4.1.12}$$

where τ is the lifetime of the initial state due to the spontaneous radiative decay. These expressions describe spontaneous emission of photons of any polarization. Since the rate for emission of a photon with a given polarization does not depend on the direction of its polarization because of an average over projections of the total angular momentum, the rate of this process for a certain polarization is one half of the total rate.

We also consider selection rules for rotational states of triatomic linear molecules. If the oscillation takes place only in the molecular axis direction, the selection rules

Table 4.2 Values of Clebsch-Gordan coefficient $C^{J'm'}_{J,m-\mu;1\mu}$ for addition of momenta and their projections J, m' and 1, μ into the momentum $J'm'$

Branch	J'	$\mu=-1$	$\mu=0$	$\mu=+1$
P	$J-1$	$\sqrt{\frac{(J-m')(J-m'+1)}{2(J+1)(2J+1)}}$	$\sqrt{\frac{(J-m'+1)(J+m'+1)}{(J+1)(2J+1)}}$	$\sqrt{\frac{(j+m')(j+m'+1)}{2(j+1)(2j+1)}}$
Q	J	$\sqrt{\frac{(J-m')(J+m'+1)}{2J(J+1)}}$	$\frac{m'}{\sqrt{J(J+1)}}$	$-\sqrt{\frac{(J+m')(J-m'+1)}{2J(2J+1)}}$
R	$J+1$	$\sqrt{\frac{(J+m')(J+m'+1)}{2J(2J+1)}}$	$-\sqrt{\frac{(J-m')(J+m')}{J(2J+1)}}$	$\sqrt{\frac{(J-m')(J-m'+1)}{2J(2J+1)}}$

for a triatomic molecule are identical to those of a diatomic molecule. We denote below a vibrational quantum number by v and a rotational momentum by j. Our goal is to determine the rate of radiative transition $w(v, j \longrightarrow v', j')$. Modeling vibrations in the CO_2 molecule by a harmonic oscillator, we obtain the selection rule for vibrational radiative transitions $v \longrightarrow v \pm 1$ for linear molecules including the CO_2 molecule. Because of a large time of radiative transitions in molecules compared with collision ones in atmospheric air, the number density of molecules N_j in a given rotational state J is determined by the Boltzmann formula

$$N_j = N_o(2J+1)\frac{B}{T}\exp\left[-\frac{BJ(J+1)}{T}\right], \tag{4.1.13}$$

where N_o is the total number density of molecules for this vibrational state, T is the gas temperature expressed in energetic units, B is the rotational constant, $BJ(J+1)$ is the excitation energy for this rotational state, and the normalized constant is taken from the condition $B \ll T$.

Being guided by the dipole character of radiation, where the rate of a radiative transition is proportional to the square of the matrix element of the dipole moment operator between transition states, one can obtain the following expression for the rate of a vibrational-rotational transition

$$w(v, J \longrightarrow v', J') = \frac{1}{\tau_{vv'}} \cdot |\langle JM|\mathbf{n}|J'M'\rangle|^2 = \frac{W(J'M')}{\tau_{vv'}} \tag{4.1.14}$$

Here $\tau_{vv'}$ is the radiative time for transition between indicated vibrational states, $\mathbf{n}$ is the unit vector directed along a molecular vibrations, J, M; J', M' are rotational momenta and their projections onto a given axis for the initial and final transition states correspondingly. This matrix element in formula (4.1.14) results from summation of the initial momentum and unit photon momentum into the momentum of a final state [3–5] and is expressed through the Clebsch-Gordan coefficient. Values of Clebsch-Gordan coefficients which are responsible for a radiative transition are given in Table 4.2 [6].

Values of the Clebsch-Gordan coefficients leads to the selection rules for radiative rotational transitions, and according to them the following transitions are possible

Table 4.3 Probabilities of radiative vibrational-rotational transitions in a given rotational state

Branch	Transition energy	$W_\perp$	$W_\parallel$
P ($J' = J-1$)	$\hbar\omega_o + 2B(J+1)$	$\frac{(J+1)(J+2)+m^2}{2(J+1)(2J+1)}$	$\frac{2J+3}{3(2J+1)}$
Q ($J' = J$)	$\hbar\omega_o$	$\frac{J(J+1)-m^2}{2J(J+1)}$	$\frac{1}{3}$
R ($J' = J+1$)	$\hbar\omega_o - 2BJ$	$\frac{J(J-1)+m^2}{2J(2J+1)}$	$\frac{2J-1}{3(2J+1)}$

$$\Delta\varepsilon = \hbar\omega_o - 2BJ,\ J' = J-1 \longrightarrow P\text{-band},$$
$$\Delta\varepsilon = \hbar\omega_o,\ J' = J \longrightarrow Q\text{-band},$$
$$\Delta\varepsilon = \hbar\omega_o - 2B(J+1),\ J' = J+1 \longrightarrow R\text{-band}, \tag{4.1.15}$$

The probability of a given final rotational state depends on relative directions of vibrational and rotational axes. These probabilities for their identical directions $W_\parallel$ and their perpendicular directions $W_\perp$ are given in Table 4.3 depending on rotational quantum numbers Jm.

Note that for an antisymmetric vibrational state, as well as for a diatomic molecule, the probability of a given final rotational state is $W_\parallel$, whereas for torsion vibrations it is equal to $(W_\parallel + W_\perp)/2$. Averaging these probabilities over the momentum projection m onto a rotational axis which varies from 0 to J, we use that the average square of the momentum projection is equal

$$\overline{m^2} = \frac{J(J+1)}{3} \tag{4.1.16}$$

This averaging leads to the following expression for each band

$$\overline{W_P} = \frac{2J+3}{3(2J+1)},\ \overline{W_Q} = \frac{1}{3},\ \overline{W_R} = \frac{2J-1}{3(2J+1)} \tag{4.1.17}$$

where indices indicate a branch. In the limit of large j these probabilities become identical and are equal to 1/3. Being guided by large rotational momenta j, we below take the probability of each branch to be 1/3. Thus, the analysis of spectroscopic properties of CO_2 molecule allows us to select vibrational-rotational radiative transitions in linear molecules which compose the molecular spectrum and determine the rates of these transitions.

4.1.3 Radiative Properties of CO_2 Molecule

In continuation of the analysis of molecular radiative properties, we consider below the carbon dioxide molecule from this standpoint. The study of radiative properties of the carbon dioxide molecule is of importance twofold. On the one hand, this is a symmetric linear molecule; therefore it is a simple example for demonstration of the general positions in radiation of molecular gases. On the other hand, carbon dioxide is of importance for the greenhouse phenomenon in atmosphere of the Earth and Venus. Therefore, radiative parameters of CO_2 molecules are required for the analysis of these phenomena. In consideration vibrational-rotational transitions of CO_2 molecules, we use the Born-Oppenheimer approximation [7] which account for a fast reaction of an electron subsystem to displacements of nuclei. As a result, molecular oscillations take place in a potential field which is formed by electrons at a given configuration of nuclei [8, 9].

Being guided by lower excitations which are associated with vibrational and rotational degrees of freedom for this molecule, we are restricted by the ground electron state with zero spin. Then the distribution of the electron number density is characterized by the axial symmetry, i.e. the electron state of this molecule is conserved as a result of turn around the molecular axis at any angle. Correspondingly, the electron state is conserved at electron reflection with respect of any plane which passes through the molecular axis.

Restricting by nuclear positions near the minimum of the potential energy surface, one can reduce the nuclear motion to their vibrations as harmonic oscillations. In the case of triatomic molecules, where three atoms are located in one line, there are 9 degrees of freedom which include three translation ones, two rotational ones for the molecule axis and four vibrational degrees of freedom. Four types of oscillations for CO_2 molecules which include the symmetric oscillation, where a distance between each oxygen atom and a central carbon atom are kept identical during this vibration, and the antisymmetric oscillation, where the distance between oxygen atoms is not changed in these oscillations along the molecular axis. In the course of the torsion oscillation (or the deformation vibration) the carbon atom moves perpendicular to the molecular axis, and because of two directions perpendicular to the molecular axis, two torsion oscillations are realized. Because the dipole moment operator is antisymmetric for molecular reflection with respect to the plane, which is perpendicular to the molecular axis and passes through the carbon nucleus, the radiative transitions involving vibrational states do not include symmetric oscillations. Only a change in antisymmetric and torsion vibrational states can lead to radiative dipole transitions.

Let the molecular axis be directed along the axis z, so that torsion oscillations occur in directions x and y. In particular, in the case of excitation of the lower torsion vibration one can compose two vibrations such that the wave function of eigen oscillation are proportional to $\exp(i\varphi)$ and $\exp(-i\varphi)$, where φ is the angle in the plane xy with respect to the axis x. In this interpretation we present torsion vibrations and rotations in the plane xy. In the case of the inversion transformation $x \leftrightarrow -x,\ y \leftrightarrow -y,\ z \leftrightarrow -z$ the torsion states are separated in even and odd substates in accordance with the property of the eigenfunction to conserve or change its sign. Namely, the torsion vibrations are even with respect to the inversion transformation are odd for the odd number of torsion vibrational quanta, and they are even, if a number of torsion excitations for this state is even.

Note that the abundance of the isotope ^{12}C in nature is 98.9%, and the abundance of the isotope ^{16}O is 99.76%. The nuclear spin of each of these isotopes is zero, so that practically all the CO_2 molecules in nature contain nuclei with zero spin. This leads to the symmetry for molecular reflection with respect to the plane which is perpendicular to the molecular axis and passes through the carbon atom. This operation is analogous to exchange by oxygen molecules which are identical because of zero nuclear spin; in addition it corresponds to reflection of the carbon atom with respect to the symmetry plane. Since the nuclear spin of the carbon atom is zero, and its electron state is conserved at this operation, we obtain the total conservation of the state of the $^{12}C^{16}O_2$ as a result of the above reflection.

Since the electron, vibrational and rotational degrees of freedom are separated, the total wave function of this molecule is the product of the electron, vibrational and rotational wave functions. Because of the symmetry of the electron state of this molecule both for inversion and reflection with respect to the symmetry plane, the vibrational and rotational states must have a certain symmetry. Because the rotational wave function for the state with the rotation momentum J of the molecule changes as $(-1)^J$ as a result of inversion [4], this gives the rotational states which can be realized for a given vibrational state. Indeed, the wave functions of symmetric and antisymmetric oscillations along the molecular axis are not changed as a result of the inversion operation, whereas due to torsion oscillations, the vibrational wave function is conserved for an even number of torsion quantum number and changes its sign for odd values of the torsion quantum number.

Thus, we obtain that in the case of the even value of the torsion quantum number, only rotational states with even values of the rotational momenta J exist, while in the case of an odd torsion quantum number, there are rotational states with odd J only realized. The CO_2 molecule is symmetric for reflection with respect to the symmetry plane, and its rotational states are separated into the even and odd states. These states are connected with the rotational momentum. Namely, the wave functions of states with even rotational numbers are conserved at the indicated operation, but the wave functions of rotational states with odd values of rotational numbers change a sign at this operation [4].

Figure 4.3 contains the spectrum for radiative vibrational-rotational transitions of the CO_2 molecule. This information is taken from the HITRAN data bank, relates to the temperature $T = 296$ K, and includes radiative transitions from the states with the rotational number $J = 16$ for even vibrational states and $J = 17$ for odd ones. The distribution function of CO_2 molecules over rotational states has the maximum at these rotational numbers for the used gas temperature $T = 296$ K. Three atoms of the CO_2 molecule lie on a line, and three types of oscillations are as follows: ν_1 is the symmetric oscillation, ν_2 is the torsion one, and ν_3 is the antisymmetric oscillation. We are restricted ourselves by lower vibrational states which give a remarkable contribution to the absorption coefficient due to vibrational-rotational transitions of the CO_2 molecule. Therefore such excited vibrational states are excluded from Fig. 4.3.

Let us formulate the selection rules for the carbon dioxide molecule being guided by those for diatomic molecules. For the strongest vibrational radiative transitions,

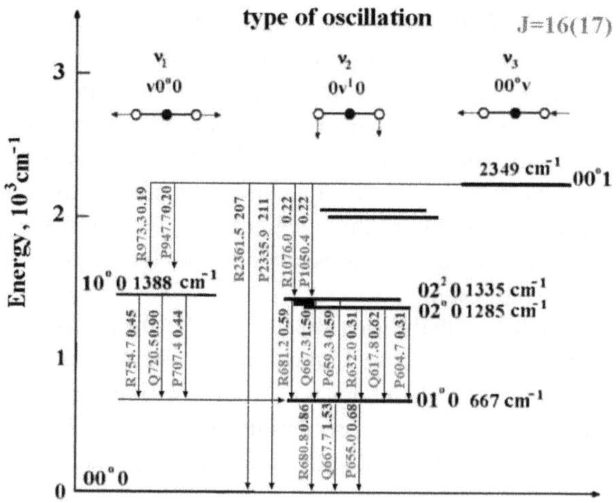

Fig. 4.3 Absorption spectrum of the carbon dioxide molecule and its radiative parameters according to HITRAN data bank [10, 11]. The energies of radiative transitions are expressed in cm^{-1} and are given in green; letters P, Q and R correspond to P, Q and R-branches of the rotational transitions correspondingly. Values of the Einstein coefficient are presented in red; they are expressed in s^{-1}. The initial rotational state for all the radiative transitions correspond to the rotational number $J = 16$, if the initial state admits only even rotational momenta, and $J = 17$ for odd rotational momenta in the initial state

the change of the vibrational quantum number is one as well as in the model of the harmonic oscillator. Note that in the case under consideration we assume vibrational excitation to be not strong, so that vibrational states of different types are separated, i.e. the total vibrational wave function is a product of wave functions for vibrational states of a different type. Evidently, if the torsion vibrational state is not changed at the radiative transitions, the selective rule is the same, as in the case of diatomic molecules. Other selection rules are in the case where the torsion vibration state is changed.

We now analyze vibrational states with a lower excited state of the torsion excitation. From two torsion states with oscillations in two perpendicular directions one can combine two rotational states around the molecular axis, and formally one can present the total molecular momentum as a sum of rotation of the molecular axis with a momentum J and torsion rotation which momentum we take to be 1/2 because of two rotational states. Correspondingly, the wave function Ψ of total rotations can be constructed from the wave function ψ of axial rotation, and wave function φ for torsion rotation. Summarizing rotation momenta, one can formally represent the total wave functions of rotations as

$$\Psi_{J+1/2} = \frac{1}{\sqrt{2}}\left(\psi_J\varphi_{1/2} + \psi_{J+1}\varphi_{-1/2}\right), \quad \Psi_{J-1/2} = \frac{1}{\sqrt{2}}\left(\psi_J\varphi_{-1/2} + \psi_{J-1}\varphi_{1/2}\right), \tag{4.1.18}$$

It should be noted that J, as the rotational number of the initial state, is an even number, at which this state is stable. Correspondingly, at torsion excitation the states with rotational numbers $J+1$ and $J-1/2$ are stable, whereas the state with rotational number J is unstable, as it follows from the symmetry of the total wave function if three atoms which are arrayed along a line [4]. If the carbon atom deviates from the line, this symmetry requirement disappears. Hence, the wave function with the rotational number J is zero only at the nuclear configuration if they form one line. As a unstable state, this state decays subsequently, but it is present in the absorption spectrum along with stable states.

We now determine the probabilities $W_P,\ W_Q,\ W_R$ for realization the $P,\ Q$ and R-branches of radiation in the case if the torsion vibrational state is changed at this transition. Assuming that the states with total rotational numbers $J+1/2$ and $J+1/2$ are formed with the equal probability, we find on the basis of the wave functions (4.1.18) the probabilities of realization of $P,\ Q$ and R branches are equal correspondingly

$$W_P=\frac{1}{4},\ W_Q=\frac{1}{2},\ W_R=\frac{1}{4}, \tag{4.1.19}$$

and the relation between the Einstein coefficients A for these branches is as follows

$$A_Q=2A_P=2A_R \tag{4.1.20}$$

4.1.4 Spectroscopic Databases

Analyzing processes in molecular gases, we use parameters of radiative transitions in molecular gases for the HITRAN data bank. Therefore we use partially notations of this data bank for molecular gases. Along with this, other data banks exist where information for various aspects of spectroscopy is given. We represent below a list of such data banks along with the HITRAN one [11].

Millimeter and Submillimeter Molecular Spectroscopy Catalog, Jet Propulsion Laboratory, USA

The *Cologne Database for Molecular Spectroscopy (CDMS)*, Universität zu Köln, Germany

GEISA Spectroscopic Database, Laboratoire de Metrorologie Dynamique, France

PNNL Vapor Phase Infrared Spectral Library, Pacific Northwest National Laboratory, USA

ExoMol, Molecular line lists for exoplanet and cool star atmospheres, University College London, UK

Ames Molecular Spectroscopic Data For Astrophysical and Atmospheric Studies, NASA Ames, USA

TheoReTs, Internet accessible information system "Theoretical Reims-Tomsk Spectral data, University de Reims, France and Institute of Atmospheric Optics, Russia

Spectroscopy and Molecular Properties of Ozone. University de Reims, France and Institute of Atmospheric Optics, Russia

NIST Wavenumber Calibration Tables from Heterodyne Frequency Measurements, National Institute of Standards and Technology, USA NIST Atomic Spectra Database, National Institute of Standards and Technology, USA

CHIANTI, An Atomic Database for Spectroscopic Diagnostics of Astrophysical Plasmas, George Mason University (USA), University of Michigan (USA), University of Cambridge (UK).

4.2 Absorption of Infrared Radiation in Gas of Linear Molecules

4.2.1 Infrared Radiation of Molecular Gas

We above have considered radiative transitions between discrete atom states, where the spectrum of absorption or emission is characterized by separate spectral lines. Each line can be splitted in a multiplet due to fine and superfine interactions inside an atom or atomic ion, but the frequency width of each multiplet is relatively small. This means that the spectrum of atoms or atomic ions consists of separate spectral lines which can be splitted in a multiplet, but at frequencies between neighboring spectral lines or multiplets it is zero. In the case of radiative transitions between electronic states of molecules, vibrational and rotational spectra apply with the electron one. As a result, an absorption line in the atomic case is transformed in an absorption band for molecular particles with an oscillating spectrum structure as a frequency function. Because of a high frequency width of an absorption band, neighboring bands may be overlapped, so that the physical picture of radiation interaction with molecular particles differs from that in the case of atoms and ions.

In order to consider the nature of vibrational-rotational transitions from the general positions of molecular spectroscopy [12–16], we below analyze the character of of absorption and emission of molecular gases. Moreover, we consider radiative transitions without change of the electron state; first we are restricted ourself by one vibrational transition. Next, for simplicity, we consider the case of linear molecules. The absorption of a molecular gas corresponds to the infrared spectrum range, and due to selection rules radiative transitions are accompanied by change of vibrational v and rotational J quantum numbers in the following way $v, J \to v', J'$ with $v' = v \pm 1$, $J' = J, J \pm 1$. Thus, according to selection rules Q-branch correspond to the rotational transition $J \to J$, P-branch refers to the transition $J \to J+1$, and R-branch relates to the rotational transition $J \to J-1$. Note that in the case of a diatomic molecule, where vibrations take place along the molecular axis, Q-branch is absent for rotational transitions.

The energy of the rotational state of the molecule E_J is [4]

$$E_J = BJ(J+1), \qquad (4.2.1)$$

Here $B = \hbar^2/2\mu$ is the rotational constant, and the energy of the rotational state is equal $\hbar\omega_J = \hbar\omega_o - 2BJ$ for P-branch with the transition $J \to J+1$ it is equal $\hbar\omega_J = \hbar\omega_o - 2BJ$, and for R- branch with the transition $J \to J-1$, the energy of the transition is $\hbar\omega_J = \hbar\omega_o + 2B(J+1)$, where $\hbar\omega_o$ is the energy difference for vibrational states of the transition.

We now consider radiative properties of a molecular gas where broadening of spectral lines is determined by collisions with molecules of a buffer gas where radiating molecules are located. Note the difference for broadening of spectral lines for atoms and molecules. In the atomic case, the interaction potential of a radiating atom in the upper state of transition with a buffer gas atom exceeds significantly that for the lower transition state; therefore broadening of an atom spectral line for the impact and quasistatic theory is determined by interaction in the upper transition state. In the case of molecular radiation, parameters of interaction of a radiating molecule and a buffer gas atom in the upper and lower states of the radiative transition are nearby, i.e. the difference of interaction potentials for the upper and lower states of the radiative transition is small.

Let us analyze this effect for molecules of CO_2 in a parent gas approximating the interaction potential of two CO_2 molecules at large distances between them by formula (2.1.41) $U(R) = -C_6/R^6$. Taking the difference of the interaction potentials for the upper and lower states of the radiative vibration transition as $\Delta U(R) = -\Delta C_6/R^6$, one can obtain instead of formula (2.1.43) for the width of the spectral line

$$\frac{\nu}{N_b} = 7.2\left(\frac{2T}{\mu}\right)^{3/10}\left(\frac{\Delta C_6}{\hbar}\right)^{2/5} \qquad (4.2.2)$$

Let us use the experimental value of the width of the spectral line for the vibrational transition 001 $\to$ 000 at 667 cm^{-1} that is equal according to measurements $\nu = 0.16 cm^{-1}$ [17–20]. Then formula (4.2.2) gives for the difference $\Delta C_6 = 32e^2a_o^5$, whereas its value is $C_6 = 118e^2a_o^5$ [1]. This shows the degree of the difference of the interaction potentials for neighboring molecular vibrational states.

One more peculiarity of spectral lines due to vibrational transitions is their restricted width for the collision mechanism of broadening. Indeed, at wings of a spectral line the distribution function a_ω according to formula (2.1.40) has the form

$$a_\omega = \frac{\nu}{2\pi\,(\omega-\omega_o)^2} \qquad (4.2.3)$$

The broadening is determined by collisions with buffer gas atoms or molecules, and ω_o is the frequency at the line center. In derivation of formula (2.1.40) for collision broadening of the spectral line we assume a collision time to be zero, and then the distribution function (4.2.3) is spreading up to infinite frequency. But according to the nature of this effect, it is necessary to restrict frequencies in this formula

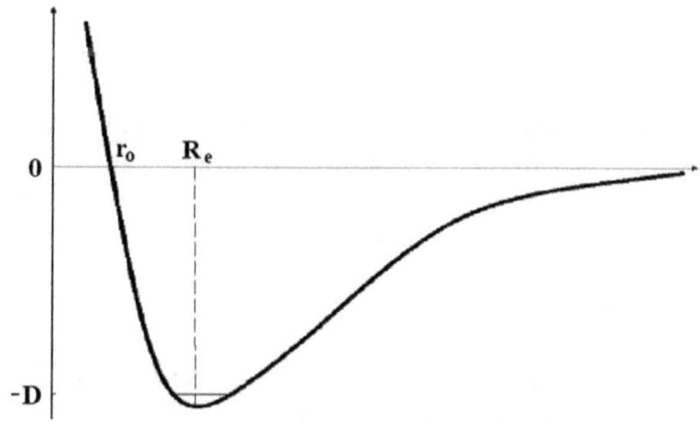

Fig. 4.4 Parameters of the Lennard-Jones interaction potential for two molecules

by $|\omega - \omega_o| \sim 1/\tau_c$, where τ_c is a duration time of the collision event. We below estimate this time for collisions of a radiating CO_2 molecule which is located in a parent gas, approximating, for simplicity, the interaction potential of two molecules of carbon dioxide by the Lennard-Jones interaction potential [21, 22]

$$U(R) = D\left[2\left(\frac{R_e}{R}\right)^6 - \left(\frac{R_e}{R}\right)^{12}\right], \tag{4.2.4}$$

Here D is the depth of the potential well for the interaction of two molecules, R_e is the equilibrium distance between molecules, which corresponds to a minimum of the interaction potential, and its parameters are given in Fig. 4.4. Though the spherically symmetric interaction potential is a crude approximation for interaction of carbon dioxide molecules, but this allows us to describe simply the effect under consideration. One can determine parameters of the interaction potential between two CO_2 molecules on the basis of the similarity law by comparison the interaction parameters of carbon dioxide molecules and inert gas atoms. Indeed, the scaling law for inert gases [23, 24] allows one to connect critical parameters of inert gases and carbon dioxide, and also their parameters near the triple point with parameters of the pair interaction potential inside these systems. In this manner, one can determine parameters of the pair interaction potential of molecules in formula (4.2.4), which in the case of carbon dioxide molecules gives $D = (38 \pm 7)$ meV and $R_e = 0.32$ nm. As is seen, the accuracy of this operation is restricted. This comparison allows us to determine the van der Waals interaction constant $C_6 = 2DR_e^6 \approx 140e^2a_o^5$ on the basis of the above scaling law, using parameters of inert gases. This exceeds the calculated value [1] by 20%.

One can define a typical collision time for two molecules of carbon dioxide τ_c as a time of approach of slow molecules from the distance R_e at the minimum interaction potential of molecules to the distance $r_o = R_e/2^{1/6}$ that corresponds to the potential wall for slow molecules at zero orbital angular momentum (see Fig. 4.3), which is equal to

$$\frac{1}{\tau_c} \equiv \Delta\omega = \frac{10}{R_e} \cdot \sqrt{\frac{2D}{m}}, \tag{4.2.5}$$

where m is the mass of the carbon dioxide molecule. Hence we obtain the following estimate for the wing width of the absorption band of carbon dioxide molecules in carbon dioxide $\Delta\omega \approx 70\,\mathrm{cm}^{-1}$. This fact restricts the flux of IR radiation emitted by carbon dioxide due to spectral line wings.

4.2.2 *Vibrational-Rotational Radiative Transitions for Diatomic Molecules*

We now determine the absorption coefficient k_ω due to optically active diatomic molecules under conditions of the thermodynamic equilibrium between vibrational and rotational molecular states. According to the selection rule, the strongest transition for a diatomic molecule within the framework of the harmonic oscillator model is $v \to v+1$ in the process of the photon absorption. Separating vibrational and rotational states, one can obtain the number density of molecules in the lower state

$$N_i = CN_v(2J+1)\exp\left[-\frac{BJ(J+1)}{T}\right],$$

Here the normalization coefficient C under the criterion $B \ll T$ is equal $C = B/T$, and N_v is the number density of molecules in the lower vibrational state of the radiation transition; it is described by vibrational and rotational states with quantum numbers vJ. We thus have for the absorption coefficient $N_v = \sum_i N_i$ where N_i is the number density of molecules in the lower transition state, and $g_i = 2J+1$ because the radiative transition takes place from a rotational state with the quantum number J. Under these conditions formula (2.2.28) gives for the absorption coefficient

$$k_\omega = N_v A_{ij} g_j \exp\left(-\frac{E_i}{T}\right)\left(\frac{\pi c}{\omega}\right)^2 a_\omega \frac{B}{T}\left[1 - \exp\left(-\frac{\hbar\omega}{T}\right)\right], \tag{4.2.6}$$

where $g_j = 2J' + 1$ is the statistical weight for the final state j with the rotational quantum number J', $E_i = BJ(J+1)$ is the rotational energy of the initial state i.

In order to reduce this expression to parameters of the HITRAN data bank which data are used below, we introduce the spectral line intensity as

$$S_{ij} = \int \sigma_\omega \mathrm{d}\nu = \int \frac{k_\omega}{N_v}\frac{\mathrm{d}\omega}{2\pi c}, \tag{4.2.7}$$

where the cross section of absorption is introduced as $\sigma_\omega \equiv \sigma_{ij} = k_\omega/N_v$, and the value ν is the reciprocal wavelength of radiation $\nu = 1/\lambda = \omega/(2\pi c)$. Using the normalization condition (2.1.1), one can obtain

$$S(\omega) = A_{ij} g_j \frac{B}{T} \exp\left(-\frac{E_i}{T}\right)\left(\frac{\pi c}{2\omega^2}\right)\left[1 - \exp\left(-\frac{\hbar\omega}{T}\right)\right] \quad (4.2.8)$$

The spectral line intensity is a convenient characteristic of vibrational-rotational transitions because this value does not depend on broadening of corresponding spectral lines.

It is necessary to compare this expression with that [25] given by the HITRAN data bank. Taking into account that the value S_{ij} has the dimensionality of length and accepting the spectroscopy unit cm^{-1} and thermal unit K as the energy units, one can present the spectral line unit of the HITRAN data bank [25] in the form

$$S_{ij} = \frac{I_a}{I_o} \frac{A_{ij}}{8\pi c \nu_{ij}^2} \frac{g_i}{Q(T)} \exp\left(-\frac{E_i}{T}\right)\left[1 - \exp\left(-\frac{\hbar\omega}{T}\right)\right] \quad (4.2.9)$$

In rewriting the HITRAN expression for the spectral line intensity, we accept that the units cm^{-1} and K to be the energy units. Next, the inertia moment is $I_a = \mu r_o^2 = \hbar^2/(2B)$ (μ is the reduced mass of nuclei, r_o is the equilibrium distance between them), the atomic value of this dimensionality is $I_o = m_e a_o^2$ (a_o is the Bohr radius), the reciprocal wavelength for this radiative transition ν_{ji} is expressed through the frequency ω as $\nu_{ji} = \omega/2\pi c$; in addition, we consider the frequency ω as a continuous variable. Next, $E_i = BJ(J+1)$ is the rotational energy, and $Q(T) = \sum_k g_k \exp(-E_k/T)$ is the partition function. Being guided by real conditions, where $T \gg B$, and only the lower vibrational state gives the contribution to the partition function, we replace the sum by integral and obtain $Q(T) = T/B$. Then formulas (4.2.8) and (4.2.9) coincide with each other. Below, considering the frequency ω as a continuous variable and being guided by large J, we have

$$g_j = 2J + 1 = \frac{\hbar(\omega - \omega_o)}{B}$$

This allows us to rewrite formula (4.2.8) in the form

$$S(\omega) = A_{ij} \frac{\lambda^2}{8\pi c} \frac{\hbar|\omega - \omega_o|}{T} \exp\left(-\frac{\hbar^2|\omega - \omega_o|^2}{4BT}\right)\left[1 - \exp\left(-\frac{\hbar\omega}{T}\right)\right], \quad (4.2.10)$$

where $\lambda = 2\pi c/\omega$ is the radiation wavelength, $\hbar\omega_o$ is the energy difference between energies of vibrational states of this transition. This expression relates both to $P-$branch and $R-$branch. We use this expression below in determination of the spectral line intensity.

Note that we consider above the frequency ω as a continuous variable. Then the spectral line intensity has the maximum at $\hbar\omega_{\max} = 2\sqrt{BT}$, and the maximum value of this quantity is equal

$$S_{\max} = A(J_{\max}) \frac{\lambda_m^2}{8\pi c\sqrt{e}} \sqrt{\frac{2B}{T}} \left[1 - \exp\left(-\frac{\hbar\omega}{T}\right)\right], \tag{4.2.11}$$

where $J_{\max}$ is the rotational momentum, λ_m is the radiation wavelength. Correspondingly, one can express the spectral line intensity $S(\omega)$ through its maximum value $S_{\max}$ as

$$S(\omega) = \sqrt{2e} S_{\max} \frac{\hbar|\omega - \omega_o|}{2\sqrt{BT}} \exp\left(-\frac{\hbar^2|\omega - \omega_o|^2}{4BT}\right) \tag{4.2.12}$$

One can compare this value with that from the HITRAN data bank [11] in the case of CO-molecules. At the temperature $T = 296$ K and for the rotational constant of the CO molecule $B = 1.93\,\mathrm{cm}^{-1}$, the maximum of the spectral line intensity corresponds to $J_{\max} = 7$; the photon energy is $\hbar\omega_7 = 2116\,\mathrm{cm}^{-1}$ for P-branch, and $\hbar\omega_7 = 2173\,\mathrm{cm}^{-1}$ for R-branch. Taking values of the Einstein coefficient $A_7 = 18.4\,\mathrm{cm}^{-1}$ and $A_7 = 17.5\,\mathrm{cm}^{-1}$ for these cases, we have the values of the spectral line intensity $S_7 = 3.87 \cdot 10^{-19}$ cm for P-branch and $S_7 = 4.56 \cdot 10^{-19}$ cm for R-branch according to data of HITRAN bank [11]. Formula (4.2.11) gives for these cases $S_7 = 4.2 \cdot 10^{-19}$ cm for P-branch and $S_7 = 3.8 \cdot 10^{-19}$ cm for R-branch. As is seen, agreement between these results take place with the accuracy of 20%.

4.2.3 Absorption Coefficient for Gas of Diatomic Molecules

We below construct the absorption coefficient k_ω for a certain band of a radiative vibrational transition of a linear molecule in the case of collision broadening of spectral lines. Then the lower and upper transition states of Fig. 1.4 are characterized by the vibrational v and rotational J quantum numbers which relate to the number of vibrational or rotational states. The selection rules extract a restricted number of transitions, and the strongest transitions for absorption of radiation are $v, J \to v+1, J+1$ for $P-$branch and $v, J \to v+1, J-1$ for $R-$branch. The absorption coefficient k_ω accounted for a transition between two states is given by formula (2.2.28) and is equal

$$k_\omega = A_{ij} N_v \frac{\lambda^2}{4} \frac{\hbar|\omega - \omega_o|}{T} a_\omega \exp\left[-\frac{\hbar^2(\omega - \omega_o)^2}{4BT}\right] \left[1 - \exp\left(-\frac{\hbar\omega}{T}\right)\right], \tag{4.2.13}$$

where λ is the radiation wavelength. We above separate vibrational and rotational states, so that in this formula N_v is the number density of molecules in the lower vibrational state v for the radiative transition $v \to v'$, and indices i and j correspond to the initial J and final J' rotational states, ω is the transition frequency, $\hbar\omega_o$ is the energy difference for vibrational transition states, c is the speed of light, A_{ij} is the first Einstein coefficient which depends weakly on the rotational number; it is taken into account a large number of rotational states which determine the absorption coefficient, and the summation in formula (4.2.13) is made over rotational states. We assume molecules in rotational and vibrational states to be under thermodynamic equilibrium with the temperature T. According to the selection rule, the vibrational number is changed by one within the framework of the harmonic oscillator for molecular vibrations. The criterion $J \gg 1$ allows one also to neglect interaction of nuclear rotation with electron momenta, i.e., $J = K$, where K is the total molecular moment.

Formula (4.2.13) joins transitions for $P-$branch with $\omega \geq \omega_o$ and $R-$branch with $\omega \leq \omega_o$. The variation energy $\hbar\omega_J$ for a given vibrational transition and the initial rotational momentum J of the molecule is given by

$$\hbar\omega_J = \hbar\omega_o - B \pm B(2J+1), \qquad (4.2.14)$$

where the sign minus relates to P absorption branch, sign plus corresponds to R-branch. From this it follows for P and R branches $\hbar\omega_J = \hbar\omega_o - B \pm 2BJ$ for large $J \gg 1$ that we use below. As is seen, the energy difference between neighboring rotational states is constant that corresponds to the Elsasser model [26], and this difference is $2B$. The photon distribution function a_ω for a certain transition has the Lorentz shape (2.1.4); in the case of the collision mechanism of broadening of spectral lines which is of the most interesting, the photon distribution function has the form

$$a_\omega = \sum_J \frac{\nu_J}{2\pi\left[(\omega - \omega_o - \omega_J)^2 + (\frac{\nu_J}{2})^2\right]}, \qquad (4.2.15)$$

where the width ν_J of spectral lines depends weakly on J.

The absorption coefficient is a harmonic function of the frequency (or almost harmonic one, if we take into account a weak frequency dependence of parameters in formulas (4.2.13) and (4.2.15)), and the period of oscillations is $2B/\hbar$. We now determine the average absorption coefficient averaged over frequencies. Accounting for the normalization condition (2.1.1) and the distance between neighboring resonance frequencies to be $2B/\hbar$, we obtain on the basis of formula (2.2.28) for the average absorption coefficient $\chi(\omega)$ which is expressed through the spectral line intensity $S(\omega)$ defined by formula (4.2.7), as

$$\chi(\omega) = \overline{k_\omega} \equiv \frac{\hbar}{2B}\int_{-\infty}^{\infty} k_\omega d\omega = \frac{2\hbar N_v \pi c}{d} S(\omega) = N_v \lambda_d S(\omega), \qquad (4.2.16)$$

Here $\lambda_d = 2\pi c\hbar/d$ is the wavelength for a photon of an energy d, which is in turn the difference of neighboring energy levels, and in the case of diatomic molecules is equal to $d = 2B$.

Finally, the value of this quantity in the case of optically active diatomic molecules is given by

$$\chi(\omega) = A_{ij} N_v \frac{\lambda^2}{8} \frac{\hbar^2 |\omega - \omega_o|}{BT} \exp\left(-\frac{\hbar^2(\omega - \omega_o)^2}{4BT}\right)\left[1 - \exp\left(-\frac{\hbar\omega}{T}\right)\right] \quad (4.2.17)$$

A general form of the absorption coefficient is

$$k_\omega = \chi(\omega)\varphi(\omega), \quad \int_{-\infty}^{\infty} \varphi(\omega)d\omega = 1 \quad (4.2.18)$$

and the function $\varphi(\omega)$ takes into account the oscillation character of the absorption coefficient. One can rewrite also the connection between the spectral line intensity and average absorption coefficient on the basis of formulas (4.2.7) and (4.2.16)

$$S(\omega) = \frac{\chi(\omega)d}{N_v \cdot 2\pi c}, \quad (4.2.19)$$

Here d is the energy difference for neighboring transitions, and for diatomic molecules we have $d = 2B$.

In order to determine $\varphi(\omega)$, we use the Mittag-Leffler theorem, that takes the form [27]

$$\sum_{k=-\infty}^{\infty} \left[(x-k)^2 + y^2\right]^{-1} = \frac{\pi \sinh 2\pi y}{y(\cosh 2\pi y - \cos 2\pi x)} \quad (4.2.20)$$

Taking now $x = \hbar(\omega - \omega_o)/B$ and $y = \hbar\nu/4B$, one can present the absorption coefficient k_ω into a given absorption band in the form (4.2.18) (for example, [28–30]) with

$$\varphi_i(\omega) = \sinh \frac{\pi\hbar\nu}{2B}\left[\cosh \frac{\pi\hbar\nu}{2B} - \cos \frac{\pi\hbar(\omega - \omega_i)}{B}\right]^{-1}, \quad (4.2.21)$$

Thus, we have that the absorption coefficient is an oscillation function of the frequency; according to formulas (4.2.18) and (4.2.21), the ratio of the neighboring maximum $k_{\max}$ and minimum $k_{\min}$ values of the absorption coefficient is

$$\frac{k_{\max}}{k_{\min}} = \frac{\cosh \frac{\pi\hbar\nu}{2B} + 1}{\cosh \frac{\pi\hbar\nu}{2B} - 1}, \quad (4.2.22)$$

In the limit of a low pressure, if $\hbar\nu \ll B$, this formula takes the form

$$\frac{k_{\min}}{k_{\max}} = \left(\frac{\pi\hbar\nu}{4B}\right)^2 \quad (4.2.23)$$

Note that in summarizing over rotational momenta we assume that the photon distribution function a_ω depends on the frequency stronger than other functions in the expression (4.2.13). Let us check it for the distribution function of molecules on energies, i.e. for the dependence of $\exp[-\hbar^2(\omega-\omega_o)^2/4BT]$, where the exponent varies through one period $2B/\hbar$ by the value $\hbar|\omega-\omega_o|/2T$. Hence, formulas (4.2.17), (4.2.18) and (4.2.21) hold true if the following criterion is fulfilled

$$\hbar|\omega-\omega_o| \ll 2T \tag{4.2.24}$$

4.2.4 Absorption Coefficient Produced by Carbon Dioxide Molecules

In consideration the spectrum of the CO_2 molecule, we are based on spectroscopy of a diatomic molecule, but take into account peculiarities of CO_2 molecules. Carbon dioxide is an important atmospheric component which gives a remarkable contribution to the greenhouse phenomenon of the Earth's and Venus atmospheres. From another standpoint, this molecule is the linear one, that allows us to analyze spectroscopic properties of gases contained carbon dioxide in the simple manner. For these reasons we analyze spectroscopic properties of a gas with carbon dioxide molecules as one of its components in detail. In evaluating the absorption coefficient k_ω of a gas with CO_2 molecules, we use the above expressions (4.2.17), (4.2.16), and (4.2.21) for diatomic molecules. In this analysis we take into account that the CO_2 molecule differs from diatomic molecules as an absorber of infrared radiation, since this molecule has only even values of the rotational momentum J or only odd ones depending on the parity of the vibration state. For this reason, the energy difference for neighboring vibrational-rotational radiative transitions is equal to $4B$ instead of $2B$, for diatomic molecules, where B is the molecular rotational constant. In addition, radiative transitions of the $Q-$branch are realized in the radiative spectrum of CO_2 molecules if the torsion vibration state changes at this transition.

In the case of a radiative vibrational-rotational transition of $P-$ or $R-$branches, one can use formulas (4.2.17), (4.2.16), and (4.2.21) for the absorption coefficient k_ω with the change the rotational constant B by $2B$. Then the absorption coefficient for P and $R-$branches is determined by formula

$$k_\omega^P = A(\omega)N_v\frac{\lambda^2}{4}\frac{\hbar|\omega-\omega_o|}{T}\exp\left[-\frac{\hbar^2(\omega-\omega_o)^2}{4BT}\right]\left[1-\exp\left(-\frac{\hbar\omega}{T}\right)\right]\frac{\sinh\frac{\pi\hbar\nu}{4B}}{\cosh\frac{\pi\hbar\nu}{4B}-\cos\frac{\pi\hbar(\omega-\omega_i)}{2B}}, \tag{4.2.25}$$

where we go from the integer rotational momentum J to a continuous quantity $\omega-\omega_o$; negative values of $\hbar(\omega-\omega_o)$ correspond to $P-$branch, while positive values of this parameter relate to $R-$branch. As is seen, the absorption coefficient as a frequency function has an oscillation structure. Figure 4.5 gives this dependence in a narrow spectrum range for P branch of the radiative vibrational transition $00^o0 \to 01^o0$ of CO_2 molecules in air at atmospheric pressure. On the basis of formula (4.2.16), this formula may be presented in the form

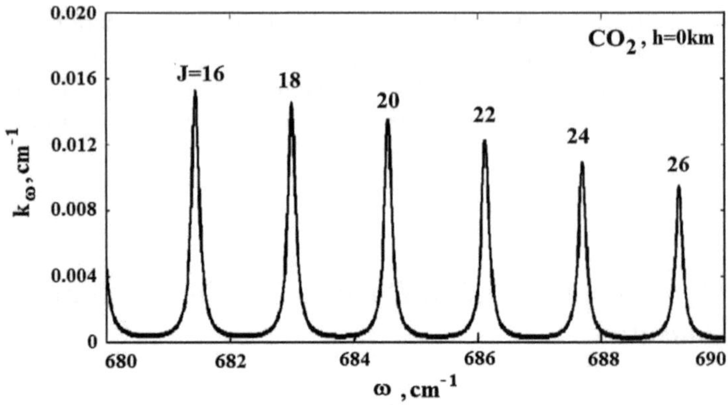

Fig. 4.5 Absorption coefficient for P branch of the 15 μm vibrational transition $00^o0 \to 01^o0$ due CO_2 molecules in air at atmospheric pressure in a narrow range of frequencies

$$k_\omega^P = N_v S_{\max} \frac{\lambda_B}{4} \cdot \frac{\hbar|\omega - \omega_o|}{2\sqrt{BT}} \exp\left[-\frac{\hbar^2(\omega-\omega_o)^2}{4BT}\right] \frac{\sinh\frac{\pi\hbar\nu}{4B}}{\cosh\frac{\pi\hbar\nu}{4B} - \cos\frac{\pi\hbar(\omega-\omega_i)}{2B}} \tag{4.2.26}$$

Figure 4.5 demonstrates this dependence for the strongest vibrational transition of CO_2 molecules in air at room temperature.

Let us consider Q−branch of radiative transitions which proceed without the change of the rotational momentum J. In this case the energy of excitation of the rotational state is given by formula (4.2.1) $E_J = BJ(J+1)$, but the transition energy or the photon energy $\hbar\omega_J$ is given by the formula

$$\hbar\omega_J = \hbar\omega_o - \Delta_J \tag{4.2.27}$$

instead of (4.2.14). In particular, on the basis of HITRAN data [11] we have for this parameter Δ_J in the case of the strongest radiative transition $00^o0 \to 01^o0$ with the energy 667 cm^{-1}, the following approximation

$$\hbar\omega_o = 667.377\,\text{cm}^{-1},\ \Delta_J = bJ + cJ^2,\ b = 2.4\cdot 10^{-3}\text{cm}^{-1},\ c = 1.0\cdot 10^{-3}\text{cm}^{-1} \tag{4.2.28}$$

Let us assume that the function $\mathrm{d}\Delta_J/\mathrm{d}J = b + 2cJ$ depends weakly on J that holds true at large J. The latter means that with taking into account that only even values of J are realized, the function $\varphi(\omega)$ according to formula (4.2.21) has the form

$$\varphi_Q(\omega) = \sinh\frac{\pi\hbar\nu}{2(b+2cJ)}\left[\cosh\frac{\pi\hbar\nu}{2(b+2cJ)} - \cos\frac{\pi\hbar(\omega-\omega_o)}{b+2cJ}\right]^{-1},\ \omega \ge \omega_o, \tag{4.2.29}$$

At moderate value of J, where the above assumption is not fulfilled, $\varphi(\omega) = 1$, whereas at large values of J, where oscillations in this function are realized, this

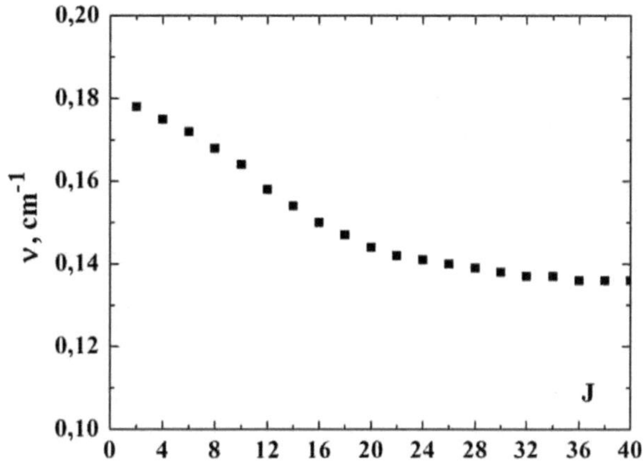

Fig. 4.6 Width of the spectral line for the vibrational transition $00^o0 \to 01^o0$ of the CO_2 molecule as a function of the rotational momentum J. This function is the same for P, Q and R-branches [10]

assumption is fulfilled. The half-width of this spectral line according to HITRAN bank data [11] is nearby to $\hbar\nu/2 = 0.07\,\mathrm{cm}^{-1}$; therefore remarkable oscillations start from the the rotational momenta $J_o = 16$, that corresponds to the maximum of the absorption coefficient.

In determination of the absorption coefficient as a function of the photon frequency ω, it is necessary to find the connection between the photon frequency and the rotational momentum J. In order to reduce the expression for the absorption coefficient to the above expressions, we take approximately $\Delta_J = cJ(J+1)$, that holds true at large J. Taking this dependence instead of formula (4.2.14), we have for the distribution function over rotational states f_J which is normalized to one ($\int f_j dJ/2 = 1$), taking into account that the rotational states with even rotational momenta exist only

$$f_J = \frac{2B(2J+1)}{T}\exp\left[-\frac{BJ(J+1)}{T}\right] = 4\sqrt{\frac{B}{T}}\sqrt{\frac{\omega-\omega_o}{\Delta\omega}}\exp\left[-\frac{\hbar(\omega-\omega_o)}{\Delta\omega}\right], \quad \Delta\omega = \frac{cT}{\hbar B} \tag{4.2.30}$$

This expression relates to positive values $\omega - \omega_o$. For the temperature at the Earth's surface $T = 288$ K, this parameter is equal $\Delta\omega = 0.051\,\mathrm{cm}^{-1}$.

We now determine the absorption coefficient on the basis of formula (2.2.28). Averaging it over oscillations, we obtain for the averaged absorption coefficient

$$\chi_Q(\omega) = A(\omega)N_v\frac{\lambda^2}{2\Delta\omega}\exp\left[-\frac{\hbar(\omega-\omega_o)}{\Delta\omega}\right]\left[1-\exp\left(-\frac{\hbar\omega}{T}\right)\right], \quad \Delta\omega = cT/(\hbar B) \tag{4.2.31}$$

Correspondingly, the absorption coefficient q_ω for Q-branch of radiative rotational transitions in accordance with formula (4.2.18) is given by

$$q_\omega = A(\omega)N_v\frac{\lambda^2}{2\Delta\omega}\exp\left[-\frac{(\omega-\omega_o)}{\Delta\omega}\right]\left[1-\exp\left(-\frac{\hbar\omega}{T}\right)\right]\frac{\sinh y}{\cosh y - \cos\frac{y(\omega-\omega_i)}{\nu}}, \quad y = \frac{\pi\hbar\nu}{2(b+2cJ)} \tag{4.2.32}$$

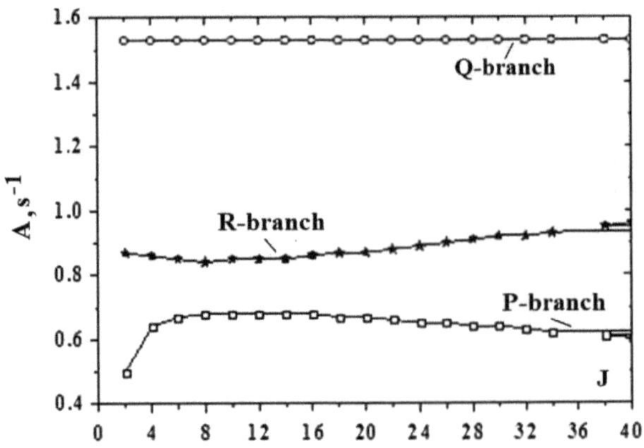

Fig. 4.7 Dependence of the Einstein coefficient A for the photon absorption process on the rotational momentum J for the vibrational transition $00^o0 \rightarrow 01^o0$ of the CO_2 molecule and for P, Q and R-branches [10]

Considering the absorption coefficient due to CO_2 molecules as a result of vibrational-rotational transitions, we assume their parameters to be independent on the rotational excitation because of the large vibrational energy compared with the rotational one. Of course, this assumption restricts the accuracy of calculations. In order to estimate this accuracy, we analyze the dependence of parameters of radiative transitions between two vibrational states on the rotational momentum J using the data of the HITRAN bank [11], restricting by the strongest transition between the ground 00^o0 and the lowest excited 01^o0 vibrational states As for the width ν of the spectral line of this transition due to collisions with air molecules at atmospheric pressure, this value is practically identical for the branches P, Q and R, but depend on the rotational number J. This dependence is given in Fig. 4.6.

Figure 4.7 contains the dependence of the first Einstein coefficient $A(\omega)$ on the rotational momentum J for the vibrational transition $00^o0 \rightarrow 01^o0$ of the CO_2 molecule in the cases of P, Q and R-branches. We consider above two cases of momentum coupling which lead to relations (4.1.17) and (4.1.19); these relations correspond to different ratios of the energy difference for neighboring rotational levels to the interaction potential between deformation oscillation and rotation. One can expect that the first case of momentum coupling holds true that leads to formulas (4.1.17) for the probability of various transition branches. Nevertheless, real values of probabilities for various branches of radiative transitions according to Fig. 4.7 data relate to the second case of momentum coupling.

The absorption coefficient due to CO_2 molecules is the sum of absorption coefficients for P, Q and R branches which describe the absorption process in different spectrum ranges. One can see that $Q-$branch is included in the total absorption coefficient which has the form of a narrow peak, while P and $R-$branches give oscillating functions in a more wide range. We now use the above formulas for the absorption coefficient for atmospheric carbon dioxide in atmospheres of the Earth and Venus. Taking the average temperature of the Earth's surface to be $T = 288$ K, we obtain that four vibrational transitions create emission of the atmosphere due to CO_2 molecules,

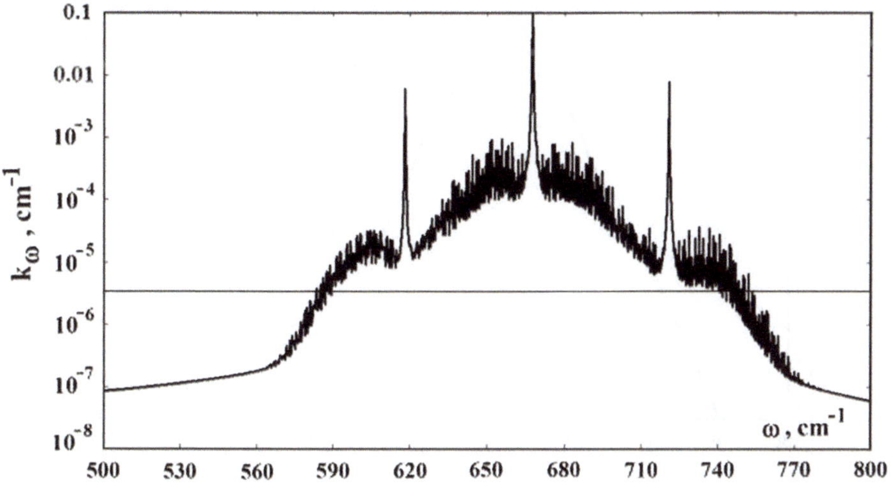

Fig. 4.8 Total absorption coefficient k_ω at the Earth's surface due to atmospheric carbon dioxide molecules for the standard atmosphere model according to formula (4.2.33) [31]

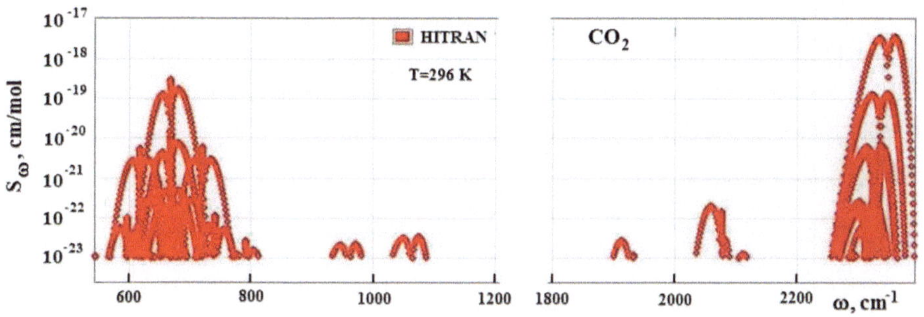

Fig. 4.9 Intensity of spectral lines due to absorption by carbon dioxide molecules in atmospheric air according data of HITRAN bank [10]

so that the absorption coefficient due to CO_2 molecules as a frequency function has the form

$$k_\omega = \sum_{i=1}^{4} k_\omega = \chi(\omega)^{(i)} \varphi^{(i)}(\omega) + q(\omega)^{(i)}, \qquad (4.2.33)$$

Here components of this expression are given by formulas (4.2.18), (4.2.21), (4.2.28), and subindex indicates the number of the vibrational transition with accounting for lower transitions of Fig. 4.3. Figure 4.8 represents the atmospheric absorption coefficient due to CO_2 molecules near the Earth surface for frequencies which determine the radiative flux to the Earth which is emitted by the atmosphere due to CO_2 molecules. We take into account an atmospheric pressure near the Earth's surface; the concen-

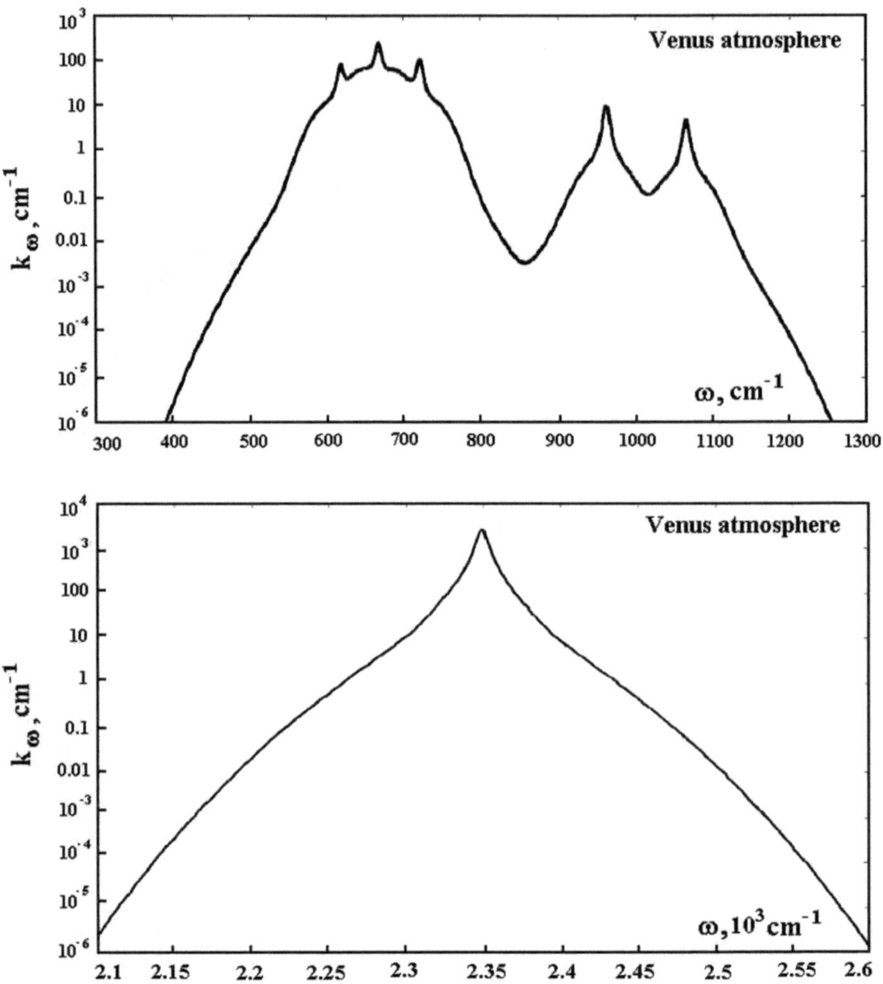

Fig. 4.10 The absorption coefficient k_ω near the surface of Venus due to carbon dioxide molecules, calculated on the basis of formula (4.2.31) [32]

tration $c = 0.04\%$ for CO_2 molecules that corresponds to their number density near the Earth surface $N_v = 1 \cdot 10^{16}\,\text{cm}^{-3}$.

As it follows from Fig. 4.8, three radiative vibrational transitions determine the absorption coefficient in the CO_2 molecule at the room temperature. This means that the radiative flux in the infrared spectrum range is determined by these three vibrational transitions which screen more weak vibrational transitions. The same conclusion follows from the analysis of the HITRAN bank data for the spectral line intensity which is presented in Fig. 4.9.

In the same manner, one can determine the absorption coefficient due to carbon dioxide molecules in the Venus atmosphere. The temperature of the Venus surface is

737 K, the pressure is 92 atm and the main part of the atmospheric gas is the carbon dioxide. This pressure leads to a large width of spectral lines, and neighboring lines are overlapped, so that $\varphi^{(i)}(\omega) = 1$. In addition, because of a high temperature, there are six vibrational-rotational transitions presented in Fig. 4.3. Therefore, the absorption coefficient is given by the following formula

$$k_\omega = \sum_{i=1}^{6} k_\omega = \chi(\omega)^{(i)} + q(\omega)^{(i)}, \qquad (4.2.34)$$

Figure 4.10 gives the frequency dependence of the absorption coefficient due to the atmospheric carbon dioxide near the Venus surface.

References

1. A.A. Radzig, B.M. Smirnov, *Reference Data on Atoms, Molecules and Ions* (Springer, Berlin, 1985)
2. G. Hancock, I.W.M. Smith, Appl. Opt. **10**, 1827 (1971)
3. A.R. Edmonds, *Angular Momentum in Quantum Mechanics* (Princeton, Princeton University Press, 1957)
4. L.D. Landau, E.M. Lifshitz, *Quantum Mechanics* (Pergamon Press, Oxford, 1965)
5. E.U. Condon, G.H. Shortley, *The Theory of Atomic Spectra* (Cambridge University Press, Cambridge, 1970)
6. https://en.wikipedia.org/wiki/Table-of-Clebsh-Gordan-coefficients
7. M. Born, J.R. Oppenheimer, Annalen der Physik **389**, 457 (1927)
8. A. Messiah, *Quantum Mechanics* (North-Holland, Amsterdam, 1960)
9. https://en.wikipedia.org/wiki/Infrared-spectroscopy
10. http://www.hitran.iao.ru/home
11. http://www.hitran.org/links/
12. G. Herzberg, *Molecular Spectra and Molecular Structure* (Princeton, Van Nostrand Reinhold, 1945)
13. M.A. El'yashevich, *Molecular Spectroscopy* (Moscow, Fizmatgiz, 1963; in Russian)
14. H.C. Allen, P.C. Cross, *Molecular Vib-Rotors; the Theory and Interpretation of High Resolution Infra-Red Spectra* (Wiley, New York, 1963)
15. G. Herzberg, *Molecular Spectra and Molecular Structure: Electronic Spectra and Electronic Structure of Polyatomic Molecules* (Van Nostrand, New York, 1966)
16. C.N. Banwell, E.M. McCash, *Fundamentals of Molecular Spectroscopy* (McGraw-Hill, London, 1994)
17. L.D. Kaplan, D.F. Eggers, J. Chem. Phys. **25**, 876 (1956)
18. J.E. Lowder, JQSRT **11**, 1647 (1971)
19. P. Varanasi, JQSRT **11**, 1711 (1971)
20. A. Levy, E. Piollet-Mariel, C. Boilet, JQSRT **13**, 673 (1973)
21. J.E. Lennard-Jones, A.E. Ingham, Proc. Roy. Soc. **106A**, 636 (1924)
22. J.E. Lennard-Jones, Proc. Roy. Soc. **107A**, 463 (1925)
23. B.M. Smirnov, Phys. Uspekhi **44**, 1229 (2001)
24. B.M. Smirnov, *Principles of Statistical Physics* (Wiley, Weinheim, 2006)
25. http://www.hitran.org/docs/definitions-and-units/
26. W.M. Elsasser, Phys. Rev. **54**, 126 (1938)
27. E.T. Whittaker, G.N. Watson, *Modern Analysis* (Cambridge University Press, London, 1940)

28. M.L. Salby, *Physics of the Atmosphere and Climate* (Cambridge University Press, Cambridge, 2012)
29. B.M. Smirnov, G.V. Schlyapnikov, Sov. Phys. Uspekhi **130**, 377 (1980)
30. B.M. Smirnov, *Plasma Processes and Plasma Kinetics* (Wiley, Weinheim, 2007)
31. B.M. Smirnov, JETP **126**, 446 (2018)
32. B.M. Smirnov, JETP **127**, 48 (2018)

Chapter 5
Elementary Radiative Processes

Abstract Parameters of elementary processes involving photons and atomic particles are evaluated. These processes include photoionization of atoms, and photorecombination of electrons and atomic ions, photodetachment of negative ions, and photorecombination of electrons and atoms, photodissociation of molecules. Photoprocesses involving highly excited atoms are analyzed. Bremsstrahlung results from collisions of electrons with atoms or ions in a weakly ionized gas. The character of emission of the solar photosphere is analyzed that consists of electron photoattachment to the hydrogen atom. The scheme of the detector of sub-millimeter radiation on the basis of highly excited atoms is described.

5.1 Radiative Transitions Involving States of Continuous Spectrum

5.1.1 Photoionization and Phodetachment of Atomic Particles

We give in Table 5.1 the list of single-photon processes involving atomic particles. Three first processes related to transitions between discrete states of atoms and molecules, are considered above. We now study the transitions between bound and continuum spectrum states induced by a photon. The first task is to obtain the expression for the cross section of photoionization of an atomic particle. In this way we modify formula (1.2.19) for the absorption rate of an atomic particle with taking into account that the frequency of an absorbed photon lies in a continuous spectrum range. We consider both the photoionization of atoms and that of negative ions. The first process (process 4 of Table 5.1) is described by a scheme

$$A + \hbar\omega \to A^+ + e,$$

while the second process (process 6 of Table 5.1) takes the form

V. Krainov and B. M. Smirnov, *Atomic and Molecular Radiative Processes*,
Springer Series on Atomic, Optical, and Plasma Physics 108,
https://doi.org/10.1007/978-3-030-21955-0_5

Table 5.1 Elementary processes of interaction between photons and atomic particles [1]

Number	Process	Scheme of the process
1	Excitation as a result of photon absorption	$\hbar\omega + A \to A^*$
2	Spontaneous radiation of an excited atom	$A^* \to \hbar\omega + A$
3	Stimulated photon emission	$\hbar\omega + A^* \to 2\hbar\omega + A$
4	Atomic photoionization	$\hbar\omega + A \to A^+ + e$
5	Photorecombination of an electron and and ion	$e + A^+ \to A + \hbar\omega$
6	Photodetachment of a negative ion	$\hbar\omega + A^- \to A + e$
7	Radiative attachment of an electron to an atom	$e + A \to A^- + \hbar\omega$
8	Photodissociation of a molecule	$AB + \hbar\omega \to A + B$
9	Photorecombination of atoms	$A + B \to AB + \hbar\omega$
10	Bremsstrahlung in electron-atom collisions	$e + A \to e + A + \hbar\omega$
11	Bremsstrahlung in electron-ion collisions	$e + A^+ \to e + A^+ + \hbar\omega$

$$A^- + \hbar\omega \to A + e.$$

Because wave functions of the continuum spectrum range are not square-integrable in the conventional sense, it is necessary to define an alternative normalization scheme for continuum spectrum wave functions. It is convenient to adopt the normalization

$$\int \psi_{\mathbf{q}}(\mathbf{r})\,\psi^*_{\mathbf{q}'}(\mathbf{r})\,d\mathbf{r} = (2\pi)^3\,\delta\left(\mathbf{q}-\mathbf{q}'\right) \tag{5.1.1}$$

where $\mathbf{r}$ is a distance between an ejected electron and the atomic core, $\psi_{\mathbf{q}}(\mathbf{r})$ is the electron wave function which determines its relative motion with the wave vector $\mathbf{q}$ in the center-of-mass frame of reference. We now consider electrons to be contained within a large space volume, which can be equal to one, since it is absent in all physical results. We below find the connection between the wave function $\psi_{\mathbf{q}}(\mathbf{r})$ and the wave function $\psi_k(\mathbf{r})$, which is normalized per unit volume according to

$$\int |\psi_k(\mathbf{r})|^2\,d\mathbf{r} = 1$$

These wave functions are coincided with the accuracy up to a factor C, that is introduced as $\psi_k(\mathbf{r}) = C\psi_{\mathbf{q}}(\mathbf{r})$. For determination C we use the normalization condition

$$\int \psi^*_k(\mathbf{r})\,\psi_m(\mathbf{r})\,d\mathbf{r} = \delta_{km},$$

or

$$\sum_m \int \psi^*_k(\mathbf{r})\,\psi_m(\mathbf{r})\,d\mathbf{r} = 1$$

If we replace an index k by the wave vector $\mathbf{q}$, and an index m by the wave vector $\mathbf{q}'$, these conditions yield

$$\sum_m \int \psi_k^*((\mathbf{r})\,\psi_m\,(\mathbf{r})\,\mathrm{d}\mathbf{r} = 1 = C^2 \int \frac{\mathrm{d}\mathbf{q}'}{(2\pi)^3} \int \psi_\mathbf{q}^* \psi_\mathbf{q}(\mathbf{r})\mathrm{d}\mathbf{r} = C^2$$

From this it follows $C = 1$. The wave function $\psi\,(\mathbf{r})$ is transformed in a plane wave at a large distance between an electron and atomic core, where interaction between them is negligible, with the accuracy up to logarithm terms in the exponent for the Coulomb interaction and with accounting for the normalization condition we have in the limit $r \to \infty$

$$\psi_\mathbf{q}\,(\mathbf{r}) \to \exp\,(i\mathbf{q}\mathbf{r})$$

We now obtain the expression for the photodetachment cross section of an atom or positive ion, or the photodetachment cross section for a negative ion. Let us rewrite expression (1.2.20) for the absorption rate with using the wave function $\psi_\mathbf{q}\,(\mathbf{r})$ for the transition matrix element instead of $\psi_j\,(\mathbf{r})$. This corresponds to a change of the matrix element square $\left|\mathbf{D}_{ij}\right|^2$ by $\left|\mathbf{D}_{o\mathbf{q}}\right|^2$, where the final-state wave function is normalized by condition (ion1). The statistical weight g_j of the final state is now given by

$$g_j = \frac{\mathrm{d}\mathbf{q}}{(2\pi)^3} = \frac{q^2 \mathrm{d}q}{8\pi^3}\mathrm{d}\Omega_\mathbf{q}, \tag{5.1.2}$$

where $\mathrm{d}\Omega_\mathbf{q}$ is the element of solid angle that characterizes the relative motion of an ejected electron. Now we employ the condition of energy conservation

$$\hbar\omega = I + \frac{\hbar^2 q^2}{2m_\mathrm{e}}, \tag{5.1.3}$$

where I is the electron binding energy for its initial state, and m_e is the electron mass. This gives

$$q\mathrm{d}q = \frac{m_\mathrm{e}\mathrm{d}\omega}{\hbar}$$

Substituting this in formula (5.1.3) and using formula (1.2.20) for the rate of a radiative transition, we obtain

$$\mathrm{d}w_i = \frac{4\omega^3 \left\|\mathbf{D}_{o\mathbf{q}}\right|^2 n_\omega}{3\hbar c^3} \frac{m_\mathrm{e} q}{\hbar \pi^3} \mathrm{d}\omega \mathrm{d}\Omega_\mathbf{q}$$

Evidently, the normalizing volume disappears from this expression. Dividing the probability $\mathrm{d}w_i$ by the photon flux according to formula (2.2.23)

$$dj = \frac{n_\omega \omega^2 d\omega}{\pi^2 c^2},$$

one can obtain the expression for the cross section of photoionization or photodetachment of the negative ion

$$d\sigma = \frac{m_e q\omega}{6\pi\hbar^2 c}\left|\mathbf{D}_{o\mathbf{q}}\right|^2 d\Omega_{\mathbf{q}} \tag{5.1.4}$$

It is convenient to analyze this result on the basis of one-electron approximation for the wave function of atomic electrons. Then a valence electron is located in a self-consistent field of the atomic core and other electrons, and the potential of the self-consistent field assumes to be spherically symmetric. In addition, transfer of a valence electron does not change the state of atomic core with other valence electrons. Using formula (5.1.4) for the cross section of this process, we characterize the state of a transferring electron by the principal quantum number n, the angular momentum number l, and the magnetic number m which is the projection of the angular momentum onto a given direction. Denoting the initial bound state by 0, let us represent the wave function of a transferring electron in the form

$$\psi_0(\mathbf{r}) = \frac{u_{nl}(r)}{r} Y_{lm}(\theta, \varphi),$$

where r, θ, φ are the spherical coordinates of the valence electron, $u_{nl}(r)$ is the radial wave function of the electron, normalized by the condition

$$\int_0^\infty u_{nl}^2(r)dr = 1,$$

and $Y_{lm}(\theta, \varphi)$ is a normalized spherical function. The final state of the liberated electron with momentum $\hbar q$ can be expressed in terms of spherical functions in the form

$$\psi_{\mathbf{q}}(\mathbf{r}) \to \exp(i\mathbf{qr}) = \frac{1}{r}\sum_{l=0}^{\infty}(2l+1)\, i^l u_{nl}(q, r)\, P_l\left(\cos\theta_{\mathbf{qr}}\right)$$

Here $\theta_{\mathbf{qr}}$ is the angle between vectors $\mathbf{q}$ and $\mathbf{r}$, and $u_{nl}(q, r)$ is the radial wave function of a free electron with an angular momentum l. Far from the atomic core, this wave function has the asymptotic form

$$u_{nl}(q, r) = \frac{1}{r}\sin\left(qr - \frac{\pi l}{2} + \delta_l\right),$$

where δ_l is the phase shift for scattering of an electron by the atomic core. This expression is valid for photodetachment of a negative ion. In the case of atomic photoionization, the phase shift δ_l depends on r.

If a released electron is fast, one can neglect the electron interaction with an atomic core, and the electron wave function in the final state has the form $\psi_{\mathbf{q}}(\mathbf{r}) \to \exp(i\mathbf{qr})$, hence

$$u_{nl}(q,r) = \sqrt{\frac{\pi r}{2q}} J_{L+1/2}(qr)$$

where $J_{L+1/2}(qr)$ is the half-integer Bessel function. A general form of the photodetachment cross section (5.1.4) in the single-electron approximation under the assumption of the spherically symmetric form of the potential of the self-consistent field for release of one valence electron has the form

$$\mathrm{d}\sigma = \frac{q\omega}{6\pi a_o c} \left|\mathbf{r}_{0\mathbf{q}}\right|^2 \mathrm{d}\Omega_{\mathbf{q}} \tag{5.1.5}$$

Here $a_o = \hbar^2/\left(m_e e^2\right)$ is the Bohr radius, and for a given transferring electron we have $\mathbf{D} = -e\mathbf{r}$. We now calculate the integral (5.1.5). Let us introduce the unit vector $\mathbf{s}$ such, that the projection of the electron momentum onto a given direction for the initial electron state is zero for this direction. Hence, since $m = 0$, the angular wave function of the initial state takes the simple form

$$Y_{lm}(\theta,\varphi) = \sqrt{\frac{2l+1}{4\pi}} P_l(\cos\theta_{\mathbf{rs}}),$$

where $\theta_{\mathbf{rs}}$ is the angle between vectors $\mathbf{r}$ and $\mathbf{s}$. On the basis of the these expressions for the wave functions, let us rewrite the above integral in the form

$$\int \left|\mathbf{r}_{0\mathbf{q}}\right|^2 \mathrm{d}\Omega_{\mathbf{q}} = \int \mathrm{d}\Omega_{\mathbf{q}} \times$$
$$\times \left| \int\int \mathbf{r}\mathrm{d}r\mathrm{d}\Omega_{\mathbf{r}} u_{nl}(r) \sqrt{\frac{2l+1}{4\pi}} P_l(\cos\theta_{\mathbf{rs}}) \sum_{l'=0}^{\infty} (2l'+1)\, i^{l'} u_{l'}(q,r)\, P_{l'}(\cos\theta_{\mathbf{qr}}) \right|^2$$

Replacing the integral square by two integrals, we have

$$\int \mathrm{d}\Omega_{\mathbf{q}} \int\int\int\int \mathrm{d}r\mathrm{d}r' \left(\mathbf{r}\mathbf{r}'\right) \mathrm{d}\Omega_{\mathbf{r}} \mathrm{d}\Omega_{\mathbf{r}'} u_{nl}(r) u_{nl}(r) \frac{2l+1}{4\pi} P_l(\cos\theta_{\mathbf{rs}})\, P_l\left(\cos\theta_{\mathbf{r}'\mathbf{s}}\right) \times$$
$$\times \sum_{l'=0}^{\infty}\sum_{l"=0}^{\infty} i^{l'-l"} (2l'+1)(2l"+1)\, P_{l'}(\cos\theta_{\mathbf{qr}})\, P_{l"}\left(\cos\theta_{\mathbf{qr}'}\right) u_{l'}(q,r)\, u_{l''}\left(q,r'\right)$$

Integrating over the solid angle $\mathrm{d}\Omega_{\mathbf{r}'}$, we use the addition theorem for Legendre polynomials

$$P_{l"}\left(\cos\theta_{\mathbf{qr}'}\right) = P_{l"}\left(\cos\theta_{\mathbf{rr}'}\right) P_{l"}\left(\cos\theta_{\mathbf{qr}}\right) + \\ + \sum_{m} \frac{(l" - m)!}{(l" + m)!} P_{l"}^{m}\left(\cos\theta_{\mathbf{rr}'}\right) P_{l"}^{m}\left(\cos\theta_{\mathbf{qr}}\right) \cos\left[m\left(\varphi_{\mathbf{r}} - \varphi_{\mathbf{r}'}\right)\right]$$

Indices at the polar angle θ indicate vectors between which it is taken, and φ is the azimuthal angle. The sum over $l"$ in the above expression is omitted due to the orthogonality of Legendre polynomials with different indices, so that

$$\int_{-\pi/2}^{\pi/2} P_l\left(\cos\theta\right) P_{l'}\left(\cos\theta\right) \sin\theta \mathrm{d}\theta = \frac{2l+1}{4\pi}\delta_{ll'}.$$

We obtain thereby

$$\int \left|\mathbf{r}_{o\mathbf{q}}\right|^2 \mathrm{d}\Omega_{\mathbf{q}} = \sum_{l'=0}^{\infty} \left|R_{ll'}\right|^2 \times \\ \times \int\int \mathrm{d}\Omega_{\mathbf{r}} \mathrm{d}\Omega_{\mathbf{r}'} \cos\theta_{\mathbf{rr}'} \left(2l+1\right)\left(2l'+1\right) P_l\left(\cos\theta_{\mathbf{rs}}\right) P_l\left(\cos\theta_{\mathbf{r}'\mathbf{s}}\right) P_l\left(\cos\theta_{\mathbf{rr}'}\right),$$

where the used notations lead to the radial matrix element

$$R_{ll'} = \int_0^{\infty} u_{nl}(r) r u_{l'}\left(q, r\right) \mathrm{d}r$$

In order to use the orthogonality of the Legendre polynomials, we introduce the following recursion relation

$$\left(2l'+1\right) x P_{l'}\left(x\right) = \left(l'+1\right) P_{l'+1}\left(x\right) + l' P_{l'-1}\left(x\right)$$

This leads to formula

$$\int \left|\mathbf{r}_{o\mathbf{q}}\right|^2 \mathrm{d}\Omega_{\mathbf{q}} = \sum_{l'=0}^{\infty} \left|R_{ll'}\right|^2 \int\int \mathrm{d}\Omega_{\mathbf{r}} \mathrm{d}\Omega_{\mathbf{r}'} \left(2l+1\right) P_l\left(\cos\theta_{\mathbf{rs}}\right) P_l\left(\cos\theta_{\mathbf{r}'\mathbf{s}}\right) \times \\ \times \left\{\left(l'+1\right) P_{l+1}\left(\cos\theta_{\mathbf{rs}}\right) P_{l+1}\left(\cos\theta_{\mathbf{r}'\mathbf{s}}\right) + l' P_{l-1}\left(\cos\theta_{\mathbf{rs}}\right) P_{l-1}\left(\cos\theta_{\mathbf{r}'\mathbf{s}}\right)\right\}$$

Using the addition theorem for the Legendre polynomials, we get finally

$$\int \left|\mathbf{r}_{o\mathbf{q}}\right|^2 \mathrm{d}\Omega_{\mathbf{q}} = \frac{(4\pi)^2}{2l+1}\left\{l\left|R_{l,l-1}\right|^2 + (l+1)\left|R_{l,l+1}\right|^2\right\}$$

We use above the conditions of the orthonormality and normalization for the Legendre polynomials. From this we have the cross section of photodetachment for a single-electron ion or atom in the form

$$\sigma_i = \frac{8\pi q\omega}{3a_o c(2l+1)}\left\{l\left|R_{l,l-1}\right|^2 + (l+1)\left|R_{l,l+1}\right|^2\right\} \tag{5.1.6}$$

It follows from the above operations, that the selection rules for radiative transitions of electrons into the continuous spectrum state are fulfilled for the single-electron atomic structure. According to the selection rules, the orbital angular momentum changes by one, i.e. the electron orbital momentum l in the initial state and the electron momentum l' of a released electron are connected by the relation

$$l' = l \pm 1 \tag{5.1.7}$$

The electron parity changes due to this radiative transition. In addition, the threshold behavior of the cross section $\sigma_i \sim q$ for photodetachment of a negative ion corresponds to the general law for the threshold electron process with a neutral particle in the final state (5.1.6); hence at a large distance from the core the electron wave function is described by the plane wave. Another threshold law corresponds to the photoionization process with the Coulomb interaction between the ejected electron and the positive ion at large separations. Note that formula (5.1.6) describes the case, where an absorbed photon has a certain polarization.

5.1.2 Two-Step Photoionization of Atoms

Laser radiation is a fine instrument which allows one to excite an atomic gas within the framework of the method of optical pumping. On the other hand, an atomic ionization with production of electrons and ions gives the possibility to analyze products of ionization in a simple manner on the basis of electric fields. Joining of these methods leads to a sensitive method to analyze gaseous systems. This is realized in the method of two-step atomic ionization. The peculiarity of atomic ionization under the action of radiation is a non-resonance character of this process. In other words, the dependence of the ionization cross section on the photon frequency depends weakly on the photon energy at the photon energies higher than the ionization threshold. Therefore, if we have a mixture of different kinds of gases, and the photon energy exceeds the ionization threshold for each component of gaseous atoms or molecules, the photoionization of each gaseous component occurs with a different efficiency. Thus, the photoionization process is not selective with respect to various gaseous components. Another situation takes place in the case of the two-step ionization approach, when the radiation includes two wavelengths. Then atoms of a certain kind are excited by radiation with the first wavelength. These excited atoms are ionized by radiation with the second wavelength, since energies of the second photons are insufficient for ionization of atoms. This method of ionization is selective, since the photon wavelength for excitation of atoms is in resonance with the excitation of this kind of atoms.

As an example, we consider the two-step ionization method of cesium atoms [2] which are located in a buffer gas. Cesium atoms are excited from its ground state into an excited state by the first laser. Then excited atoms are ionized by photons of the second laser. Formed ions are detected by standard methods of detection of charged atomic particles. As a result, it is possible to fix several atoms in the volume which contains about of 10^{19} atoms of a buffer gas. This is equivalent to discover of an element which area is less than 1 cm^2 on the Earth surface!

The same concept is used in the optohalvanic method [3–5] to analyze the sort of atoms in gases. In this case a weak gas discharge is created in a gas, so that an electric current passes through this weakly ionized gas. This gas is irradiated by a beam from a tunable laser, so that the current strength varies depending on the radiation wavelength. Peaks at certain radiation wavelength testify about the presence of atoms or molecules of a given sort in a gaseous mixture, and the value of these peaks allows one to restore the concentration of these atoms or molecules. In addition, this action of radiation on the discharge electric current may occur due to various radiative transitions that improve the accuracy of this method which give the possibility to determine the presence of admixtures in gaseous mixture with the concentration up to 10^{-10}. In particular, this method is used for detection of gunshot residues on hands [6].

5.1.3 Photodetachment of Hydrogen Negative Ion

We now calculate the cross section for photodetachment of a negative ion with a valence s-electron and photodetachment of the hydrogen negative ion as a particular process of this type. We are based on formula (5.1.6), and since existing negative ions have two electrons, i.e. valence electrons fill the shell $(s^2\,{}^1S)(H^-)$, we use the factor 2 in formula (5.1.6) for $l = 0$

$$\sigma_i = \frac{16\pi q\omega}{3a_o c} R_{01}^2 \tag{5.1.8}$$

Our goal now is to calculate the matrix element in this formula. Extracting a weakly bound electron of a negative ion and denoting its binding energy as $E_A = \hbar^2\gamma^2/(2m_e)$, one can obtain the Schrödinger equation for the electron radial wave function $u_0(r)$ in the form

$$\ddot{u}_0 = \gamma^2 u_0$$

In obtaining this equation, we neglect the interaction between a valence electron and the atom core because of the short-range interaction between them. Thus for a short-range interaction between an electron and remaining atom, the electron wave function, normalized by the condition $\int_0^\infty u_0^2(r)\mathrm{d}r = 1$ and obtained as the solution of the Schrödinger equation, is

$$u_0(r) = \sqrt{2\gamma}\exp(-\gamma r)$$

This expression is violated at electron-atom distances of the order of a_o where other electrons are located, but it is small compared with an electron size $\sim 1/\gamma$. Therefore in a basic region of location of a weakly bound electron one can present its wave function as

$$u_0(r) = B\sqrt{2\gamma}\exp(-\gamma r)$$

The parameter B is determined by the electron behavior in the region of location of other electrons; for the negative hydrogen ion this parameter is equal $B = 1.16$ [7]. In the limit $\gamma \to 0$ this parameter has the limit $B = 1$.

The radial wave function of a released electron with $l = 1$ is the first harmonic in the expansion of the wave function as the plane wave $\exp(i\mathbf{pr})$ over Legendre polynomials

$$u_l(q, r) = \sqrt{\frac{\pi r}{2q}} J_{3/2}(qr) = \frac{1}{q}\left(\frac{\sin qr}{qr} - \cos qr\right)$$

One can obtain the following expression for the radial matrix element on the basis of this wave function

$$R_{01} = \int_0^\infty u_0(r) r u_1(q, r) = \frac{2q\sqrt{2\gamma}}{\left(\gamma^2 + q^2\right)^2}$$

From this we obtain the expression for the cross section of photodetachment of the negative ion under consideration

$$\sigma_{\mathrm{det}} = \frac{4\pi n B^2}{3}\frac{e^2}{\hbar c}\frac{\gamma v_e^3}{\omega^3}, \tag{5.1.9}$$

where $n = 2$ is the number of electrons in the negative hydrogen ion, so that process can lead to release of each electron; $v_e \hbar q/m$ is the velocity of the released electron. The parameter B in this expression takes into account the fact that the electron wave function is cut off at distances $r \sim a_o$ from the nucleus. This factor is equal to one in the limit $\gamma \to 0$, and $B = 1.16$ for the negative hydrogen ion. The energy of an absorbed photon is equal

$$\hbar\omega = \frac{\hbar^2}{2m_{\mathrm{e}}}\left(\gamma^2 + q^2\right)$$

Note that the expression (5.1.9) for the cross section of photodetachment of a negative ion with valence s-electrons is based on the assumption that a size $\sim 1/\gamma$ of the negative ion is large compared with the region of interaction of a transferring electron with a remaining neutral atom. Therefore it holds true at small binding energies of an electron in a negative ion. The expression obtained is not valid for large values of q because the main contribution to the matrix element R_{01} follows from electron-nucleus distances $r \sim 1/q$, and these distances must be large compared with atomic sizes. According to formula (5.1.9), the maximum of cross section is observed at $\hbar\omega = 2E_A$ where the cross section equals $\sigma_{\max} = 0.37$ Å^2 for the hydrogen negative ion. The threshold of this process corresponds to the photon energy $\hbar\omega = E_A = 0.754$ eV, i.e., to the wavelength $\lambda = 1.64\ \mu$m. According to formula (5.1.9), the maximum cross section, that is observed at the wavelength $\lambda = 0.82\ \mu$m, is equal to $4.2 \cdot 10^{-17}$ cm^2. Figure 5.1 represents the photodetachment cross section for the hydrogen negative ion as a function of the wavelength according to formula (5.1.9) and also experimental data. In the limit $q \gg \gamma$ (but, nevertheless, $q \ll 1/a_o$), formula (5.1.9) gives for the photodetachment cross section the following asymptotic form

$$\sigma_{\text{det}} = \frac{64\pi}{3}\frac{e^2}{\hbar c}\frac{\gamma}{q^3}, \quad q \gg \gamma$$

On contrary, near the threshold, where $q \ll \gamma$, formula (5.1.9) gives

$$\sigma_i = \frac{64\pi}{3}\frac{e^2}{\hbar c}\frac{q^3}{\gamma^5}, \quad q \ll \gamma$$

The dependence $\sigma_i \sim q^3$ reflects the fact that the released electron is found in p-state (not an s-state) which wave function $u_1(r)$ at small distances from the nucleus $r \ll 1/q$ decreases proportional to r.

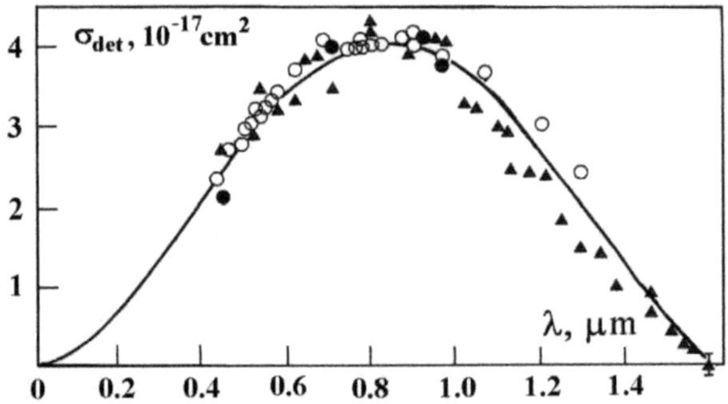

Fig. 5.1 Cross section of photodetachment of the hydrogen negative ion $\hbar\omega + H^- \to H + e$ [7]. Solid curve corresponds to formula (5.1.9), signs relate to experimental data

5.1.4 *Photoionization of Hydrogen Atom*

We now determine the cross section of photoionization of the hydrogen atom in its ground state on the basis of a general formula (5.1.8) that in this case of the one valence electron has the form

$$\sigma_i = \frac{8\pi q\omega}{3a_o c} R_{01}^2,$$

and our goal is calculation of the matrix element R_{01}. For simplicity, we use in this operation the atomic system of units $e^2 = m_e = \hbar = 1$. The radial wave function of an electron with $l = 1$ and the momentum q in the continuum spectrum of a Coulomb field is equal

$$u_1(q,r) = \frac{2}{3}\sqrt{\frac{q\left(1+q^2\right)}{1-\exp\left(-2\pi/q\right)}} r^2 \exp(iqr) F\left(\frac{i}{q}+2, 4, 2iqr\right)$$

Here $F(\alpha, \gamma, z)$ is the confluent hypergeometric function. Taking the radial wave function of the hydrogen atom in the ground state $u_0(r) = 2r\exp(-r)$, one can obtain the matrix element R_{01} as

$$R_{01} = \int_0^\infty 2r\exp(-r) u_1(q,r) \mathrm{d}r$$

To evaluate this integral, we use formula

$$\int_0^\infty \exp\left(-\lambda z\right) z^{\gamma-1} F\left(\alpha, \gamma, kz\right) \mathrm{d}z = \Gamma(\gamma)\lambda^{\alpha-\gamma}\left(\lambda - k\right)^{-\alpha}$$

This gives

$$R_{01} = \frac{8\sqrt{2\pi}}{\sqrt{q(1+q^2)^5}} \frac{\exp\left(-\frac{2}{q}\arctan\frac{1}{q}\right)}{\sqrt{1-\exp\left(-\frac{2\pi}{q}\right)}}$$

From this formula, the photoionization cross section of the hydrogen atom from the ground state is equal (the Stobbe formula)

$$\sigma_i = \frac{2^9\pi^2}{3c} \frac{\exp\left[-\frac{4}{q}\arctan\left(\frac{1}{q}\right)\right]}{\left(1+q^2\right)^4\left[1-\exp\left(-\frac{2\pi}{q}\right)\right]} \tag{5.1.10}$$

Let us consider now limiting cases of formula (5.1.10). For small q this formula leads to the threshold dependence for the photoionization cross section

$$\sigma_i = \frac{2^9\pi^2}{3e^4}\alpha a_o^2 = 6.3 \cdot 10^{-18}\mathrm{cm}^2, \tag{5.1.11}$$

where $\alpha = e^2/\hbar c$ is the fine structure constant, and we use now usual units, rather than the atomic ones. As it follows from formula (5.1.11), the threshold cross section is a finite value. This follows from the behavior of the wave function of a slow electron, when it is located in the Coulomb field of the nucleus. Then the dipole matrix element tends to infinity near the threshold as $\sim \mathrm{q}^{-1/2}$, and the matrix element at the threshold $q \ll 1$ is equal

$$R_{01} = \frac{8}{e^2}\sqrt{\frac{2\pi}{q}} \tag{5.1.12}$$

In formula (5.1.12), as well as in formula (5.1.11), $\mathrm{e} = 2.718$, is the base of the natural logarithm. Note that in the absence of the Coulomb field, i.e. if a released electron interacts with a neutral atomic particle, we have $R_{01} \sim q$ at $q \ll 1$, that leads to the threshold cross section $\sigma_i \sim q^3$. Thus, the Coulomb interaction leads to a specific electron behavior near the charged nucleus. Next, for large $q \gg 2\pi$, when the photon energy exceeds significantly the ionization potential I of ground-state hydrogen atom, that is

$$\hbar\omega \approx \frac{\hbar^2 q^2}{2m_e} \gg 4\pi^2 I \approx 40I$$

then formula (5.1.10) gives

$$\sigma_i = \frac{256\pi}{3}\frac{\alpha a_o^2}{(qa_o)^7} \approx \frac{256\pi}{3}\alpha a_o^2\left(\frac{I}{\hbar\omega}\right)^{7/2} = \sigma_o\left(\frac{I}{\hbar\omega}\right)^{7/2}, \tag{5.1.13}$$

where $\sigma_o = 5.48 \cdot 10^{-17}\mathrm{cm}^2$, and the ionization potential of the hydrogen atom e-quals

$$I = \frac{m_e e^4}{2\hbar^2} = 13.6\,\mathrm{eV}$$

As it follows from formula (5.1.13), the photoionization cross section of the hydrogen atom as a function of the photon energy decreases at large photon energies faster than the photodetachment cross section for the negative ion, where $\sigma_{\mathrm{det}} \sim q^{-3}$. If we take in formula (5.1.13) $\hbar\omega = I$, one can expect that the photoionization cross section would be close to its threshold value (5.1.11). In fact, it differs from the threshold value by the factor $\mathrm{e}^4/2\pi$, which is about an order of magnitude. This means a very extended transition from the range $q \ll$ to the range $q \gg 1$. The transition to the asymptotic dependence (5.1.13) is very slowly due to a large numerical factor in the above criterion $\hbar\omega \gg 40I$. A decrease of the photoionization matrix element and cross section according to a power law for $q \gg 1$ follows from the singularity of the Coulomb potential at $r = 0$. In fact, the matrix element R_{01} for $q \gg 1$ is the Fourier component for large values of the argument. Without the singularity in the interaction potential, the Fourier component would have an exponential dependence rather than a power dependence on the argument. Formula (5.1.13)) is valid up to values of the frequency $\hbar\omega \sim m_e c^2$, where relativistic effects become essential. These effects are not taken into account in above consideration, and this follows from the general expression (5.1.10), according to which the photoionization cross section has maximum at the threshold. If the photon energy $\hbar\omega$ increases, the cross section decreases first as $\omega^{-8/3}$, and then as $\omega^{-7/2}$ in accordance with formula (5.1.13).

We now determine the angular distribution of the ejected photoelectrons at the photoionization of the ground state of the hydrogen atom. According to formula (1.2.18), the rate of photoionization and, correspondingly, the photoionization cross section with ejection of a released electron into the solid angle element $d\Omega_{\mathbf{q}}$ is equal

$$d\sigma_i \sim (\mathbf{ns})^2 \, d\Omega_{\mathbf{q}},$$

where $\mathbf{n}$ is the unit vector in the direction of emission of the ejected photoelectron, and $\mathbf{s}$ is the unit vector along the direction of polarization of the absorbed photon. One can use the normalization condition according to which the integration over the solid angle in this expression must give the total cross section (5.1.10). One obtains

$$d\sigma_i = \frac{3}{4\pi}\sigma_i \, (\mathbf{ns})^2 \, d\Omega_{\mathbf{q}} \tag{5.1.14}$$

As it follows from formula (5.1.14), the most part of released electrons are emitted in the direction close to that of the polarization vector of the incident radiation. The probability to emit in the perpendicular direction, including that of propagation of the incident radiation, is zero. Next, if the incident radiation is unpolarized, (5.1.14) must be averaged over polarization directions $\mathbf{s}$ at a fixed direction $\mathbf{n}_0$ of the wave vector of the incident radiation. Then, from the relations $\cos\theta_{\mathbf{ns}} = \sin\theta_{\mathbf{nn_0}} \cos\varphi_{\mathbf{s}}$ and $\langle\cos^2\varphi_{\mathbf{s}}\rangle = 1/2$, we have

$$d\sigma_i = \frac{3}{8\pi}\sigma_i \, [\mathbf{n}, \mathbf{n}_0]^2 \, d\Omega_{\mathbf{q}} \tag{5.1.15}$$

If we consider photoionization of an atom K-shell which contains two electrons, it is necessary to double all the above cross sections. Applying the above formulas for the cross section of photoionization to hydrogen like ions or to electrons of the K-shell, it is necessary to change the Bohr radius a_o to a_o/Z, where Z is the nuclear charge. Then according to formula (5.1.11) we have $\sigma_i \sim Z^{-2}$ near threshold, while from formula (5.1.13) it follows that $\sigma_i \sim Z^5$ for large photon energies.

On the basis of the above results, let us analyze a general character of the electron angular distribution in the photoionization process, using single-electron approximation. It follows from formula (1.2.18), that the photoionization cross section is determined by the matrix element

$$\int \psi^*_{\mathbf{q}}(\mathbf{r})\,(\mathbf{r}\cdot\mathbf{s})\,\psi_{nlm}(\mathbf{r})\,d\mathbf{r} \tag{5.1.16}$$

Here, the vector $\mathbf{s}$ characterizes the polarization of the incident radiation, and $\mathbf{q}=\mathbf{p}/\hbar$ is the wave vector of the released electron, so that $\mathbf{p}$ is its momentum. Let us take the direction $\mathbf{q}$ as the quantum axis for the quantum numbers nlm of the initial state of a valence electron which is released in the process under consideration. The differential cross section should be averaged over all projections $m=-l,\ldots l-1,l$ of the angular momentum of the released electron. If the shell nl is filled only partially, this averaging is correct until we neglect external fields, for example, the field of the crystal cell where this atom is located. Let us denote by ϑ the angle between vectors $\mathbf{s}$ and $\mathbf{q}$; this angle determines the averaged cross section $\langle d\sigma_i\rangle$. An average over m is similar to an average over all directions of the quantum axis $\mathbf{r}$. We label as ϑ' the angle between vectors $\mathbf{r}$ and $\mathbf{q}$. On the basis of the addition theorem for trigonometric functions we have

$$\mathbf{r}\cdot\mathbf{s}=r\left(\cos\vartheta\cos\vartheta'+\sin\vartheta\sin\vartheta'\cos\varphi\right), \tag{5.1.17}$$

where ϑ and ϑ' are polar angles of these vectors, and φ is the difference of their azimuthal angles in a given frame of reference. Substituting this expression into formula (5.1.16), taking its square and averaging the result over the angle φ, one can obtain that the term $\cos\vartheta\cos\vartheta'\cdot\sin\vartheta\sin\vartheta'\left\langle\cos\varphi'\right\rangle=0$. Next, averaging of squares of each terms in formula (5.1.16) over the angle ϑ', one can obtain the following form of the photoionization cross section

$$\langle d\sigma_i\rangle=\left(\alpha+\beta\cos^2\vartheta\right)d\Omega_{\mathbf{q}} \tag{5.1.18}$$

The fact that the photoionization cross section does not contain terms of the odd power of $\cos\vartheta$ means that that it does not depend on the sign of the momentum of a released electron, or on the sign of $\mathbf{q}$, that can be obtained from general considerations. Indeed, a change of the sign of $\mathbf{q}$ is analogous to a change of the sign of $\mathbf{r}$ in the wave function $\psi^*_{\mathbf{q}}(\mathbf{r})$. But the operation $\mathbf{r}\to-\mathbf{r}$ leads to multiplication of the wave function ψ_{nlm} by a factor $(-1)^l$. Since the photoionization cross section is a quadratic function of the dipole matrix element, the cross section does not change as the result of such a transformation.

If the initial state nl is an S-state, i.e. ($l=0$), the angular dependence of the photoionization cross section is simplified. In this case, it is convenient to expand the wave function of the final state over the Legendre polynomials, which are functions of the angle ϑ' between the vectors $\mathbf{r}$ and $\mathbf{q}$, that is

$$\psi_{\mathbf{q}}(\mathbf{r})=\sum_{l=0}^{\infty}u_l(q,r)P_l\left(\cos\vartheta'\right) \tag{5.1.19}$$

The wave function of the initial state does not depend on angles. Since the function (5.1.19) does not depend on the angle φ', the second term of formula (5.1.17) is zero due to the relation

$$\int_0^{2\pi} \cos\varphi' \mathrm{d}\varphi' = 0$$

Thus, the dipole matrix element is proportional to $\cos\vartheta$, and the cross section has the form

$$\mathrm{d}\sigma_i = \beta \cos^2\vartheta \mathrm{d}\Omega_{\mathbf{q}} \tag{5.1.20}$$

5.1.5 Radiation of Solar Photosphere

As an example where the photoprocesses under consideration are of importance, is emission of the Sun. This process proceeds in the solar photosphere, and radiation results from the process

$$e + \mathrm{H} \longleftrightarrow \mathrm{H}^- + \hbar\omega \tag{5.1.21}$$

Analyzing radiation of the solar photosphere in the visible spectral range, we give basic parameters for the plasma of the solar photosphere [8–10], as well as the processes occurring in it. The main component of the photosphere are hydrogen atoms, whose concentration in this region is approximately 92%, so that further we neglect the presence of other components in the photosphere. Radiation of the Sun in the visible spectrum range occurs due to the photoattachment of electrons to hydrogen atoms, and to the opposite process—the photodecay of a negative hydrogen ion which determines absorption of photons when they try to leave the solar atmosphere. Due to a relatively large number density of hydrogen atoms in the photospheric plasma, a local thermodynamic equilibrium is maintained with respect to the formation of charged atomic particles

$$e + \mathrm{H} \leftrightarrow 2e + \mathrm{H}^+; \;\; e + \mathrm{H}^- \leftrightarrow 2e + \mathrm{H} \tag{5.1.22}$$

Considering the equilibrium properties of the photospheric plasma, we have the following relations between the number densities of hydrogen atoms N_{H}, electrons N_{e}, protons N_{p}, and negative ions N_- with taking into account the condition of plasma quasineutrality

$$N_{\mathrm{e}} = N_{\mathrm{p}} \ll N_{\mathrm{H}} \tag{5.1.23}$$

Due to the local thermodynamic equilibrium, the number density of charged particles and hydrogen atoms in the photospheric plasma of Sun are connected by the Saha relations [11]

$$N_e = (N_o N_H)^{1/2} \exp\left(-\frac{J}{2T}\right), \quad N_- = \frac{N_H^{3/2}}{4N_o^{1/2}} \exp\left(-\frac{\varepsilon_1}{T}\right), \quad N_o = \left(\frac{m_e T}{2\pi\hbar^2}\right)^{3/2} \qquad (5.1.24)$$

Here T is the plasma temperature, $J = 13.605$ eV is the ionization potential of the hydrogen atom, $\varepsilon_1 = J/2 - EA = 6.048\,\text{eV}$, where $EA = 0.754\,\text{eV}$ is the electron affinity to the hydrogen atom. Formula (5.1.24) gives $N_o = 4.3 \cdot 10^{14}\text{cm}^{-3}$ at the average plasma temperature $T = 5777\,\text{K}$. We use the NASA data for the parameters of the solar photosphere [10], namely, the pressure at the bottom of the photosphere is 125 mbar, and at the top of the photosphere it is 0.87 mbar. Accordingly, the temperature of the photosphere is 6600 K at the bottom and 4400 K at the top, and the photosphere thickness is 500 km. Taking into account the local thermodynamic equilibrium in the photosphere plasma, we obtain for the number density of atomic particles near the photosphere bottom $N_H = 1.4 \cdot 10^{17}\text{cm}^{-3}$, $N_e = 4.6 \cdot 10^{10}\text{cm}^{-3}$, and $N_- = 1.7 \cdot 10^{13}\text{cm}^{-3}$, whereas at the photosphere top these parameters are $N_H = 1.5 \cdot 10^{15}\text{cm}^{-3}$, $N_e = 1.6 \cdot 10^{7}\text{cm}^{-3}$, and $N_- = 6.4 \cdot 10^{7}\text{cm}^{-3}$.

Note that the surface of the Sun consists of convective cells—granules of a size of the order of 1000 km. A plasma from the inner regions of the Sun falls into the photosphere region as a result of convection, so that in the central regions of cells, in which the plasma penetrates from deeper regions of the Sun, the temperature is higher than that in regions between cells [8, 12]. This means that the solar surface is non-uniform, and the above parameters of the photosphere characterize their mean values.

We also have that the gradient of the number density of hydrogen atoms is not connected with the gravity. Assuming, according to the NASA data [10], the thickness of the photosphere to be equal to $\Delta = 500\,\text{km}$, we have for the temperature gradient

$$\frac{dT}{dz} = 4.4\,\text{K/km} \qquad (5.1.25)$$

In addition, by approximating the dependence of the number densities of hydrogen atoms N_H and negative hydrogen ions N_- on the altitude h by

$$N_H(h) = N_H(0)\exp(-h/\Lambda_H), \quad N_-(h) = N_-(0)\exp(-h/\Lambda_-), \qquad (5.1.26)$$

we obtain from the above NASA data for the scale of the change of the number density of atoms and negative hydrogen ions $\Lambda_H = 110\,\text{km}$, $\Lambda_- = 40\,\text{km}$.

Next, let us compare these data with parameters which follow from the analysis of the photospheric radiative properties. Indeed, absorption of a photon is determined by the photodecay process of a negative hydrogen ion. In the simplest version, in calculating the matrix element from the dipole moment between the states of the radiative transition, one can neglect the interaction between a transferring electron and atomic core, that leads to the following expression for the photodecay cross section of the negative hydrogen ion [7] which is in accordance with results of the experiments [13, 14]

$$\sigma_{ph}(\omega) = \sigma_{max} \cdot \frac{8\omega_o^{3/2}(\omega - \omega_o)^{3/2}}{\omega^3} \tag{5.1.27}$$

Here ω is the photon frequency, the threshold energy of this photoprocess is $\hbar\omega = EA = 0.754\,\text{eV}$, and the maximum cross section $\sigma_{max} = 4 \cdot 10^{-17}\text{cm}^2$ corresponds to the photon energy $\hbar\omega_{max} = 2\hbar\omega_o = 1.51\,\text{eV}$ or the photon wavelength $\lambda = 0.8\,\mu\text{m}$.

Let us determine the absorption photosphere parameters. At the wavelength $\lambda = 0.8\,\mu\text{m}$, where the absorption is maximal, we have from formula (5.1.24) for the number density of negative ions $N_- = 1 \cdot 10^{10}\text{cm}^{-3}$. From this it follows on the basis of the NASA data that the photosphere bottom is located at an altitude of $h = 300\,\text{km}$ with respect to the photosphere bottom for a frequency of the maximum absorption cross section. At this altitude, the number density of negative hydrogen ions is approximately $N_- = 1 \cdot 10^{10}\text{cm}^{-3}$, which corresponds to the number density of hydrogen atoms $N_H = 9 \cdot 10^{15}\text{cm}^{-3}$ and the number density of electrons (protons) $N_e = 4 \cdot 10^8\text{cm}^{-3}$ at this altitude. Obviously, this altitude characterizes the temperature, which corresponds to radiation at a photon wavelength of $\lambda = 0.8\,\mu\text{m}$. For other wavelengths, radiation is formed in deeper photosphere layers and corresponds to higher temperatures. Therefore, this altitude $h = 300\,\text{km}$ determines the position of the photosphere top. The position of the photosphere bottom is determined by the width of the radiation band, which is referred to the solar photosphere.

The above analysis holds true if the parameter α in formula (2.2.37) is small. According to the NASA data [10] $\mathrm{d}T/\mathrm{d}h = 4.4$ K/km, and the scale of the number density change for negative ones as absorbing particles is $\Lambda_- = 40$ km. This gives for a small parameter (2.2.37) under typical conditions $\alpha = 0.07$, i.e. the description of the photosphere radiation based on this criterion is valid.

5.1.6 Photoprocesses Involving Atmospheric Molecules

Processes of photoionization and photodissociation of atmospheric molecules N_2 and O_2 determine principal properties of upper layers of the Earth's atmosphere. Hard solar radiation is absorbed in the upper atmosphere, and this leads to ionization of air molecules and atoms, as well as to dissociation of atmospheric molecules. As a result, the content of the upper atmosphere and their electric properties are determined by these processes which we consider below partially. The main part of the solar radiation is created by the solar photosphere and relates to the visible spectrum. Alongside with that, solar radiation contains also photons in ultraviolet and

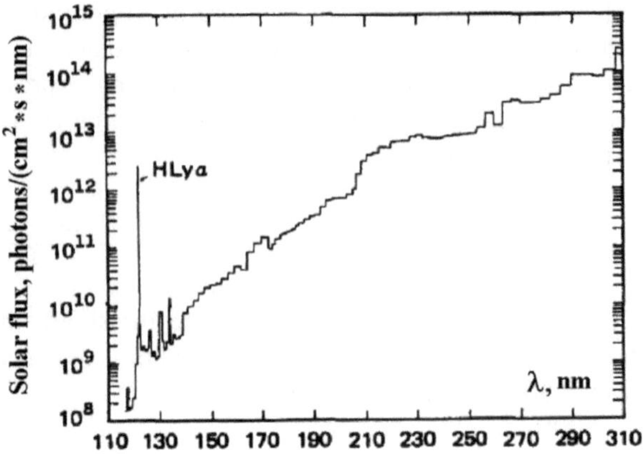

Fig. 5.2 Average flux of photons in the ultraviolet and vacuum ultraviolet spectrum ranges [15, 16]

vacuum ultraviolet spectrum ranges, and the photon flux in these spectrum ranges is given in Fig. 5.2. These photons are able to ionize and dissociate atmospheric molecules.

As a result, electrons and ions are produced, and electrons lead to the negative charge of the atmosphere at altitudes more than 80 km. Figure 5.3 connects altitudes at which the solar radiation of a given wavelength is absorbed, with atoms or molecules which are responsible for this absorption. The wavelength 102.5 nm corresponds to the ionization threshold of oxygen molecules (the ionization potential is equal to 12.88 eV) as a result of the process

$$O_2 + \hbar\omega \rightarrow O_2^+ + e. \tag{5.1.28}$$

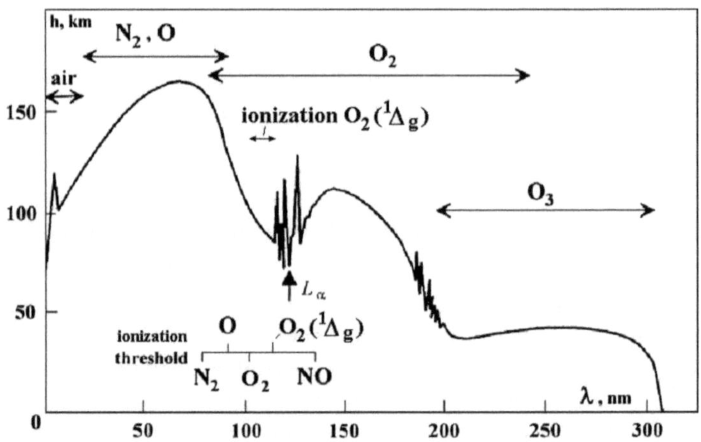

Fig. 5.3 Altitudes where hard and ultraviolet radiation is absorbed and atmospheric atoms and molecules which are responsible for this absorption [17]

The wavelength 98.0 nm corresponds to the ionization threshold of nitrogen molecules (the ionization potential is equal to 15.58 eV) in the process

$$N_2 + \hbar\omega \to N_2^+ + e \tag{5.1.29}$$

Hard solar radiation which causes ionization of atmospheric air, is produced by the solar corona, which, in turn, is formed in inner regions of the Sun which are sources of plasma fluxes. These fluxes are produced as a result of reconnection of magnetic lines of force. Plasmas of the solar corona are characterized by the electron temperature of the order of 100 eV and by the number density in the range of $(10^8$–$10^{10})\,\mathrm{cm}^{-3}$. In contrast to the solar photosphere with stable parameters of its ionized gas, parameters of plasmas in the solar corona are varied sharply in time. During the solar flare, the electron number density in the solar corona can be increased by several orders of magnitude. These fluxes are averaged in time, or they correspond to conditions of the standard atmosphere.

Let us return to Fig. 5.3 and establish the connection between the atmosphere altitude where photons of a given frequency are absorbed, and the absorption cross section of these photons by the atmosphere molecules. It is important that the atmosphere is nonuniform. Then the main contribution into absorption of photons with a given frequency is determined by altitudes at which the optical depth is estimated as

$$u(h) = \Lambda_N \sigma_\omega N(h) \sim 1,$$

where σ_ω is the absorption cross section at a given photon frequency by atmospheric molecules, $N(h)$ is the number density of absorbing molecules at altitude h (Fig. 5.4).

It is possible to clarify this expression, investigating penetration of the solar radiation through the atmosphere. Indeed, equation for the intensity I_ω of the incident solar radiation of a given frequency $I_\omega(h)$ at altitude h has the form

$$\frac{dI_\omega(h)}{dh} = -I_\omega(h)\sigma_\omega N(h) \tag{5.1.30}$$

Here for simplicity we consider the normal incidence of the solar radiation upon the Earth surface. If an incident angle θ is less than $\pi/2$, one can substitute the optical density u of the atmospheric layer by the quantity $u/\sin\theta$. The solution of the equation (5.1.30) has the form

$$\ln\left[\frac{I_\omega(h)}{I_0}\right] = -\Lambda_N \sigma_\omega N(h), \tag{5.1.31}$$

where $I_0 = I_\omega(\infty)$. It follows from this expression that the probability of survival for photons of a given frequency at altitude h of the atmospheric layer is

$$P(h,\omega) = P(x) = \frac{I_\omega(h)}{I_0};\ x = \left[\frac{h - h_0(\omega)}{\Lambda_N}\right],$$

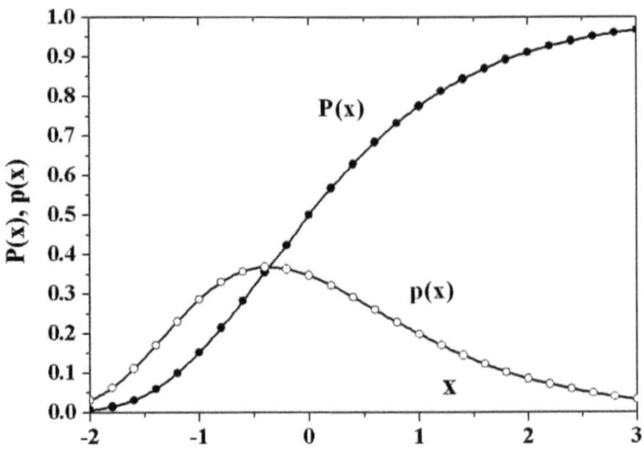

Fig. 5.4 Dependence of the surviving probability of solar photons $P(x)$ and its derivative $p(x)$ on the reduced altitude $x = [h - h_0(\omega)]/\Lambda_N$ [18]

where x is the reduced height of the layer. Thus one obtains for the probability of photon at a given frequency

$$P(x) = P(h, \omega) = \exp\left[-u(h_0)\exp(-x)\right] \qquad (5.1.32)$$

The altitude h_0 of the layer can be determined from the relation for the optical depth $u(h_0) = 1$. Then relation (5.1.32) can be rewritten in the form

$$P(x) = P(h, \omega) = \exp\left[-\exp(-x)\right] \qquad (5.1.33)$$

The dependence of the photon survival probability $P(x)$ on the reduced layer altitude x is represented in Fig. 5.5. It is seen that more than a half of photons are absorbed at altitudes from $h_0 + 0.6\Lambda_N$ $[P(x) = 0.578]$ to $h_0 - 0.6\Lambda_N$ $[P(x) = 0.162]$. Along with $P(x)$, Fig. 5.5 contains also its derivative

$$p(x) = \frac{dP(x)}{dx} = P(x)\exp(-x)$$

The maximum of this quantity corresponds to the maximum rate of photon absorption at a given frequency, so that the the maximum number of absorbed photons per unit time and unit layer depth is equal to $0.378I_0/\Lambda_N$. It is obviously that the maximum rate of photon absorption at a given frequency occurs at altitudes, where $dp(x)/dx = 0$; according to Fig. 5.5 it corresponds to the condition $x = 0$. Thus, the condition of the maximum absorption rate has the form

$$u_0(h_0) = \Lambda_N \sigma_\omega N(h_0) = 1 \qquad (5.1.34)$$

As it follows from (5.1.34), the density of absorbing particles determines the optimal altitude for maximum photon absorption at a given frequency. Therefore, basing on the cross section of photon absorption at a given frequency, one can determine the

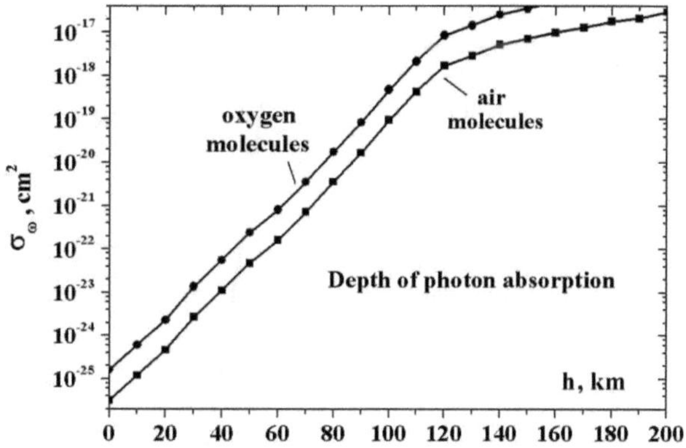

Fig. 5.5 Dependence of the photon absorption cross section at a given frequency on the altitude of the atmospheric layer, where the absorption rate has the maximum. The absorption cross sections for nitrogen and oxygen molecules assume to be identical. The maximum absorption rate of photons of a given frequency corresponds to the condition $dp(x)/dx = 0$ [18]

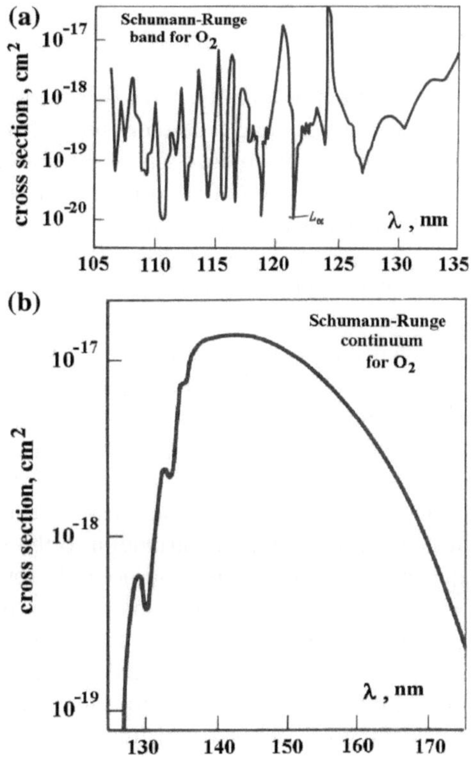

Fig. 5.6 Photoabsorption cross section of the oxygen molecule in the range of the Schuman–Runge band (**a**) and in the range of Schuman–Runge continuum (**b**) [17, 19, 20]

altitude of atmospheric layers where the main absorption takes place. Figure 5.6 gives the absorption cross section as a function of the altitude, where the maximum rate of absorption is observed.

Note that the ionization potential of the oxygen atom as one of the atmospheric air components is 13.61 eV that corresponds to the wavelength of $\lambda = 91.1$ nm. The ionization potential of the nitrogen atom is 14.54 eV that corresponds to the

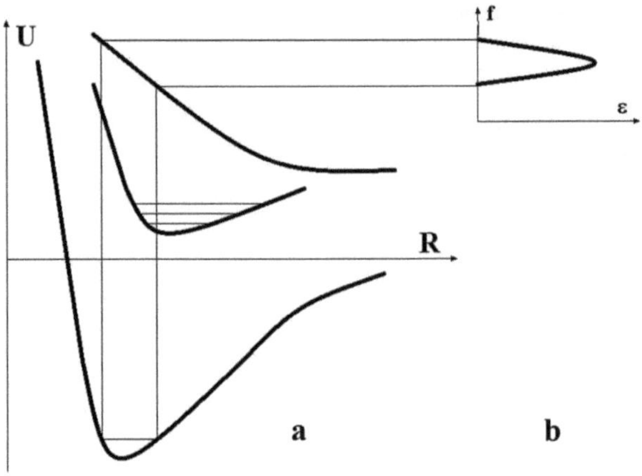

Fig. 5.7 Positions of molecular potential curves which are responsible for photodissociation of the molecule. Radiative transitions into a bound molecular state give a discrete spectrum of this process, whereas transitions in a repulsive molecular term lead to a continuous spectrum of this process

wavelength of $\lambda = 85.3$ nm. These components are responsible for the ionization in the upper atmosphere at altitudes more that 100 km. Let us consider photoionization of atmosphere air on the basis of the above results. According to observations, the maximum ionization of atmospheric air occurs at altitudes about of 160 km, where number densities of nitrogen molecules and oxygen atoms are equal approximately to $5 \cdot 10^{10}\text{cm}^{-3}$. According to Fig. 5.2 data, this takes place in the range of wavelengths of 60–90 nm. It follows from Fig. 5.5 that the photon flux for this range of wavelengths is of the order of $I_0 = 2 \cdot 10^9\text{cm}^{-2}\,\text{s}^{-1}$. In addition, from data of Fig. 5.6 one can find that the photoionization cross section at these heights is estimated as $\sigma_\omega \sim 1 \cdot 10^{-17}\text{cm}^{-2}$. In addition, according to data of Fig. 5.3 the photoionization rate at these altitudes is $0.328 I_0/\Lambda_N \sim 400\ \text{cm}^{-3}\,\text{s}^{-1}$ due to transitions in helium atoms and multicharge oxygen ions with the charge +4.

Let us consider photodissociation in the Earth atmosphere. The above analysis connects the photoabsorption cross section and the altitude of the atmosphere layer which gives the main contribution to the rate of photon absorption at a given frequency. This analysis is valid not only for ionization of the atmosphere molecules and atoms by solar radiation, but also for photodissociation of molecules which occurs due to the electron excitation of molecules and subsequent decay of the excited molecule. If the excited molecule is decayed by means of tunneling transition of atoms, then the spectrum of molecular photoabsorption has a discrete structure. The broadening of each spectral line refers to an appropriate vibration-rotation state. If the molecular excitation occurs to the repulsing term, the subsequent decay of atoms results in the continuum spectrum of photoabsorption. These possibilities are represented in Fig. 5.7.

Both mechanisms are realized at the photodissociation of the molecular oxygen; it contains the photoprocess in the range of the Schuman–Runge band (Fig. 5.4a) and the Schuman–Runge continuum (Fig. 5.4b). These processes are described by the scheme

$$\hbar\omega + O_2 \to O(^1D) + O(^1P) \tag{5.1.35}$$

and lead to production of excited oxygen atoms $O(^1D)$ which are important in the atmospheric chemistry. These processes are responsible for production of atomic oxygen at altitudes more than 120 km. At lower altitudes up to 40 km oxygen atoms are produced due to photon absorption at wavelengths from 200 to 250 nm in the Herzberg continuum; according to Fig. 5.6 the photoabsorption cross section is of the order of 10^{-24}–10^{-23} cm^2. This radiation does not penetrate into more lower atmosphere layers since it is absorbed by ozone at larger altitudes (See Fig. 5.2).

Alongside with nitrogen and oxygen, ozone is of importance for absorption the solar radiation. The most strong absorption by ozone molecules takes place in the Hartley band and continuum at photon wavelengths below 320 nm. The dependence of the absorption cross section on the wavelength in this spectral region is presented in Fig. 5.8. In addition, a weak absorption takes place in the Chappinc band for photons with wavelengths of 400–600 nm. Ultraviolet solar radiation is absorbed by ozone in the range of wavelengths 200–300 nm at altitudes around of 40 km (See Fig. 5.2 and it does not reach the Earth surface. The ozone absorption in the Hartley band takes place according to the scheme

$$\hbar\omega + O_3 \to O^1D + O_2(a^1\Delta_g) \tag{5.1.36}$$

Thus, photodecay of atmospheric ozone molecules induced by solar radiation leads to production of atoms and molecules which determine the chemical activity of the upper atmosphere. The quantum yield of excited oxygen atoms at the ozone photodissociation as a function of the radiation wavelength confirms the process of the molecular ozone photoabsorption according to the channel (5.1.36) (See Fig. 5.9).

The number density of ozone molecules as an altitude function in the atmosphere depends significantly on the geographic coordinate, season and time of day. We average over these parameters within the frame of the standard atmosphere [16]. Approximately 97% of ozone molecules are located at altitudes between 15 and 55 km; the maximum number density of the ozone molecules $5 \cdot 10^{12}$cm^{-3} corresponds

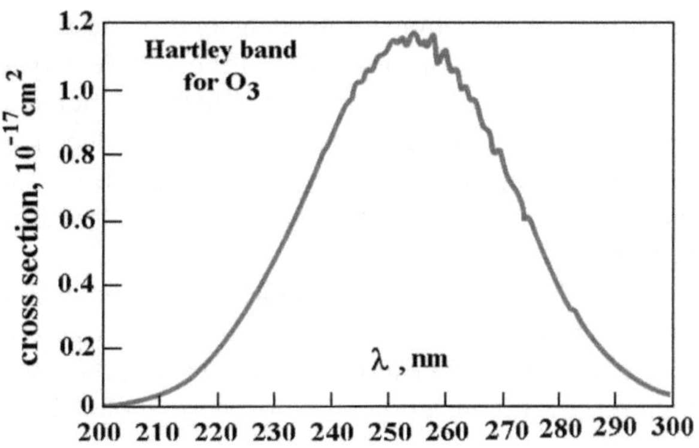

Fig. 5.8 Photoabsorption cross section by the ozone molecule in the Hartley band [17, 19–21]

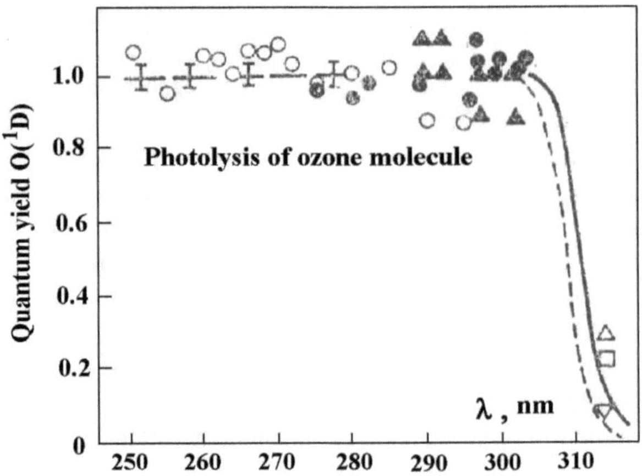

Fig. 5.9 Quantum yield of excited oxygen atoms in $O(^1D)$ states at the photodissociation of the ozone molecule. Various symbols correspond to different experimental data [18]

to altitudes 20–25 km, while the maximum concentration of the ozone molecules $c_{max} \sim 10^{-5}$ refers to altitudes about 35 km. The total ozone molecule density per unit area of the Earth is

$$\int N(O_3)dh = (2-3)\cdot 10^{19}\,cm^{-2},$$

where $N(O_3)$ is the number density of ozone molecules. As it follows from the photoabsorption cross section by the ozone molecule represented in Fig. 5.8, the optical thickness of the atmosphere is $u_\omega = 200 - 300$ for radiation with the wavelength of 250 nm. Absorption by ozone molecules in the Hartley band prevents penetration of the main part of solar UV radiation to the Earth surface. Other processes of electron excitation of the atmosphere molecules are given in [22, 23].

Excited atoms and molecules produced in the reaction described by the scheme (5.1.36) are of importance for chemistry of the upper atmosphere. Excited atoms $O(^1D)$ disappear effectively in collisions with molecules and atoms of the upper atmosphere, in spite of a low density of the atmosphere air. Oppositely, excited molecules $O_2(a^1\Delta_g)$ are formed accumulated at altitudes of 40–80 km, where ozone is distributed. Variations of the density of metastable molecules $O_2(a^1\Delta_g)$ are relatively high because of variations of the ozone density as the source of these molecules and due an unstable flux of solar radiation which can be absorbed in higher atmosphere layers. Figure 5.10 demonstrates the dependence of the number density of metastable molecules $O_2(a^1\Delta_g)$ on the altitude of atmospheric layers. It is seen, that the number density of these molecules depends sharply on an incident angle of solar radiation.

Figure 5.11 contains the dependence of the absorption cross section by excited oxygen molecules $O_2(a^1\Delta_g)$ at wavelengths where this process leads to molecular ionization. These data allow one to determine the optical thickness of the atmosphere with respect to ionization of atmospheric molecules $O_2(a^1\Delta_g)$

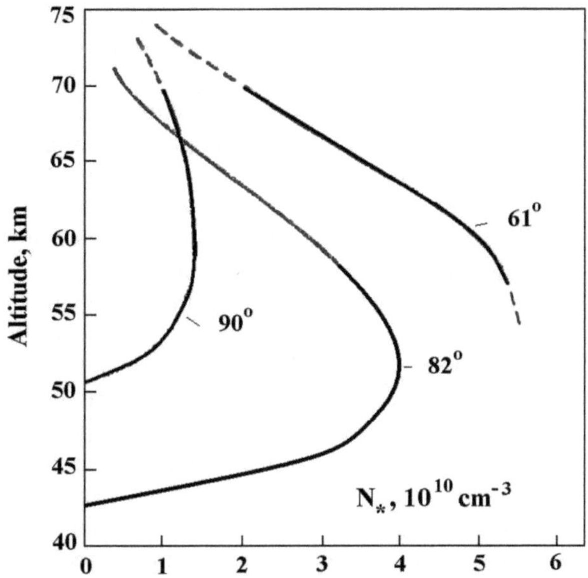

Fig. 5.10 Number density of metastable molecules $O_2(a^1\Delta_g)$ in the day atmosphere as a function of the altitude at various values of the incident angle of solar radiation [17, 24]

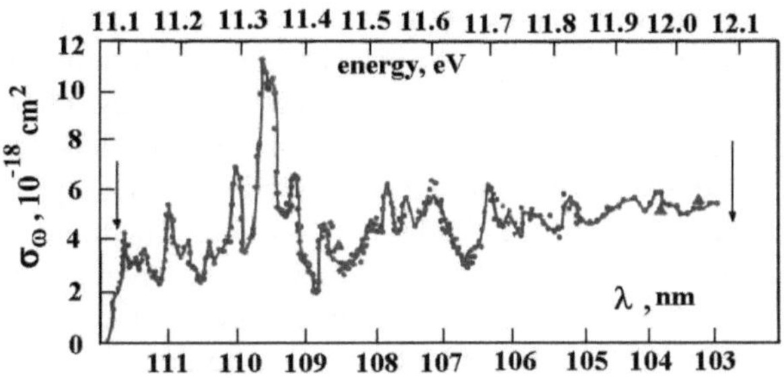

Fig. 5.11 Absorption cross section by metastable molecules $O_2(a^1\Delta_g)$ in the range of vacuum ultraviolet radiation range [17, 25]

$$u_\omega = \sigma_\omega \int \frac{N_*(h)\mathrm{d}h}{\cos\theta} \tag{5.1.37}$$

Data of Fig. 5.10 allows one to find that the number density of metastable molecules $O_2(a^1\Delta_g)$ per unit area of the Earth surface lies in the range

$$\int \frac{N_*(h)\mathrm{d}h}{\cos\theta} \approx (0.3-2)\,10^{17}\,\mathrm{cm}^{-2}$$

In accordance with Fig. 5.11 data, the photoionization cross section of the excited oxygen molecules $O_2(a^1\Delta_g)$ is of the order of $\sigma_\omega \sim 5\cdot 10^{-18}\mathrm{cm}^2$, so that the optical thickness of the atmosphere with respect to absorption by excited oxygen molecules is of order of $u_\omega \sim 0.2-1$. Thus, excited molecules in the metastable state $O_2(a^1\Delta_g)$ can give a remarkable contribution to ionization of atmospheric air in the above mentioned range of wavelengths (in addition to the NO molecules) in spite of a smallness of the photon energy for ionization of molecules in the ground state (See Fig. 5.2).

Let us consider finally ionization of tropospheric air by solar and cosmic gamma-radiation. It was shown above that photons are absorbed in atmosphere mainly in a region where the optical thickness is of the order of one. One can use this for absorption of any radiation by a nonuniform gas. Now we use this conception for cosmic radiation assuming at some frequencies that some part of this radiation penetrates into troposphere. According to contemporary understanding of the problem of cosmic rays penetrated in the Earth atmosphere, high-energy nucleons should be charged and the main part of nucleons are protons. Protons must have a high energy in order to overcome the Earth's magnetic field. In particular, in order to reach the Earth's surface in the equator region, the proton energy should excess 14 GeV. We consider below the possibility for photon components in cosmic ray and estimate the photon energy which can produce ionization in the troposphere.

Let us determine the absorption cross section at altitudes of 11–15 km where according to Fig. 5.12 the ionization rate of the atmosphere by cosmic radiation is maximum. It is obvious that the process of photoionization of air molecules starts from ionization of K-shell electrons of nitrogen and oxygen molecules. This process continues with participation of secondary electrons. The ionization potentials of helium-like ions of nitrogen, oxygen and argon which take part in this process, are 667, 871 and 4121 eV, respectively. Thus, photons which are responsible for ionization of the troposphere air, should have a large energy.

Let us find the optimal cross section for absorption of X-ray photons which produce ionization at altitudes of 11–15 km. The temperature at these altitudes is about 200 K, the number density of air molecules for the standard atmosphere model is equal $N = 7.6 \cdot 10^{18}\text{cm}^{-3}$ (the altitude of 11 km) and $N = 4.0 \cdot 10^{18}\text{cm}^{-3}$ (the altitude of 15 km). In the latter case the number density of molecular nucleons is twice larger. Thus, the optimal absorption cross section is $1.6 \cdot 10^{-25}\text{cm}^{-3}$ at the altitude of 11 km, and $3.0 \cdot 10^{-25}\,\text{cm}^{-3}$ at the altitude of 15 km. The photon absorption cross

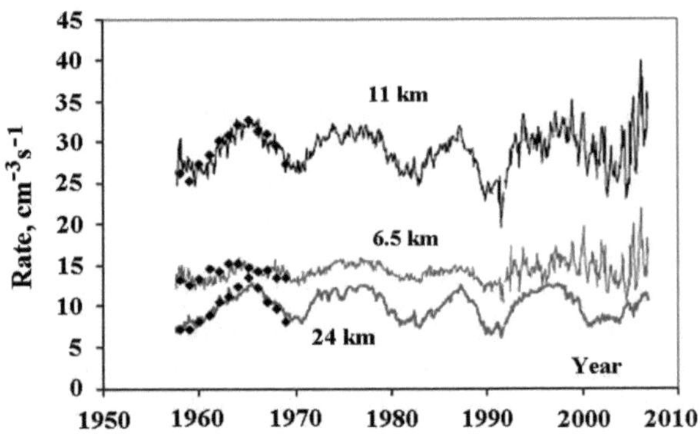

Fig. 5.12 Ionization rates of atmospheric air by cosmic rays at various altitudes [26, 27]

section involving the hydrogenlike ion at high photon energies $\hbar\omega$ compared to the ionic ionization potential J is given by the Stobbe formula

$$\sigma_\omega = \frac{\sigma_o}{Z_{\text{eff}}^2}\left(\frac{J}{\hbar\omega}\right)^{7/2} \ ; \ \hbar\omega \gg J \tag{5.1.38}$$

Here we take into account that ionization can result from an ejection of each electron from K-shell, and $Z_{\text{eff}} = Z - 0.75$ is the effective charge the nucleon and valence electrons for the ejected electron, Z is the nuclear charge, $\sigma_o = 0.55 \cdot 10^{-16}\text{cm}^2$. It should be noted that the nuclear concentrations c_i of basic air components, i.e. nitrogen, oxygen and argon, are 0.79, 0.20 and 0.005 respectively, while the relative contributions of these components to the rate of ionization of tropospheric air according to formula (5.1.38) is 0.58, 0.28 and 0.14, respectively. On the basis of formula (5.1.38) one can obtain the following equation for the altitude h_o at which the maximum rate of air ionization is observed [18]

$$\frac{2\Lambda_N\sigma_o \mathrm{N}(h_o)}{(\hbar\omega)^{7/2}} \cdot \left[\frac{2c(\mathrm{N}_2)J(\mathrm{N}_2)^{7/2}}{Z_{\text{eff}}^2(\mathrm{N}_2)} + \frac{2c(\mathrm{O}_2)J(\mathrm{O}_2)^{7/2}}{Z_{\text{eff}}^2(\mathrm{O}_2)} + \frac{2c(\mathrm{Ar})J(\mathrm{Ar})^{7/2}}{Z_{\text{eff}}^2(\mathrm{Ar})}\right] = \ln(2) \tag{5.1.39}$$

Here $\mathrm{N}(h_o)$ is the total number density of air molecules. From this it follows that the photon energy 2.4 and 2.0 keV corresponds to the maximum of the rate of air ionization at altitudes of 11 km and 15 km, respectively. These energies correspond to wavelengths of 5.1 and 6.1 Å.

5.2 Photoprocesses Involving Rydberg States

5.2.1 Photoexcitation and Photoionization of Rydberg Atoms

Properties of highly excited atoms or Rydberg atoms are determined by the Coulomb interaction of a weakly bound electron with the charge of a nucleus [28]. Therefore the states of a highly excited atom exhibit the same behavior as an excited hydrogen atom. In particular, the electron binding energy for a highly excited atom has the simple form $\varepsilon_n = -1/\left(2n^2\right)$ within an accuracy of the order of $1/n^3$. These parameters are expressed through atomic units $\hbar = m_e = e^2 = 1$ which are used throughout this section. We first find the dependence of the absorption cross section for the transition of an atom to a Rydberg state on the principal quantum number n and its connection with the photoionization cross section of this atom near threshold. Indeed, taking the cross section of atom photoexcitation according to formula (2.2.24), expressing the rate of the radiative process $1/\tau$ through the oscillator strength and using the asymptotic dependence (1.3.36) for the oscillator strength for transition in a highly excited state, we have for the cross section σ_n of formation of a highly excited state from a bound electron state

$$\sigma_n \sim \frac{a_\omega}{\omega^2 \tau} \sim \frac{\omega_{n0}^3 |\mathbf{D}_{n0}|^2}{\omega^2} a\left(\omega - \omega_{n0}\right) \sim f_{n0} a\left(\omega - \omega_{n0}\right),$$

that is,

$$\sigma_n \sim \frac{C}{n^3} a\left(\omega - \omega_{n0}\right)$$

Here $a\left(\omega - \omega_{n0}\right)$ is the distribution function for absorbed photons of frequency ω, and the frequency difference between the initial and final (Rydberg) state is given by

$$\omega_{n0} = -\frac{1}{2n^2} - \varepsilon_o,$$

where ε_o is the binding energy for the initial atom state.

The width of the distribution function is determined by the mechanism of spectral line broadening. In the limiting case, where the spectral line width exceeds significantly the frequency difference for neighboring levels, this photoexcitation cross section coincides with the photoionization cross section. Indeed, in this case the discrete spectrum of excited electrons appears to an incident photon to be the same as a continuous spectrum, and the behavior of weakly bound electrons and low-energy free electrons is similar in the field of the atomic core. Thus we have

$$\sigma_i = \sum_n \sigma_n = C \sum_n \frac{a\left(\omega - \omega_{n0}\right)}{n^3} \tag{5.2.1}$$

Note that in this photoionization cross section σ_i contains transitions in smoothed Rydberg states with formation of a slow electron with the orbital angular momentum of a weakly bound electron in the Rydberg state. For determination of the normalization constant C we replace summation in formula (5.2.1) by integration, so that

$$\sigma_i = C \int a\left(\omega + \varepsilon_o + \frac{1}{2n^2}\right) d\left(\frac{1}{2n^2}\right)$$

From the normalization condition (2.1.1) one can obtain $\sigma_i = C$. Because the photoionization cross section is constant near threshold, the relation $\sigma_i = C$ holds true for any relation between the width of the spectral line and the the energy difference between neighboring excited states of the Rydberg atom. Thus, we have the following relation between the cross section σ_n of photoexcitation of a highly excited atom with the principal quantum number n (at a given orbital momentum l) and the cross section σ_i of atom photoionization near the threshold for the same value of l

$$\sigma_n = \frac{\sigma_i}{n^3} a\left(\omega - \omega_{n0}\right) \tag{5.2.2}$$

Let us estimate the limiting binding energy of a weakly bound electron, where formula (5.2.2) holds true. Let us assume that an atom is excited by a laser with a narrow spectral line. If a gas where exciting atom is located, is rareness, the width of the absorption spectral line is determined by the Doppler broadening mechanism. Selective excitation of states with the principal quantum number n is realized, if the width of the absorption spectral line is less than the energy difference for the neighboring levels. In this case, according to formula (2.1.14) the width of the Doppler broadening is

$$\Delta\omega_D = \omega_{n0} \cdot \sqrt{\frac{T}{Mc^2}},$$

where T is the gas temperature, and M is the atomic mass. The possibility of selective excitation of Rydberg levels is determined by the condition that the width of the absorption line is small compared to the frequency difference for the transition to neighboring levels of the same symmetry, that for the neighboring Rydberg levels is equal n^{-3}. This condition has the form

$$n^3 < \frac{c}{\omega_{n0}\sqrt{\frac{T}{Mc^2}}} \sim c\sqrt{\frac{M}{T}} \tag{5.2.3}$$

This condition gives at the room temperature $n < 100$. In this estimation we assume that photons excited the state with the principal quantum number n, corresponds to the optical and neighboring spectral ranges. One can change the method of generation of Rydberg atoms by using the stepwise way, so that the final stage of this process is excitation from the Rydberg state with a principal quantum number n' to a given state with the principal quantum number n. In this case instead of the criterion (5.2.3) one can obtain

$$n^3 < \left(n'\right)^2 c\sqrt{\frac{M}{T}} \tag{5.2.4}$$

In particular, taking $n' = 100$, one can obtain according to criterion (5.2.4) for the Doppler broadening of spectral lines that it is possible the selective production of Rydberg levels even with $n \sim 1000$.

Let us determine the dependence of the photoexcitation cross section on the photon frequency near the photoionization threshold for the Lorentz broadening of spectral lines. Now the width of spectral lines Γ in the frequency distribution function of absorbed photons $a\left(\omega - \omega_{n0}\right)$ may be both larger and smaller than the frequency difference n^{-3} for transition in neighboring Rydberg levels. According to formula (5.2.2) the cross section for excitation of the Rydberg level with the principal quantum number n is equal for the Lorentz frequency distribution function of photons

$$a\left(\omega - \omega_{n0}\right) = \frac{1}{2\pi}\frac{\Gamma}{\left(\omega - \omega_{n0}\right)^2 + \Gamma^2/4}$$

The photoabsorption cross section for a given frequency ω may be represented as a sum of the excitation cross sections σ_n over all Rydberg levels

$$\sigma_a(\omega) = \sum_n \sigma_n = \sigma_i \frac{\Gamma}{2\pi} \sum_n \frac{1}{n^3} \frac{1}{(\omega - \omega_{n0})^2 + \Gamma^2/4} \tag{5.2.5}$$

The main contribution to the sum in (5.2.5) can be written in terms of effective values n^* (which are not necessary integers) for which $\omega_{n^*} = \omega$, and the summarizing value has the maximum. Expanding the transition frequency ω_{n0} near the maximum n^* in the Taylor series, we have

$$\omega_{n0} - \omega = \left(n - n^*\right) \frac{d\omega_{n0}}{dn} |_{n=n^*} = \frac{n - n^*}{n^{*3}} \tag{5.2.6}$$

Substituting this expansion in formula (5.2.5), we have

$$\sigma_a(\omega) = \sigma_i \frac{\Gamma}{2\pi n^{*3}} \sum_{n=1}^{\infty} \left[\frac{(n - n^*)^2}{n^{*6}} + \frac{\Gamma^2}{4} \right]^{-1} \tag{5.2.7}$$

Because of the strong convergence of the series (5.2.7), one can extend the lower limit of the summation range, so that $-\infty < n < \infty$. According to the Mittag–Leffler theorem [29]), we have

$$\sum_{n=-\infty}^{\infty} \frac{1}{(n - x)^2 + y^2} = \frac{\pi \sinh(2\pi y)}{y \left[\cosh(2\pi y) - \cos(2\pi x)\right]}$$

On the basis of this relation one can rewrite formula (5.2.7) as

$$\sigma_a(\omega) = \sigma_i \frac{\sinh\left[(\pi(n^*)^3\Gamma\right]}{\cosh\left[\pi(n^*)^3\Gamma\right] - \cos(2\pi n^*)} \tag{5.2.8}$$

The parameter n^* in this equation may be expressed through the frequency of the absorbed photon as

$$n^* = \frac{1}{\sqrt{2|\varepsilon_o + \omega|}}; \omega < |\varepsilon_o|$$

Figures 5.13 and 5.14 show the dependence of σ_a on the frequency ω for $\varepsilon_0 = -1/2$ (the ground state of hydrogen atom) and $\Gamma = 0.05$ a.u.

Let us consider the limiting cases of formula (5.2.8). If $\Gamma >> (n^{-*})^3$ and neighboring levels are overlapped, then $\sigma_a(\omega) = \sigma_i$. In this case photoabsorption corresponds to photoionization of an atom. In the opposite limiting case $\Gamma \ll n^{-*3}$, we have

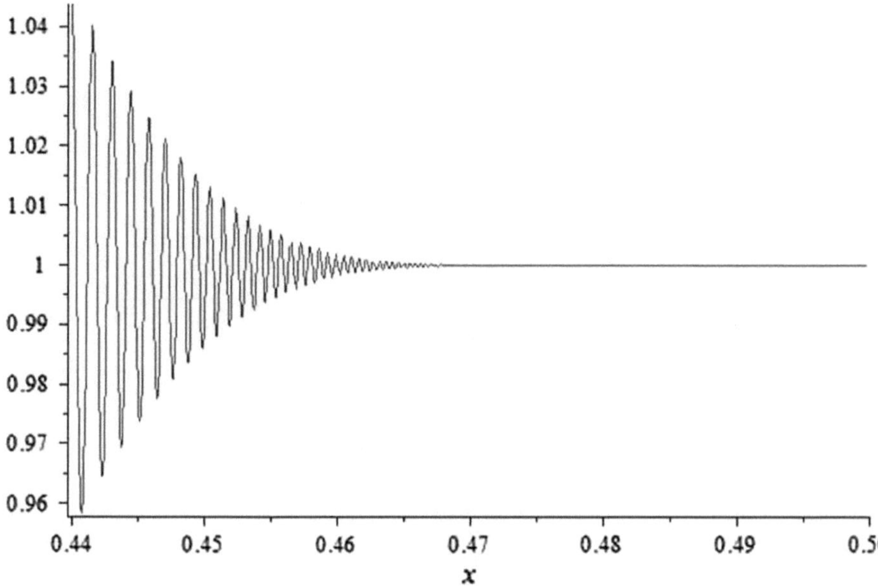

Fig. 5.13 Dependence of the ratio σ_a/σ_i on the frequency ω of the absorbed photon for the case $\varepsilon_0 = -1/2$ (ground state of hydrogen atom) and $\Gamma = 0.05$ a.u. according to formula (5.2.8). The range of frequencies of the absorbed photon is 0.44 – 0.5 a.u. [30]

$$\sigma_a(\omega) = \sigma_i \frac{\pi(n^*)^3\Gamma}{1 - \cos(2\pi n^*) + \left[\pi(n^*)^3\Gamma\right]^2/2} \tag{5.2.9}$$

As the value n^* varies, the photoabsorption cross section σ_a oscillates from the minimum value

$$\sigma_a^{\min} = \sigma_i \frac{\pi(n^*)^3\Gamma}{2} \ll \sigma_i$$

at half-integer values of n^* up to the maximum value of the photoabsorption cross section

$$\sigma_a^{\max} = \sigma_i \frac{2}{\pi(n^*)^3\Gamma} \gg \sigma_i$$

for integer values n^*. The ratio of these values is equal

$$\frac{\sigma_a^{\max}}{\sigma_a^{\min}} = \left[\frac{2}{\pi(n^*)^3\Gamma}\right]^2 \gg 1$$

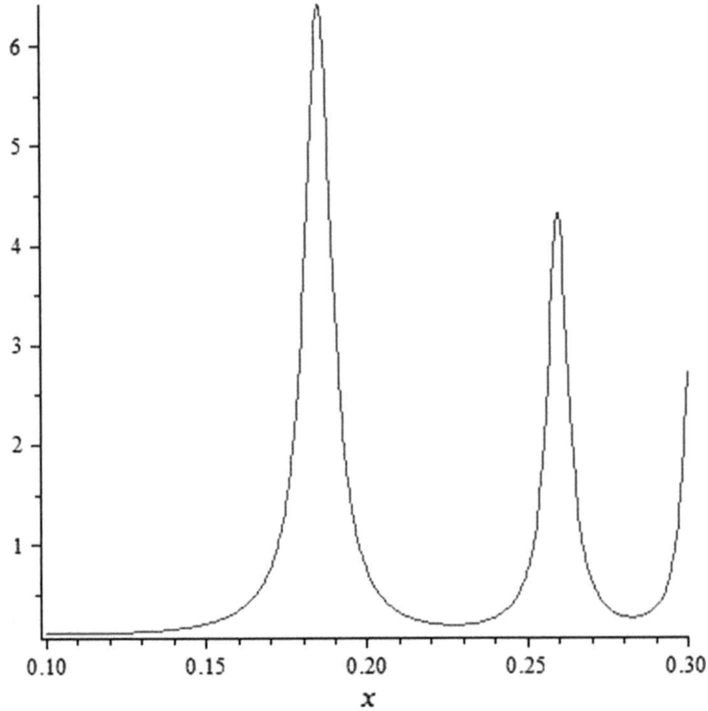

Fig. 5.14 Dependence of the ratio σ_a/σ_i on the frequency ω of the absorbed photon for $\varepsilon_0 = -1/2$ (ground state of hydrogen atom) and $\Gamma = 0.05$ a.u. according to formula (5.2.8). The range of frequencies of the absorbed photon is $0.1 - 0.3$ a.u. [30]

The value $\sigma_a^{\max}$ corresponds to resonance excitation of a discrete level with a principal quantum number $n = n^*$, where n^* is an integer. In the vicinity of the integer n, using $n^* = n + (n^* - n)$, one can obtain

$$1 - \cos\left(2\pi n^*\right) = 2\pi^2\left(n^* - n\right)^2$$

which according to formula (5.2.9) gives for the photoabsorption cross section

$$\sigma_a(\omega) = \frac{\sigma_i}{n^3}\frac{1}{2\pi}\frac{\Gamma}{(\omega - \omega_{n0})^2 + \Gamma^2/4} \tag{5.2.10}$$

As is seen, this relation coincides with (5.2.2). Next, integrating over all frequencies ω, one can obtain the sum rule

$$\int \sigma_a(\omega)\mathrm{d}\omega = \frac{\sigma_i}{n^3} \tag{5.2.11}$$

5.2.2 Photoionization of Rydberg Atoms

We now evaluate the photoionization cross sections of atoms from Rydberg states near the threshold. We first determine the oscillator strength between the nearby Rydberg states n, l and the states $n', l \pm 1$, where $\Delta n = n' - n \ll n, n'$. As the first step of this procedure we determine the matrix element of the atomic dipole moment operator between given highly excited states. Because both initial and final states are quasi-classical one, the correspondence principle of quantum mechanics can be invoked, which gives these matrix elements as the Fourier components from classical coordinates for the corresponding time-dependent values. In the case of bound states of an electron located in the Coulomb field of the atomic core, where the principal quantum number of the electron is n and the orbital angular momentum quantum number is l, it is convenient to parameterize the rectangular electron coordinates $x(t)$ and $y(t)$ for its elliptic motion in the xy plane in terms of a variable ξ, such that

$$x = n^2 (\cos \xi - \varepsilon);\; y = n^2\sqrt{1-\varepsilon^2} \sin \xi;\; t = n^3 (\xi - \varepsilon \sin \xi)$$

This describes the electron motion of an ellipse trajectory with an eccentricity ε given by

$$\varepsilon = \sqrt{1 - \left(\frac{l}{n}\right)^2}$$

Let us calculate first the matrix element of the y coordinate

$$y_{nn'} = \frac{1}{T}\int_0^T y(t) \exp\left(\frac{i\Delta n t}{n^3}\right) dt,$$

where $T = 2\pi n^3$ is the period of motion. This integral is equal to

$$y_{nn'} = \frac{in^2\sqrt{1-\varepsilon^2}}{\varepsilon \Delta n} J_{\Delta n} (\varepsilon \Delta n)$$

In the same fashion, the matrix element of the x coordinate is found as

$$x_{nn'} = \frac{n^2}{\varepsilon \Delta n} J'_{\Delta n} (\varepsilon \Delta n)$$

In the above expressions, J and J' denote the Bessel function and its derivative, respectively. The evaluation of the integrals is accomplished through an integration by parts and the use of the integral representation of the Bessel function

$$J_n(z) = \frac{1}{2\pi}\int_0^{2\pi} \exp\left[i\left(n\xi - z\sin\xi\right)\right] d\xi$$

The operator $x + iy$ corresponds to decreasing the magnetic quantum number by one. In the classical case, the orbital angular momentum then also decreases by one. In the same manner, the operator $x - iy$ corresponds to increasing the orbital angular momentum by one. Thus, we have

$$(x \pm iy)_{nl;n'l\pm 1} = \frac{n^2}{\Delta n}\left[J'_{\Delta n}\left(\varepsilon \Delta n\right) \pm \frac{\sqrt{1-\varepsilon^2}}{\varepsilon} J_{\Delta n}\left(\varepsilon \Delta n\right)\right]$$

According to formula (1.3.1), the oscillator strength for the transition $nl \to n'l \pm 1$ with $n, n' \gg 1$ and $\Delta n \ll n, n'$, is given by

$$f\left(nl \to n'l \pm 1\right) = \frac{2}{3}\omega_{n'n}\left|\frac{1}{\sqrt{2}}\left(x \pm y\right)_{nl \to n'l\pm 1}\right|^2,$$

that is,

$$f\left(nl \to n'l \pm 1\right) = \frac{n}{3\Delta n}\left[J'_{\Delta n}\left(\varepsilon \Delta n\right) \pm \frac{\sqrt{1-\varepsilon^2}}{\varepsilon} J_{\Delta n}\left(\varepsilon \Delta n\right)\right]^2 \tag{5.2.12}$$

Summarized this expression over the final electron angular momentum $l' = l \pm 1$, one can obtain the total oscillator strength for transition between states with a given values of the principal quantum numbers $n \to n'$

$$f(n \to n') = \frac{2n}{3\Delta n}\left[J'^2_{\Delta n}\left(\varepsilon \Delta n\right) + \frac{1-\varepsilon^2}{\varepsilon^2} J^2_{\Delta n}\left(\varepsilon \Delta n\right)\right] \tag{5.2.13}$$

As it follows from formula (5.2.13), in the case $n \gg 1$ the oscillator strength $f\left(nl \to n'l'\right) \sim n$ (sum rule). The oscillator strength decreases strongly with an increase of Δn, as it follows from the properties of semiclassical matrix elements.

Let us determine the average oscillator strength for transition between Rydberg states with principal quantum numbers n and n' under the assumption that the initial substates of a given Rydberg state are equally populated. In this case, averaging the oscillator strength (5.2.13) over all the available substates for the initial state with principal quantum number n, we obtain for the average oscillator strength

$$f\left(n \to n'\right) = \frac{1}{n^2}\sum_{l=0}^{n-1}(2l+1)\, f\left(nl \to n'l'\right)$$

This formula takes into account the $(2l+1)$-fold degeneration for a state with the orbital angular momentum quantum numbers l, and the total number of substates for a principal quantum number n is n^2. Using a large value of n, one can change summation by integration, take the variable ε instead of l, and introduce the relation $n^2\varepsilon \mathrm{d}\varepsilon = -l\mathrm{d}l$, which follows from the definition of $\varepsilon^2 = 1 - l^2/n^2$. This gives

$$f\left(n \to n'\right) = \frac{4n}{3\Delta n}\int_0^1 \varepsilon\left[J'^2_{\Delta n}\left(\varepsilon\Delta n\right) + \frac{1-\varepsilon^2}{\varepsilon^2}J^2_{\Delta n}\left(\varepsilon\Delta n\right)\right]\mathrm{d}\varepsilon$$

In order to evaluate this integral, let us use the property of the Bessel functions

$$z\left[J'^2_p\left(z\right) + \left(\frac{p^2}{z^2} - 1\right)J^2_p\left(z\right)\right] = \frac{\mathrm{d}}{\mathrm{d}z}\left[zJ_p\left(z\right)J'_p\left(z\right)\right],$$

that gives

$$f\left(n \to n'\right) = \frac{4n}{3\left(\Delta n\right)^2}J_{\Delta n}\left(\Delta n\right)J'_{\Delta n}\left(\Delta n\right) \qquad (5.2.14)$$

Expression (5.2.14) may be simplified for a large difference of the principal quantum numbers $\Delta n = n' - n$ that allows us to use the asymptotic form of the Bessel function. In this case we have for the Bessel function and its derivative

$$J_{\Delta n}\left(\Delta n\right) = \frac{1}{2\pi}\int_{-\infty}^{\infty}\exp\left(\frac{i\Delta n\xi^3}{6}\right)\mathrm{d}\xi = \frac{1}{2\pi\sqrt{3}}\left(\frac{6}{\Delta n}\right)^{1/3}\Gamma\left(\frac{1}{3}\right);$$

$$J'_{\Delta n}\left(\Delta n\right) = \frac{1}{2\pi}\int_{-\infty}^{\infty}\xi\exp\left(\frac{i\Delta n\xi^3}{6}\right)\mathrm{d}\xi = \frac{1}{2\pi\sqrt{3}}\left(\frac{6}{\Delta n}\right)^{2/3}\Gamma\left(\frac{2}{3}\right)$$

On the basis of these relations we have from formula (5.2.14)

$$f\left(n \to n'\right) = \frac{2n}{3\pi^2\left(\Delta n\right)^3}\Gamma\left(\frac{1}{3}\right)\Gamma\left(\frac{2}{3}\right)$$

We also use the property of the gamma functions

$$\Gamma\left(x\right)\Gamma\left(1-x\right) = \frac{\pi}{\sin\left(\pi x\right)}$$

This gives the following expression for the averaged oscillator strength

$$f\left(n \to n'\right) = \frac{4n}{3\pi\sqrt{3}\left(\Delta n\right)^3}; \ n \gg 1 \qquad (5.2.15)$$

The above expressions for the oscillator strength allow us to determine the photoionization cross section for the Rydberg states of an atom near the ionization threshold. We are based on the above concept that in the limit of a broad spectral line the photoabsorption cross section coincides with the photoionization cross section near the ionization threshold. On the basis of formula (2.2.24) one can represent the photoabsorption cross section in the form

$$\sigma_a = \frac{2\pi^2}{c} f\left(n \to n'\right) a\left(\omega - \omega_{n'n}\right)$$

Using formula (5.2.15) for the oscillator strength under the criterion $\Delta n \gg 1$, one can rewrite this formula

$$\sigma_a = \frac{8\pi n}{3\sqrt{3}c\left(\Delta n\right)^3} a\left(\omega - \omega_{n'n}\right)$$

Taking into account the relation $\Delta n = \omega n^3$, we have from this

$$\sigma_a = \frac{8\pi}{3\sqrt{3}c\omega^3 n^8} a\left(\omega - \omega_{n'n}\right) \qquad (5.2.16)$$

In order to evaluate the photoionization cross section σ_i, we use relation (5.2.11) for connection between the cross sections σ_i and σ_a. Integrating the cross section (5.2.16) over ω and taking into accounting the normalization condition (2.1.1) for the photon frequency distribution function $a\left(\omega - \omega_{n'n}\right)$, one can obtain the Kramers formula [31] for the photoionization cross section

$$\sigma_i = \frac{8\pi}{3\sqrt{3}c\omega^3 n^5} \qquad (5.2.17)$$

This expression means that the final Rydberg state n' is located in the spectrum range where the width of the distribution function $a\left(\omega - \omega_{n'n}\right)$ is large compared the frequency difference for neighboring levels. If the ionization potential of the final state is small compared to that of the initial state, that is $\omega = 1/\left(2n^2\right)$, then formula (5.2.17) gives

$$\sigma_i = \frac{64\pi}{3\sqrt{3}} \frac{e^2}{\hbar c} n a_o^2 \qquad (5.2.18)$$

In this formula we have returned from atomic units to the usual ones, and a_o is the Bohr radius. Thus the photoionization cross section for a highly excited atom near the ionization threshold, which is the maximum photoionization cross section as a frequency function, varies proportional to n, whereas the area of a highly excited atom is proportional to $n^4 a_o^2$.

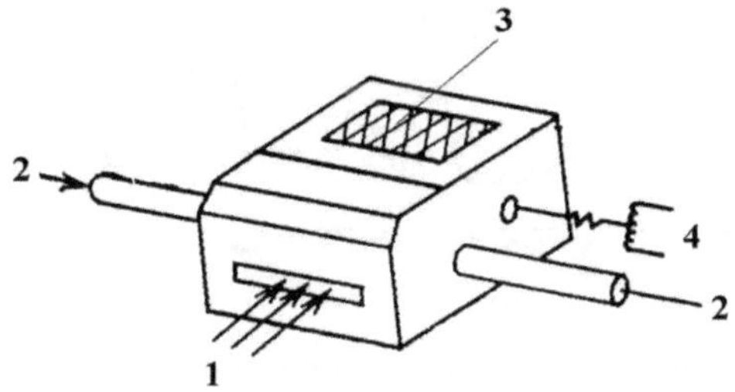

Fig. 5.15 Scheme of the detector of submillimeter radiation of the basis of highly excited atoms; 1—beam of sodium atoms, 2—laser radiation for production of highly excited atoms in the $22D$-state, 3—window for thermal radiation; 4—electric field for ionization of atoms in $23P$-state [32]

5.2.3 Rydberg Atoms in Detector of Submillimeter Radiation

Radiation transitions including highly excited atoms are elements of the various sensitive physical devices and methods. As a demonstration of this, we below consider detector of submillimeter radiation [32] which scheme is presented in Fig. 5.15. Then, a beam of sodium atoms passes through a gap into a camera which temperature is 14 K. This beam is excited by two tunable lasers; one laser has the wavelength of $\lambda = 589$ nm that causes the transition $3^2S_{1/2} \rightarrow 3^2P_{1/2}$ from the ground state to the resonantly excited state of the sodium atom. Radiation of the wavelength of 485 nm of the second laser creates transition from the resonantly excited state of the sodium atom to the Rydberg $22D$-state. Figure 5.16 represents positions of atom levels which participate in the work of this device.

It is of principle the selectivity in detection of highly excited atoms in the state $23P$. Let us introduce the critical field strength $E_{\rm cr}$ of the electric field as $E_{\rm cr} = E_o/\left(16n^4\right)$, where $E_o = 5.1 \cdot 10^9\ \mathrm{V/cm}$ is the atomic field strength. A typical flight time for excited atoms with the principal quantum number $n = 30$ between capacitor plates having the field strength $E \sim E_{\rm cr}$ is $\tau \sim 10^{-6}$ s. If this field varies so

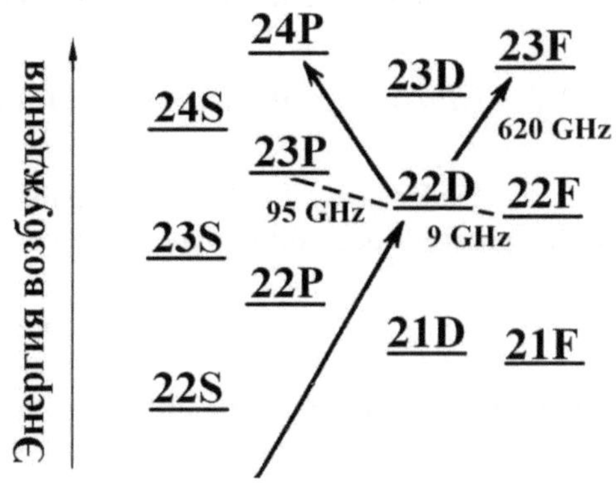

Fig. 5.16 Spectrum of highly of the excited sodium atom which partakes in the work of the detector of the submillimeter radiation

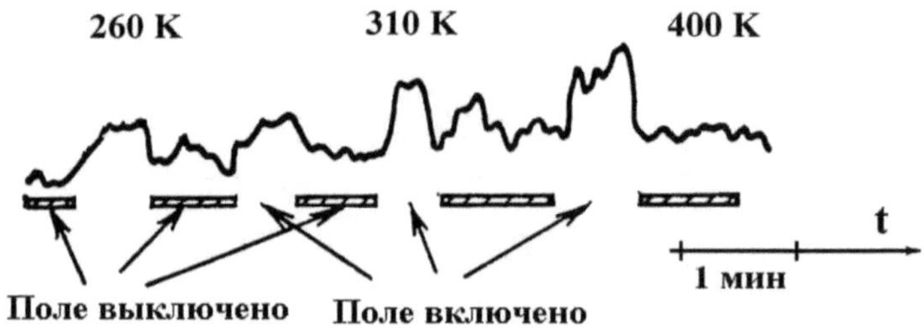

Fig. 5.17 Dependence of the output signal on the temperature of the source of thermal radiation [32]

that $(E_{cr} - E)/E_{cr} = 4.5\% - 4.9\%$, the decay probability of highly excited atoms during the flight through the region of the electric field varies from 20 to 80%. For decay of highly excited atoms with the principal quantum number $n = 31$ the quantity $(E_{cr} - E)/E_{cr}$ should be about of 12%.

Let us present typical parameters of the device under consideration. Under used conditions [32], approximately 0.1% of atoms from the incident beam are excited to the $22D$-state< so that the flux of highly excited atoms in the incident beam is of order of 10^{11} s^{-1}. The radiation lifetime of atoms in the state Na($22D$) is approximately 10^{-5} s, and about of 5% of excited atoms achieve the region of the electric field, where they are detected. In order to analyze the thermal radiation, one uses the $22D \to 23P$ transition of frequency 95 GHz. The electric field ionizes atoms in the $23P$-state, but it does not influence atoms in the $22D$-state. The rate of the transition $22D \to 23P$ produced by the thermal radiation is about of 1300 s^{-1}. Therefore at this temperature the rate for production of atoms in the $23P$-state is about of 10^5s^{-1}. The efficacy of detection of the excited atoms is 0.003, that is, about of 300 s^{-1} atoms are detected, if the temperature of a thermal source is 300 K. The character of measurements on the basis of this device is represented in Fig. 5.17.

5.3 Photoprocesses Involving Free Electrons

5.3.1 Character of Photorecombination of Atomic Particles

Let us consider general properties of photorecombination of atomic particles x and Y that proceeds according to the scheme

$$X + Y \to XY + \hbar\omega$$

Taking into account that this process is inverse with respect to the photoionization process, one can use the principle of detailed balance, that connect the cross sections of photorecombination σ_r and photoionization σ_i in the following manner

$$\frac{\sigma_r}{\sigma_i} = \frac{g_r}{g_i} \frac{j_{\rm ph}}{j_{\rm par}} \tag{5.3.1}$$

Here, g_r, g_i are the statistical weights of initial and final states, $j_{\rm ph}$ and $j_{\rm par}$ are the fluxes of photons and constituent atomic particles if one photon is located in a given volume. The statistical weights of states in the initial and final channels are equal

$$g_i = g_X g_Y \frac{4\pi q^2 {\rm d}q}{(2\pi)^3}; \ g_r = g_{XY} \cdot 2\frac{4\pi k^2 {\rm d}k}{(2\pi)^3},$$

where g_X, g_Y, g_{XY} are the statistical weights of indicated, $\mathbf{k}$ is the photon wave vector, and $\mathbf{q}$ is the wave vector of the relative motion of the particles. The photon flux if one one photon is located in an unit volume, is $j_{\rm ph} = c$, and the particle flux in this case is $j_{\rm par} = v$, where $v = \hbar q/\mu$, and μ is the reduced mass of particles X and Y, and the dispersion relation gives for the photon wave vector $k = \omega/c$. This leads to the following relation between the photorecombination and photoionization processes

$$\sigma_r = 2\frac{k^2}{q^2}\frac{g_{XY}}{g_X g_Y}\sigma_i \tag{5.3.2}$$

If one of particles in the photorecombination process is an electron $\mu = m_{\rm e}$, one can make the estimation for the ratio

$$\frac{k}{q} \sim \frac{\omega}{c\sqrt{2m_{\rm e}\varepsilon/\hbar^2}} \sim \frac{e^2}{\hbar c} \sim 10^{-2},$$

From this it follows that $\sigma_r < \sigma_i$, since $k < q$. In particular, if the photon frequency corresponds to the visible spectrum range, difference between the photorecombination and photodetachment (photoionization) cross sections is roughly four orders of magnitude.

Let us consider the photorecombination process involving a slow electron and ion according to the scheme

$$e + A^+ \to A^* + \hbar\omega, \tag{5.3.3}$$

The main contribution to the photorecombination cross section gives transitions to a Rydberg state with principal quantum number $n \sim e^2/\hbar v$, where v is the velocity of an incident electron. Below we take the statistical weights of an electron and ion to be $g_e = g_i = 1$ since the transition under consideration is not connected with electron and ion spins. The statistical weight of the atomic state is $g_a = n^2$, that is the number of substates of the degenerated Rydberg state with the principal quantum number n. Substituting the Kramers formula (5.2.17) for the photoionization cross section in formula (5.3.2), one obtains the Kramers formula for the photorecombination cross section involving a slow electron

$$\sigma_r = \frac{16\pi}{3\sqrt{3}} \frac{1}{v^2 n^3 \omega c^3} \tag{5.3.4}$$

We use here atomic units. In order slow electrons could be considered as classical particles, the condition $e^2/(\hbar v) \gg 1$ should be fulfilled.

The Kramers's formulas have a simple form and hence may be use for small values of n as estimations. In particular, if employing formula (5.2.18) for $n = 1$, one can obtain for the photoionization cross section of the hydrogen atom in the ground state with using atomic units

$$\sigma_i = \frac{64\pi}{3\sqrt{3}c},$$

whereas the exact value according to formula (5.1.11) this cross section is equal

$$\sigma_i(\text{exact}) = \frac{2^9\pi^2}{3e^4c},$$

where $e = 2.72$. The ratio of these cross sections is

$$\frac{e^4}{8\pi\sqrt{3}} \approx 1.25,$$

that is, these cross sections are nearby to each other. The accuracy of the semiclassical cross section increases with an increase of n.

As for the frequency dependence of the photoionization cross section near the threshold, the semiclassical and exact results for $n = 1$ are also similar. According to formula (5.2.17) we have $\sigma_i \sim \omega^{-3}$, while the accurate result according to formula (5.1.10) is $\sigma_i \sim \omega^{-8/3}$. The accuracy of semiclassical formulas decreases with an increase of photon frequency ω (and correspondingly the electron velocity v). Let us compare the cross section of photoionization at large electron velocities v (and, respectively, the photon frequencies ω). Then the Born approximation holds true, and the the photoionization cross section far from the threshold is given by formula (5.1.13). Taking the ratio of the semiclassical (Kramers) and Born cross sections for the hydrogen atom in the ground state $n = 1$, we have with using atomic units for large photon frequencies ($\omega \gg 1$)

$$\frac{\sigma_i\,(\text{Kramers})}{\sigma_i\,(\text{exact})} = \frac{\sqrt{\omega}}{2\sqrt{6}} > 1.$$

As is seen, the semiclassical approximation is not applicable far from the threshold.

5.3.2 *Bremsstrahlung Processes Involving Electrons*

We now study radiative processes resulted from the interaction of free electrons with atoms and ions, i.e. the bremsstrahlung process involving electrons (this term follows from the German word "die Bremsstrahlung"). In this case scattering of an electron on an atom or ion results from its interaction with a center of force. Let at the beginning the electron wave vector be $\mathbf{q}$, and scattering on a center of force be $\mathbf{q}'$, and the scattering process is accompanied by photon emission of a frequency ω. Because of the law of energy conservation, we have

$$\hbar\omega = \frac{\hbar^2}{2m_e}\left(\mathbf{q}^2 - \mathbf{q}'^2\right) \tag{5.3.5}$$

Using formula (1.2.20) for the rate of spontaneous radiation ($n_\omega = 0$) and normalized the electron wave function in the initial and final states in the standard manner, one can obtain for the bremsstrahlung rate

$$dw_b = \frac{4\omega^3}{3\hbar c^3}\left|\mathbf{D}_{\mathbf{qq}'}\right|^2 \frac{d\mathbf{q}}{(2\pi)^3}$$

where the summation is fulfilled over final photon states. Here the dipole moment equals $\mathbf{D} = -e\mathbf{r}$ where $\mathbf{r}$ is the electron coordinate. We neglect displacement of the nucleus during this process.

The flux density of incident electrons is

$$\mathbf{j} = \frac{\hbar}{2m_e i}\left(\psi_{\mathbf{q}}^*\nabla\psi_{\mathbf{q}} - \psi_{\mathbf{q}}\nabla\psi_{\mathbf{q}}^*\right) = \frac{\hbar\mathbf{q}}{m},$$

since far from the scattering center, the electron wave function is $\psi_{\mathbf{q}} = \exp(i\mathbf{qr})$. In addition, the energy conservation law (5.3.5) gives

$$d\omega = \frac{\hbar q' dq'}{m_e}$$

Dividing the transition rate on the incident flux density, we have for the cross section of bremsstrahlung

$$\frac{d\sigma_b}{d\omega} = \frac{e^2 m_e^2 \omega^3 q'}{6\pi^3\hbar^3 c^3 q}\int \left|\mathbf{r}_{\mathbf{qq}'}\right|^2 d\Omega_{\mathbf{q}'}, \tag{5.3.6}$$

where $d\Omega_{\mathbf{q}'}$ is the element of solid angle into which an electron is scattered in collision. Let us simplify formula (5.3.6) assuming the system to be spherically symmetric. This allows us to expand the electron wave function over the Legendre polynomials in the form

$$\psi_{\mathbf{q}}(\mathbf{r}) = \frac{1}{r}\sum_{l=0}^{\infty} i^l (2l+1) P_l(\cos\theta_{\mathbf{qr}}) u_l(q,r).$$

The same expansion takes place for the wave function $\psi_{\mathbf{q}'}(\mathbf{r})$ of the final state. Let us evaluate first the integral

$$I = \int \left|\mathbf{r}_{\mathbf{qq}'}\right|^2 \mathrm{d}\Omega_{\mathbf{q}'}$$

that with accounting for the above expansions has the form

$$I = \sum_{l,l',p,p'} R_{ll'}R_{pp'}(2l+1)(2l'+1)(2p+1)(2p'+1)\times$$
$$\times\int \cos\theta_{\mathbf{rr}'} P_l(\cos\theta_{\mathbf{qr}}) P_{l'}(\cos\theta_{\mathbf{q}'\mathbf{r}}) P_p(\cos\theta_{\mathbf{qr}'}) P_{p'}(\cos\theta_{\mathbf{q}'\mathbf{r}'}) \mathrm{d}\Omega_{\mathbf{r}}\mathrm{d}\Omega_{\mathbf{r}'}\mathrm{d}\Omega_{\mathbf{q}'}$$

Here indices of the angle θ indicate vectors between them this angle is taken, and $\mathrm{d}\Omega$ is the element of solid angle for the direction shown by the subscript vector. The radial matrix element has the form

$$R_{ll'} = \int_0^{\infty} u_l(q,r) u_{l'}(q',r) r\mathrm{d}r$$

The final cross section of bremsstrahlung will be expressed through these matrix elements.

We first integrate over the solid angle $\mathrm{d}\Omega_{\mathbf{q}'}$ with using the addition theorem for the Legendre polynomials

$$P_{p'}(\cos\theta_{\mathbf{q}'\mathbf{r}'}) = P_{p'}(\cos\theta_{\mathbf{q}'\mathbf{r}}) P_{p'}(\cos\theta_{\mathbf{rr}'}) +$$
$$+ 2\sum_m \frac{(p'-m)!}{(p'+m)!} P_{p'}^m(\cos\theta_{\mathbf{q}'\mathbf{r}}) P_{p'}^m(\cos\theta_{\mathbf{rr}'}) \cos(m\varphi_{\mathbf{q}'})$$

Taking into account the orthogonality property of the Legendre polynomials, we obtain from this

$$I = 4\pi\sum_{l,l',p} R_{ll'}R_{pp'}(2l+1)(2l'+1)(2p+1)\times$$
$$\times\int \cos\theta_{\mathbf{rr}'} P_l(\cos\theta_{\mathbf{qr}}) P_{l'}(\cos\theta_{\mathbf{q}'\mathbf{r}}) P_p(\cos\theta_{\mathbf{qr}'}) \mathrm{d}\Omega_{\mathbf{r}}\mathrm{d}\Omega_{\mathbf{r}'}$$

Integrating over the solid angle $d\Omega_{\mathbf{r}'}$, we use the recurrence relation for the Legendre polynomials

$$(2l'+1)\,xP_{l'}(x) = (l'+1)\,P_{l'+1}(x) + l'P_{l'-1}(x),$$

as well as the above addition rule for the Legendre polynomials and the orthogonality property. As a result, we obtain

$$I = 16\pi^2 \sum_{l,l',p} R_{ll'}R_{pp'}\,(2l+1)\times$$
$$\times\int P_l\left(\cos\theta_{\mathbf{qr}}\right)P_p\left(\cos\theta_{\mathbf{qr}}\right)\left[(l'+1)\,\delta_{p,l'+1} + l'\delta_{p,l'-1}\right]d\Omega_{\mathbf{r}}$$

or

$$I = 64\pi^3 \sum_{l=0}^{\infty}\left[lR_{l,l-1}^2 + (l+1)\,R_{l,l+1}^2\right]$$

This expression may be rewritten in the form

$$I = 64\pi^3 \sum_{l=0}^{\infty}(l+1)\left[R_{l,l+1}^2 + R_{l+1,l}^2\right]$$

Substituting this expression in formula (5.3.6), we obtain the cross section of bremsstrahlung for an electron scattered by a spherical atomic particle

$$\frac{d\sigma_b}{d\omega} = \frac{32m_e\omega^3q'}{3\hbar c^3a_oq}\sum_{l=0}^{\infty}(l+1)\left[R_{l,l+1}^2 + R_{l+1,l}^2\right] \tag{5.3.7}$$

One can simplify this expression on the basis of additional assumptions. Let us determine the cross section for emission of bremsstrahlung photons of long wavelengths, if the energy of the emitted photon is small compared to that of the electron, that is, $\hbar\omega \ll q^2/2m_e$. Let us use the asymptotic expression for the radial wave function of a scattered electron far from the scattering center, that has the form

$$u_l\,(q,r) = \frac{1}{q}\sin\left(qr - \frac{\pi l}{2} + \delta_l\right),$$

where δ_l is the partial scattering phase. In this case we assume that the main contribution to the the radial matrix elements $R_{l,l+1}$ that is responsible for bremsstrahlung, is determined by large distances between a scattered electron and the scattering center. Then we obtain

$$R_{l,l+1} = \frac{1}{qq'} \int\limits_0^\infty \sin\left(qr - \frac{\pi l}{2} + \delta_l(q)\right) \sin\left(q'r - \frac{\pi(l+1)}{2} + \delta_{l+1}(q')\right) r\mathrm{d}r$$

Excluding from this expression a strongly oscillating term, because $|q - q'| \ll q$, we obtain

$$R_{l,l+1} = \frac{1}{2q^2} \int\limits_0^\infty \cos\left[(q - q')\,r + \frac{\pi}{2} + \delta_l(q) - \delta_{l+1}(q)\right] r\mathrm{d}r,$$

where we replace q' by q if it is possible. As is seen, the main contribution to the integral comes from $r \sim 1/(q - q')$. One can evaluate this integral by twice integration by parts and excluding strongly oscillating terms. As a result, we obtain

$$R_{l,l+1} = \frac{\sin\left[\delta_l(q) - \delta_{l+1}(q)\right]}{2q^2(q - q')^2}$$

Evaluating the matrix element $R_{l+1,l}$ in the same manner, one can transform formula (5.3.7) for the cross section of bremsstrahlung in the case of emission of long wave photons to the form

$$\frac{\mathrm{d}\sigma_b}{\mathrm{d}\omega} = \frac{32 m_e \omega^3}{3\hbar c^3 a_o} \frac{1}{2q^4(q - q')^4} \sum_{l=0}^{\infty} (l+1) \sin^2\left(\delta_l(q) - \delta_{l+1}(q)\right)$$

Introducing the electron energy $\varepsilon = \hbar^2 q^2/(2m_e)$ and the photon frequency $\omega = \hbar q(q - q')/m_e$, one can obtain finally under the criterion $\hbar\omega \ll \varepsilon$

$$\frac{\mathrm{d}\sigma_b}{\mathrm{d}\omega} = \frac{16\hbar e^2}{3m_e^2 c^3 \omega} \sum_{l=0}^{\infty} (l+1) \sin^2\left(\delta_l(q) - \delta_{l+1}(q)\right) \qquad (5.3.8)$$

Note that if this cross section is integrated over ω, one can obtain the logarithm divergence. This means that the perturbation theory is violated at low frequencies.

Let us estimate the minimum emitted frequency $\omega_{\min}$ above which the expression (5.3.8) holds true. The probability w of the photon emission must be small within the framework of the perturbation theory, and emission of a single photon must be more probable than emission of two photons and more. However, at small photon frequencies, $\omega \to 0$, this is not true, since the number of emitted photons becomes infinitely large. According to formula (5.3.8) one can give the estimate for the probability to emit with a frequency above ω as the ratio of the cross section σ_b of bremsstrahlung to the scattering cross section

$$w \sim \left(\frac{m_e c}{\hbar}\right)^2 \sigma_b \sim \frac{e^2}{\hbar c} \ln \frac{\varepsilon}{\hbar\omega}$$

Therefore the criterion

$$\frac{e^2}{\hbar c} \ln \frac{\varepsilon}{\hbar\omega_{\min}} < 1, \text{ or } \hbar\omega_{\min} > \varepsilon \exp\left(-\frac{\hbar c}{e^2}\right) = \varepsilon \exp(-137)$$

gives the limit of the applicability of formula (5.3.8) for the bremsstrahlung cross section that is based on the perturbation theory.

We now determine the bremsstrahlung cross section for scattering of a fast electron by an ion on the basis of formula (5.3.6). Let us replace the matrix element of the electron coordinate by the matrix element of the electron momentum, using $\mathbf{p}_{\mathbf{qq}'} = i m_e \omega \mathbf{r}_{\mathbf{qq}'}$. For determination the matrix element $\mathbf{p}_{\mathbf{qq}'}$, we employ the wave function in the momentum space. Because of a large electron velocity, we use perturbation theory for the electron wave function. This gives

$$\psi_{\mathbf{q}}(\mathbf{q}_1) = \delta(\mathbf{q} - \mathbf{q}_1) + \frac{2m_e}{\hbar^2 \left(q^2 - q_1^2\right)} V\left(\mathbf{q} - \mathbf{q}_1\right),$$

where $V\left(\mathbf{q} - \mathbf{q}_1\right)$ is the Fourier transform of the interaction potential. One can use the same procedure for the wave function of the electron in the final state. Within the framework of first-order perturbation theory with respect to the interaction, one can represent the matrix element of the momentum operator in the form

$$\begin{aligned} \mathbf{p}_{\mathbf{qq}'} &= \int \frac{2m_e\hbar}{\hbar^2 \left(q'^2 - q_1^2\right)} V\left(\mathbf{q}' - \mathbf{q}_1\right) \mathbf{q}\delta\left(\mathbf{q}' - \mathbf{q}_1\right) d\mathbf{q}_1 + \\ &+ \int \frac{2m_e\hbar}{\hbar^2 \left(q^2 - q_1^2\right)} V\left(\mathbf{q} - \mathbf{q}_1\right) \mathbf{q}'\delta\left(\mathbf{q}' - \mathbf{q}_1\right) d\mathbf{q}_1 \\ &= \frac{\mathbf{q} - \mathbf{q}'}{\omega} V\left(\mathbf{q} - \mathbf{q}'\right), \end{aligned}$$

where

$$\hbar\omega = \frac{\hbar^2}{2m_e}\left(q^2 - q'^2\right)$$

From this, we obtain the expression for the cross section of bremsstrahlung in the Born approximation

$$\frac{d\sigma_b}{d\omega} = \frac{e^2 m_e^2 \omega^3 q'}{6\pi^3 \hbar^3 c^3 q} \frac{1}{m_e^2 \omega^2} \frac{1}{\omega^2} \int \left|\mathbf{q} - \mathbf{q}'\right|^2 \left|V\left(\mathbf{q} - \mathbf{q}'\right)\right|^2 d\Omega_{\mathbf{q}'}$$

In the case of electron scattering on the nucleus of a charge Z, the interaction potential between an electron and ion is $V = -Ze^2/r$, where Z and its Fourier transformation is

$$V\left(\mathbf{q}-\mathbf{q}'\right) = -\frac{4\pi Ze^2}{|\mathbf{q}-\mathbf{q}'|^2}$$

On the basis of this expression, one can determine the integral

$$\int \left|\mathbf{q}-\mathbf{q}'\right|^2 \left|V\left(\mathbf{q}-\mathbf{q}'\right)\right|^2 \mathrm{d}\Omega_{\mathbf{q}'} = 16\pi^2 Z^2 e^4 \int \frac{\mathrm{d}\Omega_{\mathbf{q}'}}{|\mathbf{q}-\mathbf{q}'|^2} =$$

$$32\pi^2 Z^2 e^4 \int\limits_{-1}^{1} \frac{\mathrm{d}x}{q^2 + q'^2 - 2qq'x} = \frac{32\pi^2 Z^2 e^4}{qq'} \ln\left|\frac{q+q'}{q-q'}\right|$$

Here θ is the angle between the vectors $\mathbf{q}$ and $\mathbf{q}'$. From this we obtain the expression for the cross section of bremsstrahlung from an electron scattered by an ion with a charge Z

$$\frac{\mathrm{d}\sigma_b}{\mathrm{d}\omega} = \frac{16Z^2}{3q^2\omega}\left(\frac{e^2}{\hbar c}\right)^3 \ln\left|\frac{q+q'}{q-q'}\right| \tag{5.3.9}$$

The validity of the Born approximation for the scattering of a particle in a Coulomb field is better the higher is the velocity of this particle. The criterion for the validity of the Born approximation has the form $Ze^2/(\hbar c) \ll 1$. Note that, as in the previous case, the cross section for emission of soft photons integrated over frequencies ω diverges logarithmically. In the other limiting case $q' = 0$, i.e. the total electron energy is transferred to a formed photon, this cross section is zero. Outside the Born approximation, both limiting cases give nonzero results.

We now consider bremsstrahlung in scattering of a slow electron on an atom. Then our task is determination of the sum of formula (5.3.7)

$$\sum_{l=0}^{\infty} (l+1)\left[R^2_{l,l+1} + R^2_{l+1,l}\right]$$

The difference $\mathbf{q}-\mathbf{q}'$ is not small now compared with $\mathbf{q}$, and we must take into account the main contribution to the matrix element $R_{l,l+1}$ comes from distances $r \sim 1/q \to \infty$. Then one can use asymptotic formulas for the electron wave functions far from the scattering center, where these wave functions are close to the wave functions of a free electron, if a force center is absent. Let us denote the matrix element as $R^{(0)}_{l,l+1}$ in the latter case in contrast to $R_{l,l+1}$ in the case of electron scattering on a force center. Because free electrons do not radiate, we have

$$\sum_{l=0}^{\infty} (l+1)\left[R^{(0)2}_{l,l+1} + R^{(0)2}_{l+1,l}\right] = 0 \tag{5.3.10}$$

Note that the maximum difference between the matrix elements $R_{l,l+1}$ and $R^{(0)}_{l,l+1}$ relates to small l. At low electron energies only $l = 0$ gives the contribution to the cross section, so that one can take $R_{l,l+1} = R^{(0)}_{l,l+1}$, $l \neq 0$. Then we have

$$\sum_{l=0}^{\infty} (l+1)\left[R^2_{l,l+1} + R^2_{l+1,l}\right] = R^2_{01} - [R^{(0)}_{01}]^2 + R^2_{10} - [R^{(0)}_{10}]^2$$

We now evaluate the matrix element $R^{(0)}_{01}$. The corresponding electron wave functions are

$$u_0(q,r) = \frac{\sin qr}{qr}, \quad u_1(q,r) = \frac{1}{q}\left(\cos qr - \frac{\sin qr}{qr}\right)$$

On the basis of these wave functions we obtain

$$R^{(0)}_{01} = \frac{1}{qq'}\int_0^{\infty} r \sin qr \left(\cos q'r - \frac{\sin q'r}{q'r}\right) dr =$$
$$= \frac{1}{2qq'}\int_0^{\infty} r\left[\sin(q+q')r - \sin(q-q')r\right] dr -$$
$$- \frac{1}{2qq'^2}\int_0^{\infty} \left[\cos(q-q')r - \cos(q+q')r\right] dr$$

Each of the four terms in this integral gives zero after integration. The contribution from $r \to \infty$ is zero because of the strong oscillations of trigonometric functions. Thus, one obtains $R^{(0)}_{0,1} = 0$. The matrix element $R^{(0)}_{10}$ can be obtained from the matrix element $R^{(0)}_{01}$ by replacing $q \leftrightarrows q'$ that gives $R^{(0)}_{10} = 0$. We can find also that all matrix elements $R^{(0)}_{l,l+1} = R^{(0)}_{l+1,l} = 0$. It follows also from formula (5.3.10) which contains the sum of positive terms.

For evaluation the matrix elements R_{01} and R_{10}, we use the asymptotic expression for the wave function u_0 for a slow electron, which is

$$u_0(q,r) = \frac{\sin(qr + \delta_0)}{qr},$$

since the integral is determined by large distances r between an electron and atom. The scattering phase δ_0 relates to scattering of s-electron on an atom, and at low electron energies this phase equals $\delta_0 = -Lq$, where L is the electron scattering length and does not depend on q. For the state $l = 1$ we use the wave function of a free electron. As a result we have

$$R_{0,1} = \frac{1}{2qq'} \int\limits_0^\infty r \left[\sin\left[(q+q')\, r + \delta_0 \right] - \sin\left[(q-q')\, r + \delta_0 \right] \right] \mathrm{d}r -$$

$$- \frac{1}{2qq'^2} \int\limits_0^\infty \left[\cos\left[(q-q')\, r + \delta_0 \right] - \cos\left[(q+q')\, r + \delta_0 \right] \right] \mathrm{d}r$$

Evaluating this integral, one obtains taking into account that $\delta_0 = -Lq \ll 1$

$$R_{0,1} = \frac{2Lq'}{\left(q^2 - q'^2\right)^2}$$

The value R_{10} may be obtained from R_{01} by replacing $q \to q\prime$ and $q' \to q$, so that

$$R_{10} = \frac{2Lq}{\left(q^2 - q'^2\right)^2},$$

and thus we obtain

$$\sum_{l=0}^{\infty} (l+1) \left[R_{l,l+1}^2 + R_{l+1,l}^2 \right] = \frac{4L^2 \left(q^2 + q'^2\right)}{\left(q^2 - q'^2\right)^4} = \sigma_e \frac{q^2 + q'^2}{\pi \left(q^2 - q'^2\right)^4},$$

where $\sigma_e = 4\pi L^2$ is the cross section of elastic scattering of a slow electron on a given atom. Substituting these expressions in formula (5.3.7), we obtain for the cross section of bremsstrahlung as a result of scattering of a slow electron on an atom

$$\frac{\mathrm{d}\sigma_b}{\mathrm{d}\omega} = \frac{32 m_e \omega^3 q'}{3\hbar c^3 a_o q} \sigma_e \frac{q^2 + q'^2}{\pi \left(q^2 - q'^2\right)^4}$$

Let us express the electron wave vectors q and q' through the energy of an emitted photon

$$\hbar\omega = \frac{\hbar^2}{2m_e} \left(q^2 - q'^2\right)$$

and the energy of the incident electron

$$\varepsilon = \frac{\hbar^2 q^2}{2m_e}$$

This gives

$$\frac{\mathrm{d}\sigma_b}{\mathrm{d}\omega} = \frac{4}{3\pi} \frac{e^2}{m_e c^3} \frac{(2 - \hbar\omega/\varepsilon) \sqrt{1 - \hbar\omega/\varepsilon}}{(\hbar\omega/\varepsilon)} \sigma_e \qquad (5.3.11)$$

The cross section of bremsstrahlung increases monotonically as the frequency ω decreases. For soft photons ($\hbar\omega \ll \varepsilon$) formula (5.3.11) gives

$$\frac{d\sigma_b}{d\omega} = \frac{8}{3\pi} \frac{e^2 \varepsilon}{m_e c^3} \frac{\sigma_e}{\hbar\omega} \tag{5.3.12}$$

This result follows also from formula (5.3.8) if we take into account only the term with $l = 0$ in the sum over l and substitute $\delta_0 = -Lq \ll 1$. Note that, as well as in the previous case, the cross section of bremsstrahlung according to formula (5.3.11) becomes zero at $\hbar\omega = \varepsilon$, if the total initial electron energy is transformed into the energy of an emitted photon.

In the classical limit we use formula (1.2.30) for the intensity of dipole radiation and within the framework of the classical physics evaluate the total electron energy that is consumed on radiation in one collision event. Introducing the impact parameter ρ of collision, one can obtain for the radiation intensity I as a result of bremsstrahlung

$$\frac{dI}{d\omega} = \hbar\omega \frac{d\sigma_b}{d\omega} = \frac{8\pi e^2 \omega^4}{3c^3} \int_0^\infty |\mathbf{r}_\omega|^2 \, 2\pi\rho d\rho, \tag{5.3.13}$$

where $\mathbf{r}_\omega$ is the Fourier component for the electron space coordinate.

The classical approximation holds true if the photon energy $\hbar\omega$ is small compared to a typical atomic energy. One can see that there is no dipole radiation in the collision of two electrons. Indeed, the operator of the dipole moment for two electrons is $e(\mathbf{r}_1 + \mathbf{r}_2) = 2e\mathbf{R}$, where $\mathbf{R}$ is the coordinate of the center of mass for the electrons. Bremsstrahlung is accompanied by a change of the electron energy in the center-of-mass frame of reference, that is, to a change of relative positions of electrons. One can see that bremsstrahlung does not lead to a change in $\mathbf{R}$. Separating space coordinates of two electrons in those related to their relative positions and coordinates of the center-of-mass, one can obtain because of the state of the center-of-mass of particles is not changed as a result of pair collisions, the matrix element of the dipole operator is zero in collisions of two electrons. Thus, the bremsstrahlung process is not realized in collision of two electrons.

We now calculate the cross section of bremsstrahlung as a result of scattering of a slow electron on an ion, if the energy of an incident electron is small compared to typical atomic energies. We use the Kramers formula for the photorecombination cross section (5.3.4) in collisions between a slow electron and ion that is expressed in atomic units and results in formation of Rydberg atomic states

$$\sigma_r = \frac{16\pi}{3\sqrt{3}} \frac{1}{v^2 n^3 \omega c^3}$$

The photorecombination process transits into the bremsstrahlung one, if the photon energy $\hbar\omega$ is less than the energy of an incident electron. In atomic units the law of conservation energy for the photorecombination and bremsstrahlung processes has the form

$$\omega = \frac{1}{2n^2} + \frac{v^2}{2},$$

so that, differenting this relation, one can obtain $d\omega = -dn/n^3$. The recombination cross section of an electron and ion with emission of a photon and electron transition to the Rydberg state with the principal quantum number n may be obtained from this if we get $dn = 1$ in this formula. If $dn = s$, one can find the photorecombination cross section through s neighboring equidistant Rydberg states. Then the Kramers formula (5.3.4) leads to the following expression for the bremsstrahlung cross section which is given in atomic units

$$d\sigma_b = \frac{16\pi}{3\sqrt{3}}\frac{d\omega}{v^2\omega c^3}$$

One can rewrite this expression for the bremsstrahlung cross section in usual units introducing in it the charge Z of the atomic ion

$$\frac{d\sigma_b}{d\omega} = \left(\frac{e^2}{\hbar c}\right)^3 \frac{16\pi\hbar^2}{3\sqrt{3}m_e^2 v^2}\frac{Z}{\omega c} \qquad (5.3.14)$$

The criteria for the validity of this formula is

$$\frac{m_e e^4}{\hbar^3} \gg \omega \gg \frac{m_e v^3}{e^2}, \qquad (5.3.15)$$

i.e. formula (5.3.14) holds true for emitted frequencies which are restricted both above $\omega_{max} = m_e e^4/\hbar^3$ and down $\omega_{min} = m_e v^3/e^2$. According to the condition $\omega \ll m_e^4/\hbar^3$, the radiative frequency is small compared to a typical atomic energy that requires the electron motion to be slow. This allows one to use the classical description for the electron. In order to explain the lower frequency limit, let us introduce the minimum distance r_{min} between an electron and ion at a given impact parameter ρ of collision where an attraction between them is maximal. This distance is determined from the condition that the Coulomb attractive potential is equal to the centrifugal electron energy, i.e.,

$$\frac{e^2}{r_{min}} \sim \frac{m_e v^2 \rho^2}{r_{min}^2}$$

This gives

$$r_{min} \sim \frac{\rho^2 \varepsilon}{e^2},$$

where $\varepsilon = m_e v^2/2$ is the electron energy far from an ion. Let us estimate the impact parameter ρ which gives the main contribution to emission of a photon with frequency ω. This frequency is of the order of $\omega \sim v_{max}/r_{min}$, where v_{max} is the maximum electron velocity that corresponds to the minimum distance between an electron and an ion. The latter velocity follows from the momentum conservation law $m_e v\rho = m_e v_{max} r_{min}$, that gives $v_{max} = v\rho/r_{min}$. From this it follows

$$\omega \sim \frac{v\rho}{r_{min}^2},$$

or

$$\omega \sim \frac{e^4 v}{\varepsilon^2 \rho^3}$$

Thus, we find

$$\rho \sim \left(\frac{e^4 v}{\varepsilon^2 \omega}\right)^{1/3}; \ r_{min} \sim \frac{\rho^2 \varepsilon}{e^2} \sim \left(\frac{e^2 v^2}{\varepsilon \omega^2}\right)^{1/3}$$

We have that formula (5.3.14) holds true under the condition $r_{min} \ll \rho$, where the centrifugal energy is much larger than the kinetic energy. Thus, we have two limits of validity of the bremsstrahlung cross section (5.3.14). The low limit of the criterion (5.3.15) gives the Coulomb electron-ion interaction at minimum approach exceeds significantly the electron kinetic energy far from the ion, that is the closest electron-ion approach r_{min} is small compared to the impact parameter ρ of collision, so that the Coulomb interaction is determined bremsstrahlung emission, as it was used above. Another criterion (5.3.15) requires the classical character of this process. Indeed, the classical description of this collision process is valid, if large collision momenta

$$l = \frac{m_e \rho v}{\hbar} \sim \left(\frac{m_e e^4}{\hbar^3}\omega\right)^{1/3} \gg 1$$

gives the main contribution to the cross section. Thus, the upper limit of the criterion (5.3.15) is the semiclassical criterion. Finally, the Kramers formula (5.3.14) for the bremsstrahlung cross section has a simple form, but this is realized in a restricted range of frequencies in accordance with criterion (5.3.15).

5.3.3 Radiation of Dissociative Air

The above results for photorecombination and bremsstrahlung in electron-ion collisions may be used for the analysis various cases of ionized gases. As an example of this, we below consider radiation of the plasma channel which is formed in atmospheric air during propagation through it a lightning electric current. Lightning is a complex phenomenon in the Earth's atmosphere [33–39], where the electric current passes between regions which has a different electric potential. The lightning phenomenon includes several stages, and from the standpoint of plasma emission we consider radiation of a plasma channel through which the electric current passes. In particular, this stage is the recurrent stroke, where discharging results from the electric current propagation through the conductive channel that is similar to a spark discharge [40]. There is a wide range of lightning parameters [33–36, 38], and we take below its average parameters, namely, the average electric field strength in the course of the thunderstorm weather $E = 200$ V/cm, during a time $\tau = 70\,\mu$s of this stage a current of $I = 30$ kA passes through the conductive channel, and a radius of a glowing channel is several cm. In addition, spectroscopic measurement far from lightning which are based on comparison the intensities of different spectral lines of atmospheric atoms and ions gives the lightning temperature 20000–30000 K. The analysis shows that the channel contains a fully ionized plasma with the number density of electrons and ions $N_e \sim 10^{18}\,\text{cm}^{-3}$. From the conductivity of this plasma at the indicated temperature it follows that the plasma radius which is provided indicated parameters is $r \approx 0.3\,\text{cm}$ [18], i.e. this plasma is nonuniform in the longitudinal direction. We will be guided below the parameters at the channel center.

In this analysis we will be guided by the above parameters ($E = 200\,\text{V/cm},\ I = 30\,\text{kA},\ \tau = 70\,\mu\text{s},\ r = 0.3\,\text{cm}$) and estimate the radiative specific energy. Note that the sound velocity inside the conductive channel is $c_s \sim 10^5$ cm/s, i.e. a time of establishment of an external pressure is $\tau_p \sim 1\,\mu$s, i.e. an identical pressure inside the conductive channel and outside it is supported in the course of the current propagation. In addition, the conductive channel gas is fully dissociated and ionized under indicated conditions. Radiation in the conductive channel is determined by photorecombination and bremsstrahlung processes involving electrons and ions of this plasma according to the scheme

$$e + A^+ \to e' + A^+;\ e + A^+ \to A(n) \tag{5.3.16}$$

Here e, e' are the incident and scattered electrons, $A(n)$ is an excited atom with the principal quantum number n. The first process corresponds to the bremsstrahlung at the electron-ion scattering, while the second one corresponds to the electron-ion photorecombination. In this analysis an electron can be considered as a classical particle, that is, a highly excited atom is produced in the course of the recombination process. We can unify both processes and use the Kramers formula (5.3.14) for their cross sections. Accordingly, the cross section $\mathrm{d}\sigma$ for scattering of a classical electron which moves with the speed v and forms in the end a bound state with the principal quantum number n is given by formula (5.3.14) that in usual units has the form

$$\frac{\mathrm{d}\sigma}{\mathrm{d}\omega} = \frac{16\pi e^6}{3\sqrt{3}m_e^2c^3v^2}\frac{1}{\hbar\omega} \tag{5.3.17}$$

This expression is valid both for the bremsstrahlung and for the recombination electron-ion process. Let us restrict by the photorecombination process which gives the main contribution to the radiation power. Then we have for the the photorecombination cross section expressed in usual units and related to formation of highly excited atom with the principal quantum number n

$$\sigma_r = \frac{16\pi e^{10}}{3\sqrt{3} m_e c^3 v^2 n^3 \hbar^4 \omega} \tag{5.3.18}$$

It is seen from this, that the main contribution to the radiation power is determined by small n. Though this expression is valid for large n we use below this formula for $n = 1$ of the excited atom, based on the above experience, that the Kramers formula leads to an error of 25% for the photoionization cross section from the ground state of the hydrogen atom.

The specific radiation power $W_{\rm rad}$ per unit volume of a uniform plasma is given by

$$W_{\rm rad} = \langle N_e N_i \hbar\omega\sigma_r \rangle = \left(\frac{e^2}{\hbar c}\right)^3 \frac{16\pi e^4}{3\sqrt{3} m_e n^3} \left\langle \frac{1}{v} \right\rangle \tag{5.3.19}$$

We consider quasi-neutral plasmas, where the number densities of electrons N_e and ions N_i are equal, i.e. $N_e = N_i$. Assuming the Maxwell distribution function of electrons over velocities, we have

$$\left\langle \frac{1}{v} \right\rangle = \sqrt{\frac{2m_e}{\pi T_e}},$$

where T_e is the electron temperature. This leads to the radiation power per unit volume as a result of the photorecombination of electrons and ions

$$W_{\rm rad} = N_e^2 \left(\frac{e^2}{\hbar c}\right)^3 \frac{16\sqrt{2\pi}}{3\sqrt{3}} \frac{e^4}{n^3 \sqrt{m_e T_e/\varepsilon_o}} = \frac{N_e^2 w_o}{n^3 \sqrt{T_e/\varepsilon_o}}, \quad w_o = \left(\frac{e^2}{\hbar c}\right)^3 \frac{16\sqrt{2\pi}}{3\sqrt{3}} \frac{e^2\hbar}{m_e}, \tag{5.3.20}$$

where $\varepsilon_o = m_e e^4/\hbar^2 = 27.2\,{\rm eV}$ is the atomic unit of the energy. Note that in the case under consideration, where the plasma located in a cylinder tube of a radius r, the radiation power per unit length $P_{\rm rad}$ with electron transition in a bound state with the principal quantum number N is equal

$$P_{\rm rad} = \frac{N_e^2 p_o r^2}{n^3 \sqrt{T_e/\varepsilon_o}}, \quad p_o = \pi w_o = \left(\frac{e^2}{\hbar c}\right)^3 \frac{16\pi\sqrt{2\pi}}{3\sqrt{3}} \frac{e^2\hbar}{m_e} \tag{5.3.21}$$

Let us return to the problem of radiation of the lightning channel. Taking typical parameters of the lightning plasma which are given above, namely, $N_e = 10^{18}\,\mathrm{cm}^{-3}$, $T_e = 20000\,\mathrm{K}$, $r = 0.3\,\mathrm{cm}$. One can obtain for $n = 1$ $w_o = 8 \cdot 10^{-32}$ $\mathrm{W \cdot cm}^3$, $W_{\mathrm{rad}} = 3 \cdot 10^5\,\mathrm{W/cm}^3$ and $P_{\mathrm{rad}} = 9 \cdot 10^4\,\mathrm{W/cm}$. For comparison, the specific power resulted from the electric current pass is $IE = 6 \cdot 10^6\,\mathrm{W/cm}$, i.e. two orders of magnitude higher than that due to photorecombination. Note that in this evaluation we assume that the radiative plasma is optically think. This is valid if the specific power of radiation P_{rad} is small compared with that P_b of a black body with the same temperature. The Stephan–Boltzmann law gives now

$$P_b = \sigma T_e^4 \cdot 2\pi r,$$

where $\sigma = 5.67 \cdot 10^{-12}\,\mathrm{W/(cm^2 \cdot K^4)}$ is the Stephan–Boltzmann constant. Using the above parameters, one can obtain on the basis of this formula $P_b = 2 \cdot 10^6\,\mathrm{W/cm}$. As is seen, the assumption about optically transparent plasma holds true. Note also that though we concentrate by radiation of a plasma columns formed in the atmosphere as a lightning channel, a similar object as well as radiative processes under consideration are realized in other systems. In particular, lamps of high pressure as sources of a power radiation of a wide continuous spectrum both in visible range and in ultraviolet ones take place use the plasma which are similar to this one.

5.4 Reflection of Radiowaves from Ionosphere

5.4.1 Reflection of Radiowaves by the Ionosphere E-Layer

The dielectric permittivity of the ionosphere related with free electrons is [41]

$$\varepsilon = 1 - \frac{4\pi N_{\mathrm{e}} e^2}{m_{\mathrm{e}} \omega^2} \tag{5.4.1}$$

Here ω is the radiowave frequency, and n_e is the electron number density. We can neglect the contribution of neutral molecules into the medium polarization. E-layer of the ionosphere corresponds to the altitudes of 90–140 km. In this region the experimental electron number density n_e is the linear function of the altitude z. Hence, we can express the dielectric permittivity in the form

$$\varepsilon(z) = \left(\frac{\omega_p}{\omega}\right)^2 \frac{z_0 - z}{l}, \tag{5.4.2}$$

where the quantity l determines the speed of increasing of electron number density with altitude. When $z = z_0$ we have $\varepsilon(z_0) = 0$. The plasma frequency $\omega_p = \sqrt{4\pi N_{\mathrm{e}}^{\max} e^2/m}$ corresponds to the maximum altitude z_0 and to the maximum electron number density $N_{\mathrm{e}}^{\max}$. At the lower boundary $z = z_1$ of E-layer $n_e = 0$ and

$\varepsilon \approx 1$. In practice we have $N_e = 2 \cdot 10^4\,\text{cm}^{-3}$ at $z = z_1 = 90$ km and $N_e = 10^5$ cm^{-3} at $z = z_{max} = 140$ km. Then it follows from formula (5.4.2) that

$$\frac{z_0 - z_1}{l} = \left(\frac{\omega}{\omega_p}\right)^2$$

In the case $\omega > \omega_p$ we have $z_0 > z_1 + l = z_{max}$ and electromagnetic waves with these frequencies penetrate through the E-layer. Oppositely, for $\omega \ll \omega_p$ we have $z_0 \to z_1$.

We now consider the electromagnetic wave which propagates along the vertical direction z. The Maxwell equation for the electric field strength $\mathbf{E}$ is

$$\frac{\partial^2 E}{\partial z^2} + \frac{\varepsilon\omega^2}{c^2} E = 0 \tag{5.4.3}$$

One can rewrite this equation with taking into account (5.4.2)

$$\frac{\partial^2 E}{\partial z^2} + \frac{z_0 - z}{d^3} E = 0 \tag{5.4.4}$$

Here $d^3 = l\left(c/\omega_p\right)^2$. The quantity d is the distance where the electric field strength E changes significantly. This quantity does not depend on the radiowave frequency ω. Formula (5.4.4) may be simplified with using new independent dimensionless variable $u = (z_0 - z)/d$

$$\frac{\partial^2 E}{\partial u^2} + uE = 0 \tag{5.4.5}$$

This is so called *Airy equation*. It has the analytic solution when $u >> 1$:

$$E(z,t) = \frac{A}{u^{1/4}} \cos\left(\frac{2}{3}u^{3/2} - \frac{\pi}{4}\right) \exp\left(-i\omega t\right) \tag{5.4.6}$$

If $u < 0$ ($z > z_0$), the solution of equation (5.4.5) exponentially decreases. Thus corresponds to reflection of the radiowave by plasma of the E-layer. The value of z_0 determines the reflection altitude. The experimental value is $z_0 = 130$ km for the frequency $\nu = \omega/2\pi = 4$ MHz. It should be noted that the plasma frequency on the upper border of the E-layer at the electron number density of $N_e = 10^5$ cm^{-3} is $\omega_p = 2 \cdot 10^7\,\text{s}^{-1}$. The thickness of the E-layer is $l = 50$ km. Hence,

$$d = \left[l\left(\frac{c}{\omega_p}\right)^2\right]^{1/3} = 0.2\,\text{km} \ll z_0$$

This small value determines the typical distance where the electric field of the radiowave exponentially decreases.

5.4.2 Reflection of Radiowaves by the Ionosphere F-Layer

The experimental electron number density in the ionosphere F-layer (140–300 km) can be approximated by the parabola

$$n_e(z) = n_e^{\max}\left[1 - a\left(\frac{z - z_m}{z_m}\right)^2\right], \qquad (5.4.7)$$

Here $n_e^{\max} = 7 \cdot 10^5\ \mathrm{cm}^{-3}$ is the maximum value of the electron number density, $z_m = 250$ km is the altitude, where this maximum is achieved, and $a = 14$. The altitude z_0 from which the radiowave is reflected, is determined from equation

$$\varepsilon\left(z_0\right) = 1 - \frac{4\pi e^2 n_e\left(z_0\right)}{m\omega^2} = 0 \qquad (5.4.8)$$

Substituting formula (5.4.7) in (5.4.8), one can obtains

$$\frac{\varepsilon_0}{1+\varepsilon_0} = a\left(\frac{z_m - z_0}{z_m}\right)^2, \qquad (5.4.9)$$

where the notation is introduced

$$\varepsilon_0 = \frac{4\pi e^2 n_e^{\max}}{m\omega^2} - 1 > 0 \qquad (5.4.10)$$

It follows from this equation that

$$z_m - z_0 = z_m\sqrt{\frac{\omega_p^2 - \omega^2}{a\omega_p^2}} \qquad (5.4.11)$$

In the case $\omega = \omega_p$ we have from formula (5.4.11) $z_m = z_0$ and according to classical mechanics the radiowave with this frequency penetrates inside the F-layer. However, it is known from quantum mechanics that the penetration coefficient in this case is equal 0.5. If $\omega < \omega_p$, penetration of a radiowave inside F-layer is exponentially small in the quantum consideration. On contrary, if $\omega > \omega_p$, the exponentially small probability of quantum reflection of the radiowave by the parabolic barrier is given by

$$w = \exp\left(-\frac{\pi\omega^2 z_m \varepsilon_o}{c\omega_p\sqrt{a}}\right) \ll 1 \qquad (5.4.12)$$

We have $z_m \sim 250$ km and $c/\omega_p \sim 5$ m, so that the exponent in formula (5.4.12) is larger than one at $\omega > \omega_p$.

References

1. B.M. Smirnov, *Physics of Ionized Gases* (Wiley, New York, 2001)
2. G.S. Hurst, M.H. Nayfeh, J.P. Young, Phys. Rev. A **15**, 2283 (1977)
3. L.J. Radziemski, R.W. Solarz, J.A. Paisner (eds.), *Laser Spectroscopy and Its Applications* (Marcel Dekker, New York, 1987)
4. V.N. Ochkin, N.G. Preobrazhensky, N.V. Shaparev, *Optphalvanic Effect in Ionized Gas* (Gordon and Breach, London, 1998)
5. V.N. Ochkin, *Spectroscopy of Low Temperature Plasma* (Wiley, Berlin, 2009)
6. C.R. Dockery, S.R. Goode, Appl. Opt. **42**, 6153 (2003)
7. B.M. Smirnov, *Negative Ions* (McGrow Hill, New York, 1982)
8. https://en.wikipedia.org/wiki/Photosphere
9. https://scied.ucar.edu/sun-photosphere
10. https://nssdc.gsfs.nasa.gov/planetary/factsheet
11. M.N. Saha, Proc. Roy. Soc. **99A**, 135 (1921)
12. D.J. Mullan, *Physics of the Sun* (CRC Press, Boca Raton, 2009)
13. S.J. Smith, D.S. Burch, Phys. Rev. **116**, 1125 (1959)
14. D. Feldman. *Zs.Naturforsch* **25**, 621 (1970)
15. M. Ackerman, in *Mesospheric Models and Related Experiments*, ed. by G. Fiocco (Springer, New York, 1971)
16. *U.S. Standard Atmosphere* (U.S. Government Printing Office, Washington, 1976)
17. M.J. McEwan, L.F. Phillips, *Chemistry of the Atmosphere* (Halsted Press, New York, 1975)
18. B.M. Smirnov, *Microphysics of Atmospheric Phenomena* (Springer, Switzerland, 2017)
19. K. Watanabe, E.C.Y. Inn, M. Zelikoff, J. Chem. Phys. **21**, 1026 (1953)
20. J.W. Chamberlain, *Theory of Planetary Atmospheres: An Introduction to their Physics and Chemistry* (Academic Press, Orlando, 1987)
21. M. Griggs, J. Chem. Phys. **49**, 857 (1968)
22. R.P. Wayne, Ann. N. Y. Acad. Sci. **171**, 199 (1970)
23. K.N. Liou, *An Introduction to Atmospheric Radiation* (Academic Press, Amsterdam, 2002)
24. M.H. Hirsch, P.N. Eisner, J.A. Slevin, Phys. Rev. **178**, 175 (1969)
25. R.E. Huffman, J. Geophys. Res. (e.a.) **76**, 1028 (1971)
26. H.V. Neher, J. Geophys. Res. **72**, 1527 (1967)
27. H.V. Neher, J. Geophys. Res. **76**, 1637 (1971)
28. T.F. Gallagher, *Rydberg Atoms* (Cambridge University Press, Cambridge, 1994)
29. E.T. Whittaker, G.N. Watson, *Modern Analysis* (Cambridge University Press, London, 1940)
30. V.P. Krainov, H.R. Reiss, B.M. Smirnov, *Radiative Processes in Atomic Physics* (Wiley, New York, 1997)
31. H.A. Kramers, Phil. Mag. **46**, 836 (1923)
32. H. Figger, G. Leuchs, R. Strauchinger, H. Walther, Opt. Commun. **33**, 37 (1980)
33. M.A. Uman, *Lightning* (McGrow Hill, New York, 1969)
34. J. Latham, I.M. Stromberg, The thunder cloud, in *Lightning*, ed. by R.H. Golde (Academic Press, London, 1977), p. 99
35. M.A. Uman, *About Lightning* (Dover, New York, 1986)
36. M.A. Uman, *The Lightning Discharge* (Academic Press, New York, 1987)
37. E.M. Bazelyan, Y. Raizer, *Lightning Physics and Lightning Protection* (IOP Publishing, Bristol, 2000)
38. V.A. Rakov, M.A. Uman, *Lightning, Physics and Effects* (Cambridge University Press, Cambridge, 2003)
39. J.R. Dwyer, M. Uman, Phys. Rep. **534**, 147 (2014)
40. E.M. Bazelyan, Y. Raizer, *Spark Discharge* (CRC Press, Roca Baton, 1997)
41. J.D. Jackson, *Classical Electrodynamics* (Wiley, New York, 1975)

Chapter 6
Photon Interaction with Clusters and Microparticles

Abstract There are various mechanisms of particle interaction with photons that cause photon absorption. In the case of dielectric particles it results from interaction of the radiation field with a particle dipole moment which is induced by this field. Absorption of infrared radiation by a particle proceeds by excitation of internal degrees of freedom that in the case of separation of the particle in molecules corresponds to molecular vibrations and rotations. The latter is of importance for radiative transitions in aerosols, i.e. in atmospheric particles. The interaction of metal particles with an electromagnetic wave takes place through an electron subsystem twofold. If the electron subsystem partakes in this interaction as a whole, the photon absorption is determined by its plasma properties. In other case, the absorption results from electron excitation of metal atoms which constitute the metal particle. In particular, from the analysis of experimental data it is shown that light absorption is described by radiative transitions in individual atoms which interact strongly with surrounding ones. In addition, electrons of the metal particles screen an electromagnetic field, and if a particle size is not too small, absorption proceeds in a thin layer near its surface, or in the skin layer.

6.1 Scattering of the Electromagnetic Wave on Atomic and Small Particles

6.1.1 Resonance Fluorescence Involving Molecules and Atoms

Interaction of an atomic particle with an electromagnetic wave results in processes of scattering of this wave and its absorption. Elementary processes of photon collisions with a molecule are presented in Fig. 6.1. They include one-photon processes—absorption and emission during photon-molecule collisions. Last three processes are two-photon processes. Rayleigh scattering is an elastic photon-molecule process,

V. Krainov and B. M. Smirnov, *Atomic and Molecular Radiative Processes*, Springer Series on Atomic, Optical, and Plasma Physics 108,
https://doi.org/10.1007/978-3-030-21955-0_6

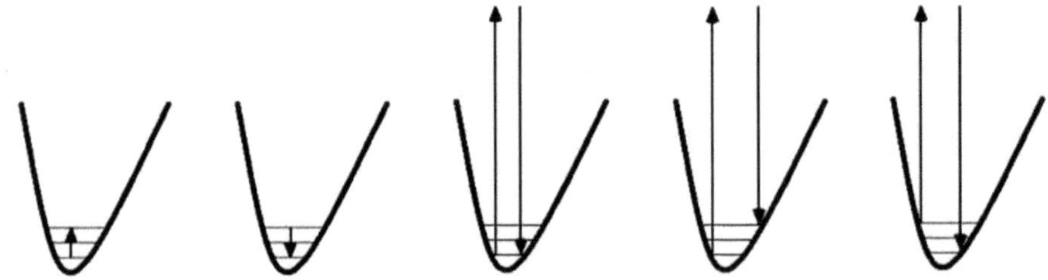

Fig. 6.1 Types of photon scattering on the molecule

other two processes in Fig. 6.1, i.e. the Stokes process and anti-Stokes process, are inelastic collisions with a loss and increase of the photon energy.

One can see the symmetry between some processes of Fig. 6.1. This symmetry is expressed through the principle of detailed balance which connects the cross sections of direct and inverse processes with each other. In particular, the symmetry between one-photon processes, i.e. between processes of emission and absorption, follows from formulas (2.2.24) and (2.2.25). We below show the connection between the Stokes process and anti-Stokes one, i.e. two-photon processes in photon-molecule scattering which are presented in Fig. 6.1.

Note that the most intense Stokes and anti-Stokes processes as a result of two-photon scattering on a molecule are realized, if the first step of these processes is the resonance photon absorption in accordance with Fig. 6.2 for molecule transitions. Similar resonance processes for atoms are shown in Fig. 6.3; they take into account that the final state include a group of levels which may be degenerated. These levels belong to states of fine or superfine structures; we characterize each level by a momentum J_0 for the initial state and a momentum J_k for the final state, so that the statistical weights for the initial state g_0 and for the final one are

$$g_0 = 2J_0 + 1, \ g_k = 2J_k + 1$$

We below determine the connection between rates of the processes of resonance fluorescence which are shown in Fig. 6.3. These processes are detailed inverse ones, and we use the cross section of photon absorption (2.2.24) for the total radiative process as a transition between electron terms, and the radiative time τ_{k0} accounts for the radiative transition between electron terms which are identical for both radiative processes in Fig. 6.3. In considering the transition states i and j of these processes as to be related to the same electron state and assuming the process to be incoherent, one can express the cross sections of the processes of Fig. 6.3 through the same parameters. Indeed, assuming the collision character of broadening of spectral lines, we have for the cross section of the direct radiative process using the formula (2.2.24)

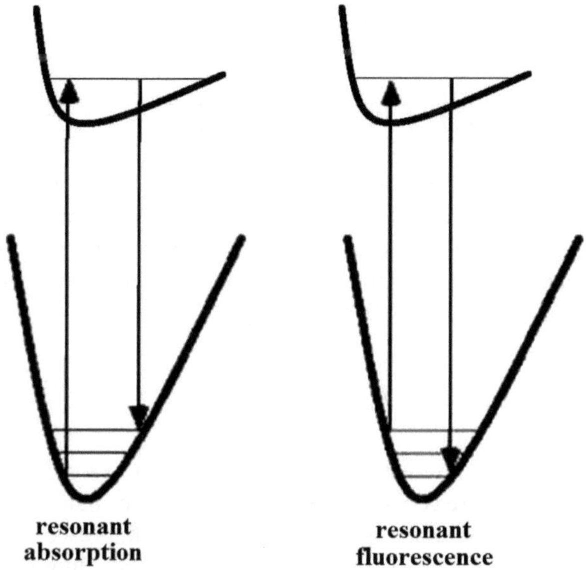

Fig. 6.2 Resonant photon scattering on a molecule, so that the first step of the radiative process is photon absorption, and the second step is photon emission

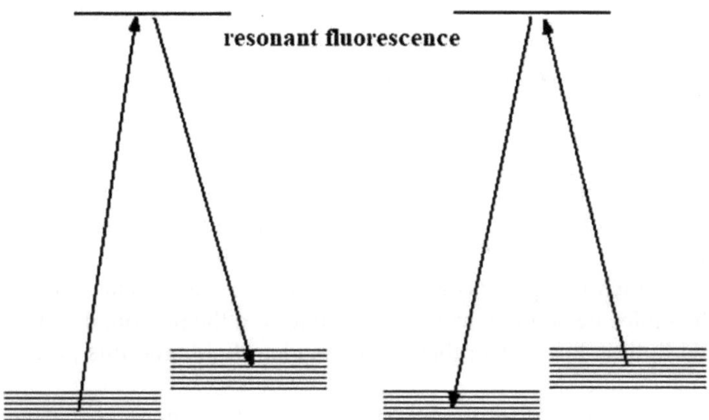

Fig. 6.3 Resonance fluorescence in photon scattering on an atom; the initial and final states include the corresponding group of levels

$$\sigma(\omega,\ i \to j) = \frac{2J_k + 1}{2(2J_0 + 1)} \frac{\pi c^2}{\omega^2} \frac{1}{\tau_{mi}\tau_{jm}} \frac{1}{\left[(\omega_{ji} - \omega)^2 + (\nu/2)^2\right]}, \tag{6.1.1}$$

where $\omega_{ji} = (E_j - E_i)/h$, so that E_j, E_i are the energies of these levels, m is an intermediate state, τ_{mi}, τ_{jm} are partial radiative lifetimes of an intermediate state m with respect to the transition into states i and j respectively. In the same manner we have for the cross section of the inverse process

$$\sigma(\omega, j \to i) = \frac{2J_0 + 1}{2(2J_k + 1)} \frac{\pi c^2}{\omega^2} \frac{1}{\tau_{mi}\tau_{jm}} \frac{1}{\left[(\omega_{ji} - \omega)^2 + (\nu/2)^2\right]}, \tag{6.1.2}$$

It should be noted that if we assume the radiation to be incoherent and therefore it is based on statistical principles, one can obtain the probability of the transition to a state of this group of levels as $1/2J + 1$. If the width of the spectral line is determined by radiative decay of the excited state, one can obtain an estimate for the fluorescence cross section $\nu \sim 1/\tau mi$

$$\sigma \sim \frac{\pi c^2}{\omega^2} \sim \pi \bar{\lambda}^2, \tag{6.1.3}$$

where $\bar{\lambda} = c/\omega$ is the photon wavelength. It is seen that this cross section does not depend on the fine structure constant $\alpha = e^2/\hbar c$ unlike the analogous non-resonance cross sections. Hence, the cross section for resonance fluorescence is much larger (in $(\omega_{ji}\tau_{ji})^2$ times) than the cross section for non-resonance fluorescence. Note also that (6.1.3) is the maximum cross section of photon absorption, for the transition of the atomic particle from the state i to the state j.

6.1.2 Raman Scattering on Atomic Particles

Raman spectroscopy is based on the elastic light scattering (see Fig. 6.1) that gives information about molecular vibrations. In the case of the Rayleigh scattering the energy exchange between incident and scattered photons is absent. But due to the interaction with the molecule, a photon can obtain vibrational quanta from the molecule, and then this process is known as anti-Stokes Raman scattering. Oppositely, if the molecule acquires vibration quanta from the photon, then the frequency of scattered light is lower than that of the incident light, and this process is called Stokes Raman scattering.

Combination scattering (or Raman scattering) [1–6] is photon scattering on atomic particles or on atomic systems at which the direction of the photon motion and, possibly, its frequency, change (Fig. 6.1). We consider here only relatively weak intensities of the electromagnetic wave, when the radiation electric field strength is assumed to be small compared to the atomic field strength. Then the process of the radiation scattering on atomic particles (or systems) is a two-photon process; various channels of this process are shown in Fig. 6.1. Participation of three and more photons is insignificant due to the weak radiation intensity. It is seen from Fig. 6.1 that there are three channels of the photon scattering during its capture by a virtual level and subsequent birth of another photon. In the case of the Rayleigh scattering the photon with the same frequency appears, while at the Stokes and anti-Stokes scattering processes, the atomic particle transfers to higher, or lower level, respectively. In the case of photon scattering on molecules, the Stokes and anti-Stokes processes allows to determine the molecular spectrum for a given symmetry of states.

We here regard photon scattering as a two-photon process in which the first photon of frequency ω_1 is absorbed and a second photon of frequency ω_0 is emitted, with the simultaneous transition of an atomic electron from the initial state i to the final state j. This process differs from fluorescence in that the final state j can differ from the initial state i (this is so-called combination scattering or Raman scattering). Consider first the case of resonant Raman scattering, where the photon frequency ω_1 is nearly the same as the excitation frequency ω_{ji} of the state i of the atomic electron. The photo-absorption distribution function is given by (2.1.40). Subsequent steps in the solution of this problem are parallel to the solution in Sect. 2.1, except that (2.1.40) should be multiplied by the factor τ_j/τ_{mj}, where τ_{mj} is the lifetime of the state m with respect to the transition into the final state j. As a result, instead of (2.1.40) we find

$$\sigma_c = \frac{2J_k+1}{2\left(2J_0+1\right)}\frac{\pi c^2}{\omega^2}\frac{1}{\tau_{mj}\tau_{mi}}\frac{1}{\left(\omega_{mi}-\omega\right)^2+\left[1/\left(2\tau_m\right)\right]^2}. \tag{6.1.4}$$

In particular, the result at the exact resonance is

$$\sigma_c^{\max} = \frac{2J_k+1}{2\left(2J_0+1\right)}\frac{4\pi c^2}{\omega^2}\frac{\tau_k^2}{\tau_{jm}\tau_{mi}}. \tag{6.1.5}$$

As should be the case, this quantity is less than the maximum value for the photoabsorption cross section. We observe that (6.1.4) is of the same order of magnitude as the resonant fluorescence cross section.

We now treat nonresonance Raman scattering. The probability of the two-photon transition induced by fields with electric field strengths E_1 and E_2 is determined by the second-order perturbation theory. According to the "Fermi golden rule" of quantum mechanics we have

$$w_{ij} = \frac{\pi E_1^2 E_2^2}{8\hbar^2}\left|\sum_m\left[\frac{\left(\mathbf{s}_2\mathbf{D}_{jm}\right)\left(\mathbf{s}_1\mathbf{D}_{mi}\right)}{\omega_{ji}-\omega_1}+\frac{\left(\mathbf{s}_1\mathbf{D}_{jm}\right)\left(\mathbf{s}_2\mathbf{D}_{ji}\right)}{\omega_{mi}+\omega_2}\right]\right|^2\delta\left(\omega_{ji}-\omega_1+\omega_2\right) \tag{6.1.6}$$

Here we take into account that the incident photon with the frequency ω_1 is absorbed, and the photon with the frequency ω_2 is emitted.

According to results of the Chap. 1 we can connect the electric field strength E of the incident electromagnetic wave with the photon number n_ω

$$E^2 = \frac{8\hbar\omega^3\delta\omega}{\pi c^3}n_\omega$$

($\delta\omega$ is the difference of frequencies of the neighboring modes). Then, substituting

$$\sum_k \to \frac{1}{\delta\omega}\int d\omega, \tag{6.1.7}$$

in (6.1.2), one obtains

$$w_{0m} = \left|\sum_k \left[\frac{(\mathbf{s}_2\mathbf{D}_{mk})(\mathbf{s}_1\mathbf{D}_{k0})}{\omega_{k0}-\omega_1} + \frac{(\mathbf{s}_1\mathbf{D}_{mk})(\mathbf{s}_2\mathbf{D}_{k0})}{\omega_{k0}+\omega_2}\right]\right|^2 \delta\left(\omega_{m0}-\omega_1+\omega_2\right)\times$$
$$\times \frac{\omega_1\omega_2 d\mathbf{k}_1 d\mathbf{k}_2}{8\pi^3\hbar^2}. \tag{6.1.8}$$

Here we suppose that in the incident photon beam there are no photons with the frequency ω_2, that is $n_{\omega_2} = 0$.

Since both absorbed and emitted photons in (6.1.8) have definite polarizations (s_1 and s_2, accordingly), then the flow density of incident photons is equal to

$$j_\omega = \frac{n_{\omega_1}\omega_1^2 d\omega_1}{2\pi^2 c^3} \tag{6.1.9}$$

Thus it follows from (6.1.8) the expression for cross section of nonresonance combination scattering

$$d\sigma_c = \frac{\omega_1\omega_2^3}{c^4\hbar^2}\left|\sum_m \left[\frac{(\mathbf{s}_2\mathbf{D}_{mk})(\mathbf{s}_1\mathbf{D}_{mi})}{\omega_{mi}-\omega_1} + \frac{(\mathbf{s}_1\mathbf{D}_{mk})(\mathbf{s}_2\mathbf{D}_{mi})}{\omega_{mi}+\omega_2}\right]\right|^2 d\Omega_2. \tag{6.1.10}$$

In the derivation of this expression we integrated (6.1.8) over frequency ω_2 of the emitted photon, using the energy conservation law for this process. The quantity $d\Omega_2$ is the solid angle of the scattered photon, and m is the index labeling the final state of the atomic electron. The energy conservation law gives $\omega_{ji} - \omega_1 + \omega_2 = 0$. In particular, when the initial and final states are the same, that is, when $j = i$ and $\omega_1 = \omega_2$, (6.1.10) gives the nonresonance fluorescence cross section.

Now we calculate the cross section for photon scattering by a free electron. We suppose that the energy of the incident photon is small as compared to the electron rest energy, that is, $\hbar\omega \ll m_e c^2$. The photon momentum is $\hbar\omega/c$. The change of photon momentum in the scattering process and the electron momentum after scattering are of the same order of magnitude: $\hbar\omega/c$ (excluding scattering through very small angles). The energy gained by the electron in the collision is of the order of $(\hbar\omega)^2 / \left(m_e c^2\right)$; it is small as compared to the rest energy, which is equivalent to the statement that the velocity increment v of the electron from the scattering is small as compared to c. The electron motion is thus nonrelativistic one. Since the change of the energy of the photon, also $\sim (\hbar\omega)^2 / \left(mc^2\right)$, is small compared to the initial photon energy $\hbar\omega$, the photon-electron scattering is quasi-elastic, which means $\omega_1 \cong \omega_2$.

We shall use (6.1.10) for the calculation of the cross section assuming that $\omega_1 = \omega_2$, and using the semiclassical approximation for the free electron states due to the semiclassical character of the initial continuum electron state i. The dipole operator of an electron is $D = -er$, where r is the electron coordinate. Equation (6.1.4) leads to

$$d\sigma_c = \frac{4\omega^4}{c^4\hbar^2}\left|\sum_m (\mathbf{s}_2\mathbf{r}_{im})(\mathbf{s}_1\mathbf{r}_{mi})\frac{\omega_{mi}}{\omega_{mi}^2-\omega^2}\right|^2 d\Omega. \tag{6.1.11}$$

We wish to evaluate the sum in (6.1.11). The coordinate axes are selected so that s_1 is along the z axis and s_2 is in the xz plane. The sum is then of the form

$$s_{2x}\sum_k \omega_{mi}x_{im}z_{k0} + s_{2z}\sum_m \omega_{mi}\left|z_{mi}\right|^2.$$

The first of the sums in this expression is zero because of the odd parity of the product $\omega_{mi}x_{im}z_{mi}$: it changes sign when $z \to -z$. The second sum can be calculated using the sum rule, (1.3.20) of the Sect. 1.3.2, for dipole transitions. The above expression then yields

$$\frac{\hbar s_{2z}}{2m_e} = \frac{\hbar(\mathbf{s}_1\mathbf{s}_2)}{2m_e} \tag{6.1.12}$$

Substituting this expression into (6.1.11) for cross section of Raman scattering gives this cross section in the form

$$d\sigma = r_e^2(\mathbf{s}_1\mathbf{s}_2)^2 d\Omega \tag{6.1.13}$$

Here the quantity

$$r_e = \frac{e^2}{m_e c^2} \tag{6.1.14}$$

is the classical electron radius. Equation (6.1.13) is called the Thomson formula. It is a purely classical result, since the Planck constant does not appear. To calculate the total cross section, we integrate (6.1.13) over the solid angle $d\Omega$. We select the polar axis of a system of spherical coordinates to lie along the polarization vector s_1 of the incident photon. The notation θ is introduced for the angle between vectors s_1 and k_2. Since vectors s_2 and k_2 are perpendicular to each other, then we find that $s_1s_2 = \sin\theta$. If the vector s_2 is normal to the plane composed from vectors s_1 and k_2 the cross section of scattering is zero, since then vectors s_1 and s_2 will be normal to each other. Hence,

$$\sigma = r_e^2\int \sin^2\theta d\Omega = \frac{8\pi}{3}r_e^2. \tag{6.1.15}$$

Equations (6.1.13) and (6.1.15) can be also obtained in the classical radiation theory by solving the Newtonian equations of motion for induced electron oscillations and considering the emission of secondary waves with the same frequency. The classical results fail when the photon energy $\hbar\omega$ is of the order of the electron rest energy mc^2 or greater. Then most of the incident photon energy is transferred to the electron, and the scattering is therefore inelastic. In this case, relativistic and quantum effects will be important simultaneously, and the electron spin will be also an essential element in the description of the scattering.

Further we calculate the low-energy scattering cross section for a photon scattering from an atom with zero angular momentum. The frequency of the photon is taken to be small compared to typical atomic frequencies, $\omega \ll \omega_{mi}$. This limit is thus opposite to that considered for a photon scattering of a free electron with $\omega \gg \omega_{mi}$. The small frequency condition allows us to simplify (6.1.10) to

$$d\sigma_c = \frac{4\omega^4}{c^4\hbar^2}\left|\sum_m \frac{(\mathbf{s}_2\mathbf{D}_{im})\,(\mathbf{s}_1\mathbf{D}_{mi})}{\omega_{mi}}\right|^2 d\Omega, \tag{6.1.16}$$

The initial state i is specified to be an S state, so that its magnetic quantum number is $M_i = 0$. The state m is therefore a P state in accordance with the dipole selection rule, and so $M_m = 0, \pm 1$. We take the axis of quantization z to lie along s_1. The vector D is along the z direction. In the opposite case, the quantity $D_{im}s_1$ vanishes. Hence we have

$$\mathbf{D}_{0k}\mathbf{s}_2 = (\mathbf{s}_1\mathbf{s}_2)\,(D_z)_{im}\,. \tag{6.1.17}$$

We now define the polarizability tensor

$$\alpha_{ij} = 2\sum_m \frac{(D_i)_{im}\left(D_j\right)_{mi}}{\hbar\omega_{mi}} \tag{6.1.18}$$

It follows from the above considerations that α_{ij} is a diagonal tensor, so that $\alpha_{ij} = \alpha\delta_{ij}$ and

$$\alpha = \frac{2e^2}{\hbar}\sum_m \frac{|z_{mi}|^2}{\omega_{mi}}. \tag{6.1.19}$$

When this result is substituted into (6.1.16), we find the scattering cross section

$$d\sigma = \frac{\omega^4\alpha^2}{c^4}\,(\mathbf{s}_1\mathbf{s}_2)^2\,d\Omega = \frac{\omega^4\alpha^2}{c^4}\sin^2\theta d\Omega, \tag{6.1.20}$$

where θ is the angle between the polarization direction s_1 of the incident photon and the direction k_2 of the wave vector of the scattered photon. After integration over the angular coordinates, we find the total cross section for photon scattering by an atom in the low photon frequency limit to be

$$\sigma = \frac{8\pi\omega^4\alpha^2}{3c^4}. \tag{6.1.21}$$

We now wish to solve the same problem by the classical approach. From (1.2.31), the intensity of scattered light is

$$I(t) = \frac{2}{3c^3}\left|\ddot{\mathbf{D}}(t)\right|^2, \tag{6.1.22}$$

where $\mathbf{D}$ is the induced dipole moment produced by the field of an electromagnetic wave with an electric field given by $\mathbf{E}\cos\omega t$. By the definition of atomic polarizability α, we have $\mathbf{D} = -\alpha\mathbf{E}\cos\omega t$. We are thus led to the intensity of scattered light expressed as

$$I(t) = \frac{2\omega^4\alpha^2}{3c^3}\mathbf{E}^2\cos^2\omega t. \tag{6.1.23}$$

To find the cross section, we should divide this quantity by the energy flux of the incident radiation. This energy flux is given by the Poynting vector $c\,[\mathbf{E}, \mathbf{H}]\,/4\pi$, where E and H are the electric and magnetic fields of the electromagnetic wave. In our case, the energy flux has the magnitude $cE^2\cos^2\omega t/4\pi$. When we define the scattering cross section as the ratio of the intensity of scattered light to the energy flux of the incident radiation, we obtain

$$\sigma = \frac{8\pi\omega^4\alpha^2}{3c^4}. \tag{6.1.24}$$

This agrees with the quantum result in (6.1.21). The advantage of the quantum-mechanical derivation is that it makes it possible to obtain the explicit expression for atomic polarizability. It is seen from the derivation that the scattering process considered is purely classical one. A classical dipole moment radiates the same frequency that is induced by the electromagnetic wave. Such scattering is called Rayleigh scattering. It is interesting that both Rayleigh scattering ($\omega \ll \omega_{ji}$) and Thomson scattering ($\omega \ll \omega_{ji}$) are purely classical phenomena. The maximum in the scattering of visible light by atoms with absorption frequencies in the ultraviolet range corresponds to the violet cut-off of the spectrum, since the scattering cross section increases very strongly with frequency: as ω^4. The limit $\omega \ll \omega_{ji}$ holds true nevertheless. This explains the blue color of the sky. Sunset is of a red color for the same reason: the strong scattering of the violet part of the solar spectrum in the direct flux of the solar rays leaves a predominance of red in the remaining part of the sunlight.

The static polarizability can be exactly calculated for the ground state of the hydrogen atom. The result is that $\alpha = (9/2)\,a_o^3$ where $a_o = \hbar^2/m_e e^2$ is the Bohr radius. Hence the cross section for the hydrogen ground state for low-energy photon scattering, with $\hbar\omega \ll \hbar^2/m_e a_o^2$, is given by

$$\mathrm{d}\sigma = \frac{81}{4}r_e^2\left(\frac{\hbar^3\omega}{me^4}\right)^4(\mathbf{s}_1\mathbf{s}_2)^2\mathrm{d}\Omega \tag{6.1.25}$$

This expression describes accurately the elastic scattering cross section from zero frequency up to the frequency of the first resonance when $\hbar\omega = \hbar\omega_{k0} = 3me^4/\left(8\hbar^2\right)$. The indices 1 and 2 refer, respectively, to the ground and first excited states of the hydrogen atom. An additional problem with accounting for degeneracy of the state i with respect to magnetic quantum numbers appears in the case of nonzero angular momentum. If the low-energy photon is scattered by an atom in an excited state, then Raman scattering occurs as well as Rayleigh scattering, with the consequent transition of the atom to a lower lying state m.

We now calculate the dependence of the intensity of induced Raman scattering on the propagation distance of the photon beam in the gas. We consider (6.1.10) for the Raman scattering cross section $d\sigma_c$ when an atomic electron makes a transition from the initial state i to the final state j. If we denote by N/Ω the density of atoms, then the quantity

$$g = \frac{N}{\Omega}\sigma_c \tag{6.1.26}$$

presents the number of photons with frequency ω_2 that is generated in a unit distance along the photon beam. The total Raman scattering cross section, σ_c, is obtained from (6.1.10) by performing the integration over the angles of the emitted photons of frequency ω_2.

We now have $n_{\omega_2} \neq 0$, since the photons transfer from an incident beam of frequency ω_1 to photons of scattered light with frequency ω_2. If we select coordinates with the z axis along the propagation direction of the incident beam, then by the requirement that each absorbed photon gives rise to a scattered photon, we have

$$n_{\omega_1}(z) + n_{\omega_2}(z) = \text{const} = n_{\omega_1}(0), \tag{6.1.27}$$

where $n_{\omega_1}(0)$ is the initial amount of photons in the incident beam. In the usual scheme of quantization, each mode of oscillation is contained in the volume Ω, so we suppose that the typical characteristic length along the z axis is much greater than $\Omega^{1/3}$. We can now write balance equations that determine the dependence of the quantities $n_{\omega_1}(z)$ and $n_{\omega_2}(z)$ on z. Equation (6.1.26) establishes the amount of photons that appears in a unit length along the photon beam, under the condition that there is one photon of frequency ω_1 and no photons of frequency ω_2. However, if at the coordinate z we have the amount $n_{\omega_1}(z)$ photons with frequency ω_1 and $n_{\omega_2}(z)$ photons of frequency ω_2 then (6.1.8) means that the amount of photons appearing in a unit length along the beam with frequency ω_2 is

$$w(z) = g n_{\omega_1}(z)\left[1 + n_{\omega_2}(z)\right]. \tag{6.1.28}$$

Hence, the balance equations are of the simple form

$$\frac{dn_{\omega_1}(z)}{dz} = -\frac{dn_{\omega_2}(z)}{dz} = -w(z). \tag{6.1.29}$$

The solution of the system (6.1.29) under the conditions (6.1.27) is elementary. We write it in the form

$$n_{\omega_2}(z) = \frac{\exp(Gz) - 1}{1 + \left[\exp(Gz)/n_{\omega_1}(0)\right]}, \tag{6.1.30}$$

where the quantity G is defined as

$$G = g\left[1 + n_{\omega_1}(0)\right] = \frac{N}{\Omega}\sigma_c\left[1 + n_{\omega_1}(0)\right]. \tag{6.1.31}$$

The quantity G is called the *increment coefficient*. We see that the amount of scattered photons increases at first linearly with z. This corresponds to the general theory developed above in this chapter. This linear dependence occurs when $Gz \ll 1$. When $Gz \sim 1$, the linear increase becomes an exponential one. Finally, when $Gz > 1$, saturation takes place, $n_{\omega_2}(z) \to n_{\omega_1}(0)$, so that all photons from the incident beam are replaced by photons in the scattered state.

The intensity of the induced Raman scattering for photons with frequency ω_2 is given by

$$I_2(z) = c\hbar\omega_2\frac{n_{\omega_2}(z)}{\Omega}, \tag{6.1.32}$$

where $n_{\omega_2}(z)$ is determined by (6.1.27). In the linear regime, (6.1.30) becomes

$$n_{\omega_2}(z) = Gz\frac{n_{\omega_1}(0)}{1 + n_{\omega_1}(0)}, \tag{6.1.33}$$

which is in good agreement with (6.1.10). If we take the volume Ω with the length z in the direction of the photon beam with frequency ω_1, where N the number of atoms in this volume, then the cross section of the volume Ω is Ω/z. We now calculate the energy flux through this volume for the photons of frequency ω_1, and obtain

$$I_1(z)\frac{\Omega}{z} = c\hbar\omega_1\frac{n_{\omega_1}(z)}{\Omega}\frac{\Omega}{z} = -Gc\hbar\omega_1\frac{n_{\omega_1}(0)}{1 + n_{\omega_1}(0)} + c\hbar\omega_1\frac{n_{\omega_1}(0)}{z}. \tag{6.1.34}$$

Using (6.1.31) for the parameter G we rewrite the first term in (6.1.34) in the form

$$-\frac{N}{\Omega}\sigma_c c\hbar\omega_1 n_{\omega_1}(0) \tag{6.1.35}$$

To calculate the cross section we divide (6.1.35) by the particle density N/Ω and by the photon flux of the incident photons $c\hbar\omega_1 n_{\omega_1}(0)$. As should be expected, we obtain σ_c, the Raman scattering cross section given in (6.1.10). In the nonlinear regime, the increment coefficient G is more useful than the cross section σ_c.

6.1.3 Rayleigh Scattering by Dielectric Particles

Scattering and absorption of solar light by small dielectric dust particles is produced by their polarization in the external electric field. There is no skin-layer inside a dielectric particle, so that the external field penetrates through the whole particle. Besides of this, the magnetic component of scattering is small compared to the electric part, due to absence of the conduction currents.

There are condensation nuclei, tiny suspended particles, either solid or liquid, upon which water vapor condensation begins in the atmosphere. There are also much smaller nuclei in the atmosphere. The discovery that the air is full of tiny particles around which water droplets may condense to create clouds was made by Scottish physicist John Aitken (1839–1919). Much smaller particles are called Aitken nuclei. They ordinarily play no role in cloud formation because they do not induce condensation unless the air is highly supersaturated with water vapor. Most condensation nuclei are produced by wave action over the oceans and by natural and man-made fires over land. When mixed with the more hygroscopic material, dust and soil particles blown into the atmosphere also are sources of nuclei. Numerous measurements provide support for the hypothesis that layers of high concentrations of Aitken nuclei near the tops of marine clouds are due to photochemical nucleation. Chemical factors support the view that Aitken nuclei are dielectric particles of sizes in the range 0.01–0.1 μm. On average, their concentration varies from less than $10^3/\mathrm{cm}^3$ over oceans to $10^6/\mathrm{cm}^3$ in urban areas. It is tentatively concluded that Aitken particles in the troposphere account for most of the sulfate in the atmosphere.

We first consider scattering of light on the dielectric dust particles. The radius of this particle r_o is assumed to be small compared to the wavelength of the incident light $\lambda = 2\pi c/\omega$, that is

$$\frac{\lambda}{r_o} \sim \frac{c}{\omega r_0} \ll 1. \tag{6.1.36}$$

We assume also that the dielectric permittivity $\varepsilon(\omega)$ is not too large, i.e. the condition

$$1 < \sqrt{\varepsilon} \ll \frac{c}{\omega r_o} \tag{6.1.37}$$

is fulfilled. Then we can solve the static problem (the Laplace equation) both inside and outside of the particle. It should be noted that the inequality (6.1.37) is fulfilled well also for polar molecules for which the dielectric permittivity ε is near unit for light range of the electromagnetic frequencies. As a result, the temporal dependence of the electrostatic potential can be neglected everywhere.

The Laplace equation for the electrostatic potential φ inside and outside of the particle is of the form

$$\varphi(r,\theta) = Cr\cos\theta;\; r \le r_o;\; \varphi(r,\theta) = (-Er + \frac{D}{r^2})\cos\theta;\; r \ge r_o \tag{6.1.38}$$

Here $a \ll \lambda, \lambda_e$ is the radius of the dust particle, and λ is the wavelength of the incident light, $\lambda_e = \lambda/\sqrt{\varepsilon}$ is the wavelength inside the particle. The quantity $\mathbf{E} = \mathbf{E}_0 \exp(-i\omega t)$ is the electric field strength of the incident light wave.

Using the continuity condition of the potential on the surface of the particle (at $r = r_o$), one obtains the connection between coefficients C and D

$$C = -E + \frac{D}{r_o^3}. \tag{6.1.39}$$

The second equation for these coefficients follows from the continuity condition for the normal component of the electric displacement

$$-\varepsilon C = E + \frac{2D}{r_0^3} \tag{6.1.40}$$

Excluding the quantity D from two last equations, one finds the value of C

$$C = -\frac{3}{\varepsilon + 2} E. \tag{6.1.41}$$

Thus, the field strength of the uniform electric field inside the small dust particle is

$$\mathbf{E}_{\text{in}} = \frac{3}{\varepsilon + 2} \mathbf{E} \tag{6.1.42}$$

The electric polarization $\mathbf{P}_{\text{in}}$ (the dipole moment of the unit volume) is also uniform everywhere inside the particle; it is equal to

$$\mathbf{P}_{\text{in}} = \frac{\varepsilon - 1}{4\pi} \mathbf{E}_{\text{in}} = \frac{3}{4\pi} \left(\frac{\varepsilon - 1}{\varepsilon + 2} \right) \mathbf{E}, \tag{6.1.43}$$

and the dipole moment of the whole dust particle is

$$\mathbf{p} = \frac{4\pi a^3}{3} \mathbf{P}_{in} = \left(\frac{\varepsilon - 1}{\varepsilon + 2} \right) a^3 \mathbf{E}. \tag{6.1.44}$$

This solution is equivalent to the well known expression for the static dipole moment of the dielectric ball in a constant electric field. The difference is only that (6.1.44) contains the dielectric permittivity $\varepsilon(\omega)$ which corresponds to the frequency of the visible light, instead of the static dielectric constant ε_{st}. In the case of polar dielectrics the quantity ε_{st} can be several decimal orders of magnitude larger than the dielectric permittivity $\varepsilon(\omega)$ in the light frequency range. Now we determine the differential cross section of scattering; it is obtained from (6.1.20) by substitution the polarizability which is equal to ratio of the dipole moment (6.1.44) by the electric field strength of the incident light wave:

$$d\sigma_s = \left(\frac{\varepsilon - 1}{\varepsilon + 2}\right)^2 \frac{\omega^4 r_o^6}{c^4} \left(1 - \sin^2 \vartheta \cos^2 \varphi\right) d\Omega. \tag{6.1.45}$$

In order to consider the non-polarized solar light, we should average $d\sigma_s$ over the angle φ :

$$\langle d\sigma_s \rangle = \left(\frac{\varepsilon - 1}{\varepsilon + 2}\right)^2 \frac{\omega^4 r_o^6}{c^4} \frac{1 + \cos^2 \vartheta}{2} d\Omega. \tag{6.1.46}$$

Integrating over the solid angle, one obtains the total cross section of the Rayleigh light scattering on a small spherical dust particle:

$$\langle \sigma_s \rangle = \frac{8\pi}{3} \left(\frac{\varepsilon - 1}{\varepsilon + 2}\right)^2 \frac{\omega^4 r_o^6}{c^4}. \tag{6.1.47}$$

This cross section is proportional to ω^4, and it is much less than the geometrical cross section of the dust particle πr_o^2, if $a \ll \lambda \sim c/\omega$. It should be noted that in the opposite limiting case $a \ll \lambda$, the cross section coincides the geometrical cross section both for dielectric and for metal dust particles.

The Rayleigh law (6.1.47) $\sim\omega^4$ explains the cyan color of the heaven at the scattering of solar light. The maximum of the Planck spectrum of solar light corresponds to the yellow color. The difference in the intensity of cyan and violet components of the Planck solar light is less than 20%. The visible cyan color of the heaven is explained by the human eye sensitivity. The cyan light is perceived by the eye better than the violet light, by more than one order of magnitude!

6.1.4 Small Dielectric Particles in Electromagnetic Field

Scattering of an electromagnet wave on a macroscopic particle is determined by electric properties of the particle; in other words, scattering results from reaction of a particle material to the radiation field. Note that the polarizability of a particle α characterizes the particle reaction of the action of the electric field. This quantity is introduced as a connection between the induced dipole moment $\mathbf{D}$ of the particle and the electric field strength $\mathbf{E}$ which creates this dipole moment, so that

$$\mathbf{D}(\omega) = \alpha(\omega)\mathbf{E}(\omega) \tag{6.1.48}$$

We first establish the connection between the absorption cross section by a particle with its polarizability. Indeed, the interaction potential between the particle and electric field is $V = -\mathbf{DE}$. From this we have for the power absorbed by the particle

$$P = -\left\langle \mathbf{E} \frac{d\mathbf{D}}{dt} \right\rangle,$$

where brackets mean the averaging over time. Taking the electric field strength of a monochromatic electromagnetic wave in the form

$$\mathbf{E} = \mathbf{E}_0 \exp(i\omega t) + \mathbf{E}_0^* \exp(-i\omega t),$$

where ω is the frequency of the electromagnetic wave, we obtain for the particle dipole moment induced by the electromagnetic wave as

$$\mathbf{D} = \alpha(\omega)\,\mathbf{E}_0 \exp(i\omega t) + \alpha^*(\omega)\,\mathbf{E}_0^* \exp(-i\omega t),$$

where $\alpha(\omega)$ is the particle polarizability.

From this it follows for the absorbed power

$$P = i\omega|E_0|^2\left(\alpha^* - \alpha\right)$$

The flux density of the electromagnetic energy is

$$J = \frac{c|E_0|^2}{2\pi}$$

Then the absorption cross section $\sigma_{\rm abs}$ by the particle as the ratio of the absorbed power density to the flux of the electromagnetic energy is equal

$$\sigma_{\rm abs} = \frac{P}{J} = 4\pi\frac{\omega}{c}\,{\rm Im}\,\alpha(\omega) \qquad (6.1.49)$$

Thus, absorption of radiation by a spherical particle is determined by its polarizability. On the other hand, this takes place because the dielectric permittivity ε in the region of particle location differs from that in a surrounding space. Hence, the dielectric permittivity of a particle matter is connected with its polarizability. Let us determine this connection for a spherical particle which radius r_o is large compared to the wavelength. One can use the Poisson's equation for the electric potential φ is $\Delta\varphi = 0$ under these conditions both inside, and outside the particle. The boundary condition for the normal component of the electric displacement near the particle surface has the form

$$\varepsilon\frac{\partial\varphi(R \to r_o - 0)}{\partial R} = \frac{\partial\varphi(R \to r_o + 0)}{\partial R}, \qquad (6.1.50)$$

where R is a distance from a particle center, r_o is a particle radius, and ε is its dielectric permittivity. An electric field induces a particle dipole moment $\mathbf{D}$ is connected with the cluster polarizability as $\mathbf{D} = \varepsilon\mathbf{E}$. This leads to the electric field potential φ outside the particle which is induced by the electric field and by the particle dipole moment

$$\varphi = -\mathbf{E}\mathbf{R} + \frac{\mathbf{D}\mathbf{R}}{R^3} \qquad (6.1.51)$$

Since the electric potential inside the cluster is restricted, the solution of the Poisson's equation $\Delta\varphi = 0$ may be represented in the form

$$\varphi = C\mathbf{E}\mathbf{R}, \tag{6.1.52}$$

where C is a numerical coefficient. This coefficient and the polarizability of the particle can be found from the condition of continuity of the electric potential φ. This gives

$$C = \frac{3\varepsilon}{\varepsilon + 2}; \ \alpha = \frac{\varepsilon - 1}{\varepsilon + 2}a^3 \tag{6.1.53}$$

This relation between the polarizability and its cluster dielectric permittivity holds true also for an alternating electric field $\mathbf{E}_0 \cos\omega t$, where it has the form

$$\alpha(\omega) = \frac{\varepsilon(\omega) - 1}{\varepsilon(\omega) + 2}a^3, \tag{6.1.54}$$

Let us consider the classical limit of scattering of electromagnet wave, where the radiation intensity as a result of scattering is given by formula (1.2.30), i.e.

$$I(t) = \frac{2}{3c^3}\left[\ddot{\mathbf{D}}(t)\right]^2 = \frac{2\omega^4}{3c^3}\alpha^2(\omega)E^2 \tag{6.1.55}$$

The cross section of scattering σ_s is the ratio of this quantity by the energy flux $E^2/4\pi$ of the incident radiation that is given by the Rayleigh formula

$$\sigma = \frac{8\pi\omega^4\alpha(\omega)^2}{3c^4} \tag{6.1.56}$$

6.2 Absorption of Radiation by Metal Particles

6.2.1 *Interaction of Metal Particles with the Electromagnetic Wave*

Interaction between an electromagnetic wave and a metal particle is determined, in the first place, by interaction with valence electrons of the metal. We first consider such an interaction with a large particle which may be considered as a metal piece. We give in Table 4.2 electron parameters which influence on the interaction of univalent metals with the electromagnetic wave. In this case valence electrons of atoms become metal valence electrons in the metal formation from atoms. Along with the frequency ω of an electromagnetic wave, one can construct two frequency parameters, namely, the plasma frequency $\omega_p = \sqrt{4\pi N_e e^2/m_e}$ and the metal conductivity Σ.

In order to understand relation between these values, we consider two cases. The first case corresponds to the interaction of visible light with a metal surface. The wavelength of the green color wave is $\lambda = 0.5\,\mu\text{m}$, and its frequency equals $\omega = 3.7 \cdot 10^{16}\,\text{s}^{-1}$. Comparing with data of Table 4.2, we have

$$\omega_p \gg \omega \gg \Sigma \tag{6.2.1}$$

Another example relates to emission from a metal surface at the temperature $T = 1000\,\text{K}$. According to the Wien law, the emission maximum corresponds to the wavelength $\lambda = 0.29\,\mu\text{m}$, and its frequency equals $\omega = 3.7 \cdot 10^{16}\,\text{s}^{-1}$, and we obtain the same relation between the above frequencies. Therefore we ignore below the plasma frequency and consider only parameters ω and Σ which determine the interaction of radiation with a metal particle.

In the case of metal objects the interaction with an electromagnetic wave proceeds through valence electrons; therefore this interaction is stronger than that involving dielectric particles. The skin effect takes place for large particles, so that valence electrons in the metal screen the field of the electromagnetic wave. Hence, the interaction occurs in the metal region near the metal surface; thus, scattering of an electromagnetic wave by a small metal particle is analogous to that in the case of a bulk metal (Table 6.1).

Hence, one can use formulas for scattering of radiation and emission by a small metal particles on the basis of that for bulk metal which are considered in detail in [8]. One can construct these formulas with using a small parameter $\alpha = \omega/2\pi\Sigma$. In this case the penetration depth δ is given by [8]

$$\delta = \frac{\lambda}{(2\pi)^{3/2}}\sqrt{\frac{\omega}{\Sigma}}, \tag{6.2.2}$$

and the particle radius $r_o \gg \delta$. The electric and magnetic fields of the electromagnetic wave decrease inside the metal as $\exp(-z/\delta)$ where z is the distance from the plane boundary inside the metal. Correspondingly, the absorption cross section σ_a of a electromagnetic wave by the metal particle is [8]

Table 6.1 Parameters of univalent metals at room temperature due to valence electrons [7]. Here ρ is the metal mass density, N_e is the number density of valence electrons in the metal, Σ is the metal conductivity, $\omega_p = \sqrt{4\pi N_e e^2/m_e}$ is the plasma frequency for electrons of the metal, so that e, m_e are the electron mass and charge correspondingly

Parameter/metal	Li	Na	K	Cu	Rb	Ag	Cs	Au
ρ, g/cm^3	0.53	0.97	0.89	9.0	1.5	10.5	1.9	19
N_e, 10^{22} cm^{-3}	4.6	2.5	1.4	8.5	1.1	5.8	0.87	5.9
Σ, 10^{16} s^{-1}	9.7	19	12	54	7.0	57	4.4	40
ω_p, 10^{16} s^{-1}	1.2	0.90	0.66	1.6	0.58	1.4	0.53	1.4

$$\sigma_a = \pi r_o^2 \kappa(\alpha), \ \kappa(\alpha) = \sqrt{\frac{\alpha}{2}}\left[\ln\left(\frac{1}{\alpha}\right) - \frac{\pi}{2} + 1\right], \ \alpha = \frac{\omega}{2\pi\Sigma} \ll 1, \tag{6.2.3}$$

where $\kappa(\alpha)$ is the gray coefficient for the metal surface. Correspondingly, the flux of thermal radiation I_ω from the particle surface is given by

$$I_\omega = I_\omega^{(0)}\kappa(\alpha), \tag{6.2.4}$$

where $I_\omega^{(0)}$ is the radiative energy flux of the black body (the Planck's radiation).

In particular, the conductivity of silver is $\Sigma = 5.5 \cdot 10^{17}\,\mathrm{s}^{-1}$ at the temperature $T = 300$ K, and $\Sigma = 6.6 \cdot 10^{16}\,\mathrm{s}^{-1}$ at the temperature $T = 2000$ K. According to the Wien's law, the optimal radiation frequency at the temperature $T = 2000$ K is $\omega_{\max} = 1.3 \cdot 10^{15}\,\mathrm{s}^{-1}$. At this temperature a small parameter is $\alpha = 3.1 \cdot 10^{-3}$, that gives for the gray coefficient $\kappa = 0.16$.

6.2.2 *Absorption of Radiation by Metal Nanoparticles*

A size of metal nanoclusters is small compared to the penetration depth for an electromagnet wave, and these particles are uniform in the interaction with radiation. Electrons of metal nanoclusters and microparticles can spread freely over the particle and interact with an electromagnetic wave as free charges. There are two ways of the behavior of the electron subsystem in this interaction. First, the electron subsystem of the particle partakes in interaction as a whole, so that collective properties of the electron subsystem determine absorption of an electromagnetic wave by a metal particle. Second, the spectrum of metal atoms usually includes radiative transitions in a visible spectral range. These spectral lines are broadening in a condensed metal due to the interaction with neighboring atoms which consist partially of ions and electrons. But a general character of radiative transitions in atoms may be conserved in the system of bound atoms. The choose between these two types of interaction involving valence electrons can be done on the basis of experimental data.

A cluster is a system of a finite number of bound atoms. We consider here the metal clusters consisting of large number of bound atoms. This cluster is a uniform particle of a spherical shape where valence electrons can freely propagate inside a cluster volume. However, a cluster size is small compared to a depth of the skin-layer in a bulk metal, and therefore the cross section for interaction of clusters with an electromagnetic wave is proportional to the number of atoms in the cluster, i.e., to the number of valence electrons.

In addition, a cluster radius r_o is small compared to the radiation wavelength λ

$$r_o \ll \lambda \tag{6.2.5}$$

Considering a metal cluster as a macroscopic system, we take the cluster polarizability to be proportional to the number of its atoms. In addition, for this metal particle the following criterion holds true

$$\omega \ll \Sigma, \tag{6.2.6}$$

where Σ is the specific conductivity of the cluster matter. This criterion allows one to reduce the problem to the stationary case, where the stationary polarizability of the spherical cluster is $\alpha = r_o^3$. Accordingly, formula (6.1.30) for the absorption cross section may be represented as

$$\sigma_{\text{abs}} = \frac{12\pi\omega}{c} \frac{\varepsilon''}{(\varepsilon' + 2)^2 + \varepsilon''^2} r_o^3 = \frac{\pi\omega}{c} r_o^3 g_{sph}(\omega); \; g_{sph}(\omega) = \frac{12\varepsilon''}{(\varepsilon' + 2)^2 + \varepsilon''^2} \tag{6.2.7}$$

Here the dielectric permittivity of the cluster matter is taken in the form $\varepsilon(\omega) = \varepsilon'(\omega) + i\varepsilon''(\omega)$ It is seen that the absorption cross section by a spherical macroscopic cluster may be estimated as

$$\sigma_{\text{abs}} \sim \frac{r_o}{\lambda} \pi r_o^2,$$

i.e., this cross section is small compared to the geometrical cross section πa^2.

We now apply the above results for metal clusters contained of a finite number of bound atoms. The absorption process is determined by valence cluster electrons. Assuming that these electrons are free, under criterion (6.2.5), the dielectric permittivity of an electron plasma is given by

$$\varepsilon(\omega) = 1 - \frac{\omega^2}{\omega_p^2} \tag{6.2.8}$$

Here ω_p is the plasma frequency or Langmuir frequency, that is given by the expression

$$\omega_p = \sqrt{\frac{4\pi N_e e^2}{m_e}},$$

where N_e is the electron number density, e and m_e are the electron charge and mass, respectively. Assuming that $\varepsilon'' \ll 1$, we transform formula (6.2.7) taking into account the expression (6.2.8) near the resonance frequency

$$\sigma_{\text{abs}}(\omega) = 2\pi \frac{\hbar\omega^2}{c} a^3 \frac{\Gamma}{\hbar^2(\omega - \omega_0)^2 + \Gamma^2} = \sigma_{\max} \frac{\Gamma^2}{\hbar^2(\omega - \omega_0)^2 + \Gamma^2}, \tag{6.2.9}$$

where ω_0 is the resonance Mie frequency,

$$\omega_0 = \frac{\omega_p}{\sqrt{3}},$$

Γ is the resonance width according to

$$\Gamma = \frac{\hbar\omega_0\varepsilon''}{6},$$

and σ_{max} is the maximum absorption cross section,

$$\sigma_{max} = 2\pi\frac{\hbar\omega^2}{\Gamma c}a^3. \qquad (6.2.10)$$

From formulas (6.2.9) and (6.2.10) it follows the integral relation

$$\int \sigma_{abs}(\omega)d\omega = \pi\sigma_{max}\frac{\Gamma}{2\hbar} \qquad (6.2.11)$$

where the resonance width is assumed to be relatively small. Within the frame of liquid drop model, the cluster radius is given by [9, 10]

$$r_o^3 = r_w^3 n,$$

where r_w is the Wigner-Seitz radius, and n is the number of cluster atoms. Under used conditions, the cluster is assumed to be a uniform particle, and the absorption cross section is proportional to the number of cluster atoms.

Though the above cluster model of a uniform spherical particle is rough, it is convenient to demonstrate the mechanisms of cluster absorption through the interaction between an electromagnetic field and valence electrons which leads to the resonance character of the absorption cross section as a function of a photon frequency. In practice, the absorption spectrum has more complex structure, and it can include several peaks. Table 4.3 contains parameters of the absorption cross sections for some metal clusters for which these cross sections can be approximated by a simple resonance dependence. Basing on the experimental data [11–14] for the cross sections of absorption by clusters consisting of Li, K and Ag atoms, one can check the validity of the plasma model for the absorption cross section of clusters. It is convenient to introduce the parameter

$$\xi = \sigma_{max}\frac{\Gamma c}{2\pi\hbar\omega_0^2 a^2}, \qquad (6.2.12)$$

that is equal to one, if formula (6.2.9) is correct.

As it follows from Table 6.2, the parameter ξ differs from one stronger than the limits of its accuracy. This means violation of the macroscopic character of absorption in accordance with formula (6.2.7) and prohibits to describe valence electrons as free ones that leads to formula (6.2.9) for the absorption cross section. Thus, the concept of the interaction of the electromagnetic wave with cluster electrons as plasma ones is violated.

One more comparison confirms this conclusion. The resonance frequency for metal clusters with one valence electron per atom which are given in Table 6.2 is equal to

Table 6.2 Parameters of the absorption cross sections for metal clusters

Cluster	$\hbar\omega_0$, eV	Γ, eV	σ_{max}/n, Å^2	ξ	β	f
Li^+_{139}	2.92	0.90	62	2.8	0.24	0.58
Li^+_{270}	3.06	1.15	120	3.2	0.30	0.73
Li^+_{440}	3.17	1.32	280	4.9	0.50	1.20
Li^+_{820}	3.21	1.10	440	3.3	0.52	0.85
Li^+_{1500}	3.25	1.15	830	3.5	0.66	0.91
K^+_9	1.93	0.22	26	2.9	0.27	0.91
K^+_{21}	1.98	0.16	88	2.9	0.52	0.96
K^+_{500}	2.03	0.28	1750	4.0	1.3	1.40
K^+_{900}	2.05	0.40	2500	4.5	1.2	1.59
Ag^+_9	4.02	0.62	8.84	2.6	0.24	0.87
Ag^+_{21}	3.82	0.56	16.8	2.1	0.26	0.64

$$\omega_0 = \omega_p/\sqrt{3} = \frac{e}{\sqrt{m_e r_W^3}}$$

The number density of valence electrons is equal to

$$N_e = \frac{3}{4\pi r_W^3},$$

where r_W is the Wigner-Seitz radius. The value of $\hbar\omega_0$ is equal to 13.5 eV for large Li clusters, to 7.1 eV for large K clusters, and to 14.7 eV for for large Ag clusters. A large difference of these data from measured values of Table 6.2 allows us to conclude that the plasma model for valence electrons in the analysis of absorption of an electromagnetic wave by metal clusters is incorrect. In addition, we show in Fig. 6.4 experimental dependencies for the absorption cross sections of some Ag clusters on the photon frequency. As it is seen, the absorption cross sections can contain both one and two resonance maxima. This is also rejects the plasma model for valence electrons.

We now consider the sum rule for a metal cluster. At fixed nuclei the absorption spectrum of the cluster consists of a finite number of spectral lines; the number of these lines is comparable with the number of cluster valence electrons. In the limit of one atom this spectrum is transformed into one or several resonance spectral lines. Let us introduce the effective oscillator strength f per one valence electron. Then the total oscillator strength is equal to nf, where n is the number of valence electrons in the cluster. Due to nuclear motion, the absorption spectrum of clusters is transformed from set of separate spectral lines into the continuous curve. However, the sum of the oscillator strengths does not change. One can expect that the effective oscillator strength f per one atom depends weakly on the cluster size, and it corresponds to

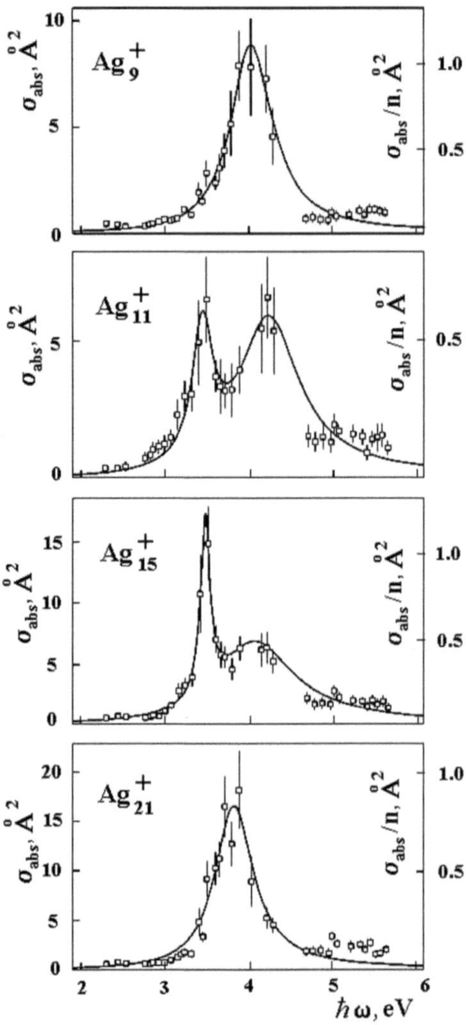

Fig. 6.4 Absorption cross sections for silver clusters [14]

the atomic value. Below we check this concept using the analysis of clusters with a plasma form of the absorption cross section which are contained in the Table 6.2.

Let us use the general expression for the cross section of photon absorption by an atomic system in the form

$$\sigma_{\rm abs}\,(i \to j) = \frac{\pi^2 c^2}{\omega^2}\frac{a_\omega}{\tau_{\rm ij}}\frac{g_j}{g_i} = \frac{2\pi^2 c^2}{m_e c} f_{\rm ij} g_k a_\omega \qquad (6.2.13)$$

Here ω is the frequency of a given electron transition between states i (the lower state) and j (the upper state); g_i, g_j are statistical weights of these transition states, τ_{ij}

is the radiative lifetime with respect to this transition, a_ω is the frequency distribution function for the radiated photons which is normalized as (2.1.1)

$$\int a_\omega d\omega = 1;$$

finally, f_{ij} is the oscillator strength for this transition. The sum rule for dipole radiation transitions of valence electrons in the spectral range including resonance transitions, is of the form

$$\int \sigma_{\rm abs}(\omega) d\omega = \frac{2\pi^2 e^2}{m_e c} nf \tag{6.2.14}$$

This sum rule is analogous to those for atoms (1.3.11).

If the absorption cross section has a resonance structure, analogously to that for clusters in Table 6.2, the integral relation (6.2.11) is also applicable. Then it follows from relations (6.2.11) and (6.2.14) that

$$f = \frac{\sigma_{\max} \Gamma m_e c}{2\pi^2 e^2 n \hbar} \tag{6.2.15}$$

The values of the effective oscillator strength for metal clusters with resonance structure of absorption are given in Table 6.2. Different values for each element are explained, in our opinion, by a restricted accuracy of used data. Average values of the oscillator strength for each cluster correspond to oscillator strengths of low-lying transitions $^2S_{1/2} \rightarrow ^2 P_{1/2} \rightarrow ^2 P_{3/2}$ of their atoms. The total oscillator strengths are equal to 0.74 for the lithium atom, 1.5 for the potassium atom, and 0.77 for the silver atom. Coincidence of the oscillator strengths for clusters with corresponding values for atoms confirms the fact that the absorption spectra of clusters can be obtained by transformations of atomic resonance lines due to the interaction between atoms and due to the nuclear motion.

In considering of the interaction of metal clusters with the radiation field, on the one hand, these clusters are uniform with respect to radiation, and on the other hand, they may be considered as macroscopic ones, where the interaction with an electromagnetic field is determined by the electron subsystem. Let us construct this cluster from n bound metal atoms and transfer an electron of one of these atoms into the resonance excited state by means of the dipole radiative transition, so that the oscillator strength of this transition is of the order of one. This excitation is spread over the cluster, and the spectrum of cluster excitation consists of n discrete spectral lines. The total spectrum of cluster excitation with accounting for atom motion has a continuous structure with one or several maxima. This form of the cluster spectrum follows also from computer simulations, and the integral absorption cross section is proportional to the number of valence electrons.

6.2.3 Emission of Metal Clusters in Hot Gases

A strong interaction between metal clusters and resonance radiation may be of importance both for radiative properties of metal clusters located in a buffer gas and for the heat balance of this system. We below determine the spectral radiation power $p(\omega)$ of a cluster at a certain temperature T. This spectral power is expressed through the absorption cross section σ_{abs} according to the Kirchhoff law as

$$p(\omega) = \hbar\omega \cdot I(\omega) \cdot \sigma_{\text{abs}}(\omega)\,, \tag{6.2.16}$$

where in accordance with formula (2.2.7) $I(\omega)$ is the isotropic flux of the black body radiation at a given frequency and temperature

$$I(\omega) = \left(\frac{\omega}{\pi c}\right)^2 \frac{1}{\exp(\hbar\omega/T) - 1} \tag{6.2.17}$$

Here $\sigma_{\text{abs}}(\omega)$ is the absorption cross section for a given cluster as a small particle. From this it follows that the spectral radiative power by a small cluster has the form

$$p(\omega) = \frac{\hbar\omega^3}{\pi^2 c^2} \frac{\sigma_{\text{abs}}(\omega)}{\exp(\hbar\omega/T) - 1} \tag{6.2.18}$$

In particular, the total radiation power of a small macroscopic particle with a radius a is [15, 16]

$$P = \int_0^\infty p(\omega)\mathrm{d}\omega = \frac{12\pi}{\hbar c} a^3 g\sigma T^5 \kappa = \frac{46\pi a^3 g\sigma T^5}{\hbar c}; \quad \frac{Ta}{\hbar c} \ll 1 \tag{6.2.19}$$

Here the quantity

$$g = \frac{\varepsilon''}{(\varepsilon' + 2)^2 + (\varepsilon'')^2}$$

is given by formula (6.2.7) and is assumed to be independent on the frequency; σ is the Stephan-Boltzmann constant, and the numerical coefficient is $\kappa = 3.83$. It is seen that the radiation power by a small macroscopic particle under equilibrium conditions is proportional to T^5, in contrast to the classical dependence $\sim T^4$ for the radiative power of a macroscopic black body surface.

According to formula (6.2.7) and (6.2.14), the absorption cross section for a small cluster is proportional to the number of cluster atoms. Hence, the specific absorption cross section, i.e., the absorption cross section by one atom, does not depend on the cluster size. Therefore, the cluster radiative power per unit volume is proportional to the number density of bound atoms. This statement does not depend on the distribution function of clusters, or of small particles over their sizes. Thus, the total radiation power for a given volume of a gas, or of a plasma, is determined

Table 6.3 The specific radiation power P_{rad}, 10^7 W/g, and the light yield for large clusters at various temperatures, expressed in lm/W and given in parentheses

Cluster	3000 K	3500 K	4000 K
Ag	0.71 (51)	1.6 (25)	3.5 (88)
K	4.0 (106)	8.6 (111)	17 (165)
Li	2.0 (51)	4.9 (80)	10 (102)
Black body	(22)	(39)	(57)

by the total number of bound atoms, and it does not depend on the size of the cluster. This general conclusion is based on the statement that absorption cross section is proportional to the number of bound atoms and it is valid both for clusters and small macroscopic particles.

We now use parameters of the absorption cross sections for lithium, potassium and silver clusters given in Table 6.2 in order to analyze numerically the radiative parameters of the plasma which contains clusters. Table 6.3 presents the specific radiation powers for clusters; they are determined by the expression

$$P_{rad} = \int \frac{p(\omega)\, d\omega}{M}, \qquad (6.2.20)$$

where M is the cluster mass. Here we take into account that the radiation power is proportional to the total mass of bound radiating atoms in clusters. Table 6.3 contains also (in brackets) the light yield of cluster radiation where the absorption cross sections for these clusters are used as model ones. The light radiation yield characterizes the efficiency of the eye perception that is given by the expression

$$\eta = \frac{\int p(\omega) V(\omega) d\omega}{\int p(\omega) d\omega}, \qquad (6.2.21)$$

where the spectral radiation power $p(\omega)$ is calculated on the basis of formula (6.2.18), and the visibility function $V(\omega)$ determines the perception of radiation by eye; this function has maximum about of 683 lm/W for the wavelength of radiation of $\lambda = 555$ nm. For comparison, Table 6.3 contains also the light yield of the black body. It is seen that clusters as light sources are better than a black body because of a more favorable radiation spectrum (a thermal infrared radiation is excluded from the radiation spectrum of the clusters). It follows from data of the Table 6.3 that at the temperature of $T = 3600$ K the averaged radiation power of the clusters is $1 \cdot 10^8$ W/g. This value is convenient for estimates.

Thus, metal clusters or small macroscopic particles which are located in a hot, or ionized gas, can be responsible for radiation of these systems. For example, this occurs in the flame where radiation is produced by small soot particles.

6.3 Absorption by Atmospheric Particles

6.3.1 Aerosols and Water Microdrops in Atmosphere

Aerosols are particles of nano-sized and micro-sized particles located in the atmosphere [17–22]. Sometimes water microdrops which form clouds and are of importance in electric and radiative properties of the Earth's atmosphere, are included in the list of aerosols [23, 24]. Below the object of our consideration will be just water drops as an important atmospheric radiator in the infrared spectrum range [25–28]. But first we glance at aerosols as small particles which are influenced by atmospheric properties.

Figure 6.5 contains various types of nanoparticles and microparticles which can be presented in atmospheric air. For comparison, nanoclusters and electric probes are added to this list, though these objects exist irrespectively the atmosphere. Aitken particles were investigated from 19th century [30–33], earlier than other atmospheric particles. They are located at high altitudes, above clouds, and their basis are radicals of sulfur compounds which result from vaporization of meteorites at high altitudes and from processes which proceed at the Earth surface or in the atmosphere at low altitudes. Since their size is below 0.1 μm, Aitken particles are also responsible for a blue sky color because shortwave photons scatter on these particles effectively. The number density of Aitken particles at altitudes 10–20 km is $10^2 - 10^4\,\mathrm{cm}^{-3}$ [34].

Aerosols at low altitudes results from processes which proceed at the Earth's surface. Sulfur SO_x and nitrogen NO_x oxides, as well as atmospheric ions, are nuclei of condensation in formation of water microdrops. In addition, sulfur and nitrogen

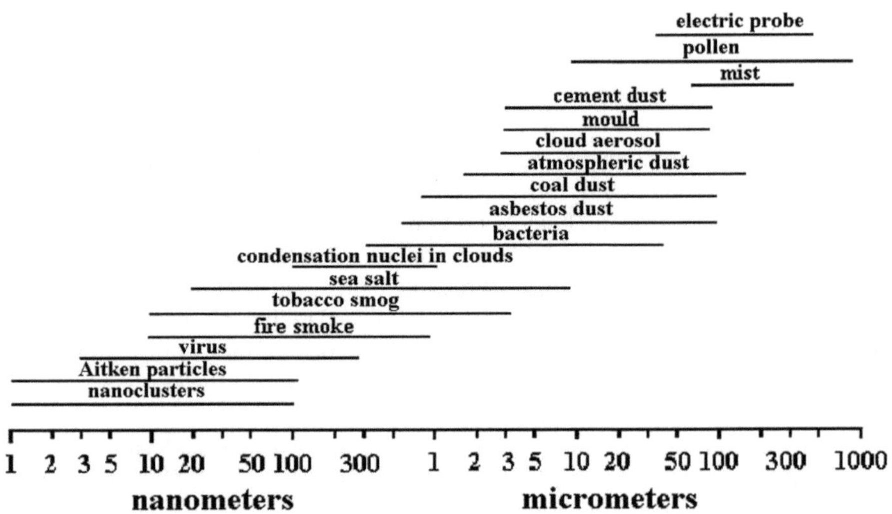

Fig. 6.5 Typical size of aerosols and microparticles located in the Earth atmosphere [29]

oxides, as well as soot particles, are formed in processes of combustion of solid and liquid organic compounds in air. A large amount of aerosols is formed as a result of volcano eruption. Some part of particles in atmospheric air goes up from the Earth's surface under the wind action. The putrefaction processes at the Earth's surface cause extraction of some aerosols and their transport to the atmosphere. In spite of the variety of aerosols located in atmospheric air in accordance with Fig. 6.5, they do not influence radiative properties of the atmosphere because of a small amount.

The Earth's atmosphere contains a large amount of atmospheric water which results from circulation of water in the nature. Water circulation consists of water evaporation from open water reservoirs at the Earth's surface and rain or snow return water back to the Earth's surface. This process establishes the amount of atmospheric water $1.3 \cdot 10^{19}$ g [35–38], and the total rate of water evaporation from the Earth's surface is $3.9 \cdot 10^{20}$ g/yr [39–42]. Precipitation of atmospheric water on the Earth's surface with uniform distribution over it gives a layer of liquid water of a thickness 2.5 cm [43]. The average concentration of water molecules in atmospheric air as approximately 0.4%, whereas near the Earth's surface the average concentration of water molecules in air is equal 1.7%, i.e. the number density of atmospheric water molecules decreases sharper with an altitude increase than that for air molecules. An average time of residence of water molecules in the atmosphere is approximately 9 days [42]. These data are the basis to analyze the atmospheric greenhouse phenomenon.

A small part of atmospheric water is found in the atmosphere in the form of water microdrops which interact effectively with infrared radiation. Near the Earth's surface, the average partial pressure of water is about 2 *Torr*, whereas the saturated water pressure is 4.7 *Torr* at the surface temperature [44, 45]. Therefore atmospheric water is in the form of a vapor consisting of free molecules mostly, and a small part of water exist in the form of aerosols, at least, at altitudes below 3 km. Aerosols are formed and exist at larger altitudes; processes with their participation [21, 26–28, 46, 47] are important for optic and electric atmospheric properties. Formation of aerosols may proceed if the partial pressure of water exceeds the saturated vapor pressure at a current atmosphere temperature which is taken from [45]. The ratio of these pressures is the air moisture. There are reliable methods of measurements of the global moisture (for example, [48, 49]) which allows one to analyze the evolution of the moisture of atmospheric air at some altitudes in time.

Atmospheric water is the main greenhouse component and includes a water vapor consisting of free water molecules and water aerosols which compose clouds. Figure 6.6 gives the absorption spectrum of liquid water. As is seen, the absorption coefficient in the visible spectral range is lower by seven orders of magnitude compared with that in the infrared spectral range. This means that water aerosols are transparent for solar radiation, and clouds are seen due to absorbed admixtures, whereas these microdrops are sources of IR radiation. Let us estimate the role of water aerosols in IR emission of the atmosphere taking the cross section of absorption of an infrared photon $\sigma = \pi r^2$ under the criterion $\lambda \leq r$, where r is a drop radius, and λ is the wave length. Correspondingly, the optical thickness u of the layer with the depth l and number density N of water microdrops is equal

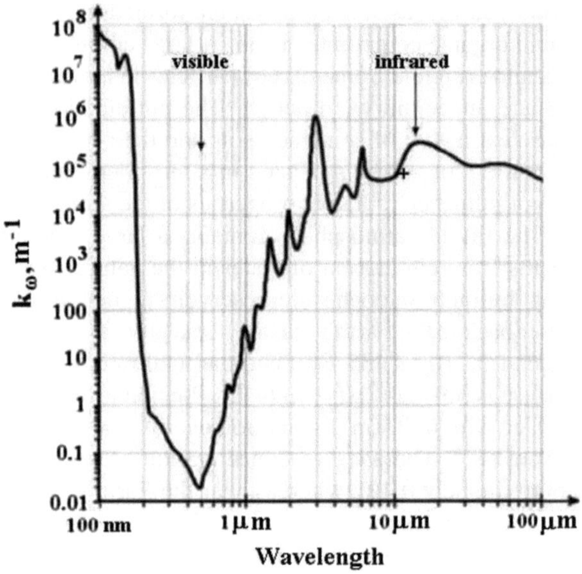

Fig. 6.6 Absorption coefficient for a bulk liquid water layer under normal conditions [50]. A cross indicates the absorption coefficient for aerosols [51]

$$u = \pi r^2 Nl, \tag{6.3.1}$$

and the water mass in aerosols per unit square of the Earth's surface m is

$$m = \frac{4}{3}\pi r^3 Nl = \frac{4ru\rho}{3}, \tag{6.3.2}$$

where ρ is the water density. One can see that the observed optical thickness of the atmosphere ($u \approx 3$) may be provided by approximately 0.2% of atmospheric water if it is located in the form of liquid microdrops-aerosols.

Let us note two types of interaction of an electromagnetic wave with a dielectric particle. Above we considered the case when this interaction results from electric properties of a particle material which is described by its dielectric constant. Another character of this interaction consists in absorption of radiation by this particle due to transitions between molecular states of this particle and also due to transitions between states owing to interaction between molecules of this condensed matter. Evidently, the absorption coefficient of Fig. 6.6 for liquid water in the infrared spectrum range is determined by such transitions.

Let us analyze also the result of experiment [51] which is given in Fig. 6.6 according to which water microdrops of an average radius $r = 10\,\mu\text{m}$ are characterized by the absorption cross section $\sigma_{\text{abs}} = \pi r^2$ for thermal atmospheric radiation. Then the optical thickness u of a layer of a thickness l is equal at the number density N of microdrops

$$u = Nl\sigma_{\text{abs}} = Nl\pi r^2$$

From the another standpoint, the mass per unit area $M = \rho l$ is

$$M = Nl \cdot \frac{4}{3}\pi r^3 \rho,$$

where ρ is the mass density of water. On the basis of these formulas, the absorption coefficient is equal to

$$k_\omega = \frac{u}{l} = \frac{3\sigma_{abs}}{4r^3}$$

Under the above considered conditions ($\sigma_{abs} = \pi r^2$) the absorption coefficient $k_\omega \sim 10^5\,m^{-1}$ (it is included in Fig. 6.6 shows the identity of the absorption coefficient for liquid water and a gas of water microdrops).

6.3.2 *Water Microdrops in Clouds*

Considering an interaction of an individual atmospheric water microdrop with atmospheric radiation and being guided by experiment [51], we take a drop radius to be $r = 10\,\mu m$. In this case the absorption cross section of thermal atmospheric radiation $\lambda \sim r \approx 10\,\mu m$ by a given microdrop is close to its geometric cross section; it is equal to $\sigma_{abs} \sim 3 \cdot 10^{-6}\,cm^2$, whereas for solar radiation the absorption cross section is equal $\sigma_{abs} \sim 10^{-11}\,cm^2$. In order to analyze radiative parameters of a microdrops in a cumulus, we take typical parameters of water microdrops in cumulus as [52–55])

$$r_o = 8\,\mu m, \; N_d = 10^3 cm^{-3}, \tag{6.3.3}$$

where r_o is the average drop radius, and N_d is an average density of water microdrops in cumulus in a thunderstorm weather. This gives for the absorption coefficient $k_\omega = \sigma_{abs} N_d \sim 10^{-3}\,cm^{-1}$, and a typical observed optical thickness $u = k_\omega L$ of cumulus corresponds to its thickness $L \sim 30$m.

One can see according to Fig. 6.6 data, that the cloud optical thickness in the visible spectrum range is equal $u \sim (10^{-4} - 10^{-3})$, i.e. a cloud is transparent for visible radiation. Hence, clouds in the course of their formation are invisible. Through a time they become visible as a result of attachment of absorbed components, in particular, a dust or optically active atoms in the visible spectrum range. Below we estimate an amount of sodium atoms in a drop which provide absorption of visible radiation. Sodium atoms are formed in a drop as a result of attachment of molecules NaCl to this drop or joining of this drop with a small particle of this salt.

We assume that injection of sodium atoms into a water microdrop leads to broadening of an absorption line, so that it is transformed in an absorption band of a width $\hbar\Delta\omega \sim 1$eV. According to formula (2.2.24) the absorption cross section may be estimated as

$$\sigma_{abs} = \frac{\lambda^2}{4\Delta\omega\tau}, \tag{6.3.4}$$

where a typical wavelength in the visible spectrum range is $\lambda \sim 0.5\,\mu$m. Taking a typical lifetime for an upper state of transition is $\tau \sim 10^{-8}$ s. This gives a typical absorption cross section $\sigma_{Na} \sim 5 \cdot 10^{-17}$cm^2. Let a drop contains n sodium atoms, and then a cloud becomes nontransparent in a visible spectrum range, if its optical thickness is

$$u = NLn\sigma_{Na} \sim 1$$

From this consideration we have $n \sim 6 \cdot 10^9$. Because an individual microdrop contains $n = 7 \cdot 10^{13}$ water molecules, one can find a typical concentration of sodium atoms in water, that is, $\sim$0.01%. This amount of sodium solved in water provides the visibility of clouds in a sky.

Though it is known that the covering of a sky by clouds is approximately 70%. But this does not mean that the other part of sky is not occupied with water microdrops. Indeed, at the first stage of the nucleation process that leads to conversion of water molecules in water microdrops transparent drops are formed, and only through a time they become visible after attachment of absorbed atoms or dust particles to microdrops.

Let us take the cross section of absorption σ_{abs} by an aerosol particle of a radius r to be [16]

$$\sigma_{abs} = \pi r^2, \quad (6.3.5)$$

if an aerosol radius is large compared to the wavelength of radiation λ. In other limiting case we have [8]

$$\sigma_{abs} \sim \frac{r^3}{\lambda}, \quad r \ll \lambda \quad (6.3.6)$$

From this one can construct the absorption cross section by an aerosol particle as a function of an aerosol size in the form

$$\sigma_{abs} = \frac{\pi r^2}{1 + C\frac{\lambda}{r}} \quad (6.3.7)$$

From the above experimental data it follows that in this frequency range formula (6.3.5) holds true with an accuracy of 20%, and the cross section of absorption of thermal radiation by aerosols of a radius (8–10) μm is $(2.5 \pm 0.5) \cdot 10^{-6}$cm^2.

This leads to the depth of a formed water layer (30–40) μm that corresponds to the concentration of atmospheric water in aerosols roughly $(1-2)$%. It is seen that water aerosols may give a remarkable contribution to atmospheric emission. This result causes alarm with respect to the climate change because transition of a small part of an atmospheric water vapor in aerosols may lead to a significant change of the atmospheric optical thickness. In particular, according to studies [56–59] cosmic rays influence on formation of aerosols and clouds in the Earth's atmosphere.

Thus, we have that the spectral radiative flux of the Earth's atmosphere may include some frequency bands which are created by vibration-rotation or rotation radiative transitions of certain components. One can introduce the effective temperature for a certain frequency band in accordance with formula (7.1.17). The optical thickness of the atmosphere starting from this layer in the direction perpendicular to the Earth surface equals to 2/3 at a given frequency, and the temperature of this atmospheric layer is the radiative temperature for emission at a given frequency. We below use this concept in the analysis of emission of atmospheric CO_2 molecules.

Note that clouds consisting of water aerosols water aerosols influence on the Earth's energetics. In accordance with Fig. 6.6, the absorption of radiation by water aerosols is strong for the infrared spectrum range, whereas water aerosols are transparent for visible radiation. But this fact corresponds to pure liquid water, and if some chemical components are dissolved in water, the situation may be changed. Hence, on the first stage of formation of water aerosols they are transparent microdrops, and subsequently after attachment of some impurities water aerosols become visible. One can demonstrate the influence of admixtures on optical properties of aerosols in the case when the sodium salt NaCl is dissolved in water of an aerosol; then the solute sodium atoms determine the absorption of visible radiation by aerosols.

Let us assume that yellow spectral line of absorption of a free sodium atom is converted in an absorbed band as a result of interaction with surrounding water molecules, if a sodium atom is located in water. Taking a width of the absorption band $\Delta\omega$ to be $\hbar\Delta\omega \sim 1$eV, one can obtain for the absorption cross section σ as a result of interaction of an electromagnetic wave with a dissolved sodium atom

$$\sigma_{\rm vis} = \frac{\lambda^2}{4} \cdot \frac{1}{\Delta\omega\tau}, \tag{6.3.8}$$

where λ is a wavelength, τ is the radiative lifetime of an excited atom which is estimated as

$$\frac{1}{\tau} = \frac{2e^2\omega^2}{m_e c^3} f g_o \tag{6.3.9}$$

Here m_e is the electron mass, f is the oscillator strength for transition between atomic states, g_o is the statistical weight of the lower transition state which is the ground electron state. Because $f \sim 1$ and $g_o \sim 1$, it follows from formula (6.3.8) an estimate for the absorption cross section

$$\sigma_{\rm vis} \sim \frac{e^2}{\Delta\omega m_e c} \tag{6.3.10}$$

This formula gives $\sigma_{\rm vis} \sim 5 \cdot 10^{-18}\,{\rm cm}^2$. Let us take for definiteness aerosols of a cumulus clouds with an average radius $r = 8\,\mu$m and the absorption cross section $\sigma \approx 2.5 \cdot 10^{-6}\,{\rm cm}^2$ for infrared radiation. This cross section may be reached as a result of absorption of dilute sodium atoms if the concentration c(Na) of these atoms in aerosols is c(Na) $\sim 0.5\%$. These conditions may be attained in reality.

6.3.3 *Atmospheric Water Microdrops as Atmospheric Radiators and Absorbers*

Water drops are effective radiators and absorbers of infrared radiation. Being located in the Earth's atmosphere, they interact effectively with infrared radiation that passes through the atmosphere. We now consider the energetic balance of an individual water microdrop which interacts with this infrared radiation through processes of emission and absorption; it interacts also with surrounding air through the thermal conductivity process. This microdrop exchange through the thermal conductivity of air; the power P_κ which the drop takes from air or transfers to it in the case of a higher drop temperature compared to the air temperature far from the drop, is equal [15, 16]

$$P_\kappa = 4\pi r_o \kappa \Delta T, \tag{6.3.11}$$

where r_o is a drop radius, κ is the air thermal conductivity, and ΔT is the temperature difference for the drop under consideration and surrounding air far from it. Here we assume that the drop radius r_o is large compared to the mean free path of molecules in air.

Considering a microdrop as a black body, we have the following energetic balance equation

$$4\pi r_o \kappa \Delta T = 4\pi r_o^2 \sigma T^4 - P_{\text{abs}}, \quad \Delta T = T - T_a, \tag{6.3.12}$$

where σ is the Stephan-Boltzmann constant, T is the drop temperature, P_{abs} is the power absorbed by drop in the form of infrared radiation, T_a is the air temperature. In accordance with Fig. 6.6, we assume that absorbed radiation is emitted at a distance λ from the drop, that is, the mean free path of an infrared photon in atmospheric air; then we obtain for the absorbed power

$$P_{\text{abs}} = 4\pi r_o^2 \int_{-1}^{1} \sigma T(\cos\theta)^4 = 4\pi r_o^2 \int_{-1}^{1} \sigma \left[T + \lambda \frac{dT}{dh} \cos\theta \right]^4 = 4\pi r_o^2 \sigma \left[T^4 + 4 \left(\lambda \frac{dT}{dh} \right)^2 T^2 \right] \tag{6.3.13}$$

Here $dT/dh = 6.5$ K/km is the temperature gradient in the atmosphere, and we take roughly $\lambda \approx 2$ km for infrared radiation. From this we find the difference between temperatures of the drop and surrounding air

$$\Delta T = \frac{4 r_o \sigma T^2}{\kappa} \left(\lambda \frac{dT}{dh} \right)^2 \tag{6.3.14}$$

It is seen that the drop temperature is higher than the temperature of surrounding air in the case $h > \lambda$ which is represented in Fig. 6.7. For a drop radius $r_o = 10\,\mu$m we take $h = \lambda = 2$ km, where the temperature is $T = 270$ K with $\kappa = 2.4\text{W/m}^2$ [45]; then we obtain $\Delta T = 1 \cdot 10^{-5}$ K. This value increases if the altitude decreases. In particular, when the microdrop reaches the Earth's surface ($h = 0$), we have instead of (6.3.14)

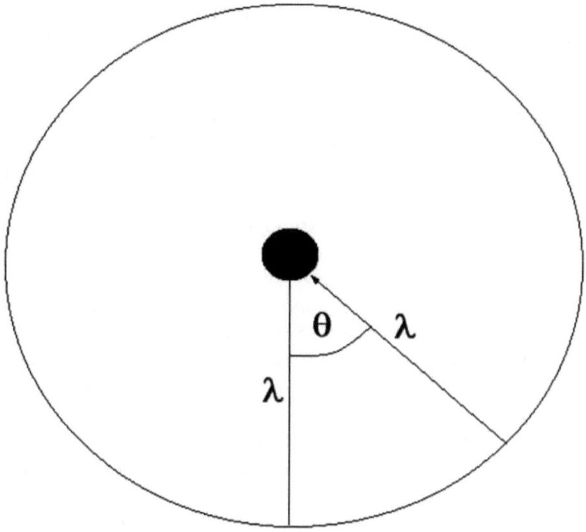

Fig. 6.7 Geometry of a microdrop located in atmospheric air and interacted with passed radiation

$$\Delta T = -\frac{2r_o}{\kappa}\sigma T^3 \lambda \frac{dT}{dh} \quad (6.3.15)$$

In this case the drop temperature is lower than that for surrounding air, and $\Delta T = -6 \cdot 10^{-3}$ K. Thus from this analysis one can conclude that if a microdrop is located in atmospheric air, its temperature coincides practically with the temperature of surrounding air.

References

1. L. Long, *The Raman Effect: The Unified Theory of Raman Scattering by Molecules* (Wiley, New York, 1977)
2. N.B. Colthrup, L.H. Daly, S.E. Wiberley, *Introduction to Infrared and Raman Spectroscopy* (San Diego, Academic Press, 1990)
3. P. Hendra, C. Jones, G. Warnes, *FT Raman Spectroscopy* (Ellis Horwood Ltd, Chichester, 1991)
4. J.R. Ferraro, K. Nakamoto, *Introductory Raman Spectroscopy* (Academic Press, San Diego, 1994)
5. R.L. McCreery, *Raman Spectroscopy of Chemical Analysis* (Wiley, New York, 2000)
6. E. Smith, G. Dent, *Modern Raman Spectroscopy. A Practical Approach* (Chichester, Wiley, 2005)
7. B.M. Smirnov, *Reference Data on Atomic Physics and Atomic Processes* (Springer, Heidelberg, 2008)
8. L.D. Landau, E.M. Lifshitz, *Electrodynamics of Continuous Media* (Pergamon Press, Oxford, 1984)
9. E.P. Wigner, F. Seitz, Phys. Rev. **46**, 509 (1934)
10. E.P. Wigner, F. Seitz, Phys. Rev. **46**, 1002 (1934)
11. C. Bréchignac, P. Cahuzac, F. Carlier, J. Leygnier. Chem. Phys. Lett. **164**, 433 (1989)
12. C. Bréchignac, P. Cahuzac, N. Kebaili, J. Leygnier, A. Sarfati, Phys. Rev. Lett. **68**, 3916 (1992)

13. C. Bréchignac, P. Cahuzac, J. Leygnier, A. Sarfati, Phys. Rev. Lett. **70**, 2036 (1993)
14. J. Tiggesburker, L. Köller, H.O. Lutz, K.-H. Meiwes-Broer, Chem. Phys. Lett. **190**, 42 (1992)
15. B.M. Smirnov, *Clusters and Small Particles Processes in Gases and Plasmas* (Springer, New York, 1999)
16. B.M. Smirnov, *Cluster Processes in Gases and Plasmas* (Wiley, New York, 2010)
17. H.R. Byers, *Elements of Cloud Physics* (University of Chicago Press, Chicago, 1965)
18. N.H. Fletcher, *The Physics of Rainclouds* (Cambridge University Press, London, 1969)
19. S. Twomey, *Atmospheric Aerosols* (Elsevier, Amsterdam, 1977)
20. H. Proppacher, J. Klett, *Microphysics of Clouds and Precipitation* (Reidel, London, 1978)
21. M.L. Salby, *Fundamentals of Atmospheric Physics* (Academic Press, San Diego, 1996)
22. B.J. Mason, *The Physics of Clouds* (Oxford Univesity Press, Oxford, 2010)
23. L.S. Ivlev, *Chemical Composition and Structure of Atmosphere Aerosols* (Izd. Leningrad State University, Leningrad, 1982). (in Russian)
24. I.V. Petryanov-Sokolov, A.G. Sutugin, *Aerosols* (Moscow, Nauka, 1989). (in Russian)
25. O. Boucher, *Atmospheric Aerosols. Properties and Climate Impacts* (Springer, Dordrecht, 2015)
26. H.R. Grim, *Plasma Spectroscopy* (Cambridge University Press, New York, 1964)
27. R.G. Flleagle, J.A. Businger, *An Introduction to Atmospheric Physics* (Academic Press, San Diego, 1980)
28. K. Friedlander, *Smoke, Dust, and Haze. Fundamentals of Aerosol Dynamics.* (Oxford University Press, Oxford, 2000)
29. B.M. Smirnov, *Fundamental of Ionized Gases. Basic Topics of Ionized Gases* (Wiley, Weinheim, 2012)
30. J. Aitken, Nature **23**(583), 195–197 (1880)
31. J. Aitken, Nature **23**(588), 311–312 (1881)
32. J. Aitken, Nature **23**(591), 384–385 (1881)
33. J. Aitken, Trans. Roy. Soc. Edinburgh **35**(1), 1–19 (1888)
34. S.H. Harris, in *Encyclopedia of Physics*, ed. by R.G. Lerner, G.L. Trigg (VCH Publisher, New York, 1990, p. 30; Weinheim, Wiley, 2005, p. 61)
35. I.A. Shiklomanov, in *Water in Crisis: A Guide to the World's Fresh Water Resources*, ed. by P.H. Gleick (Oxford University Press, Oxford, 1993), pp. 13–24
36. I.A. Shiklomanov, J.C. Rodda (eds.), *World Water Resources at the Beginning of the Twenty-First Century* (Cambridge University Press, Cambridge, 2003)
37. R.W. Healy, T.C. Winter, J.W. Labaugh, O.L. Franke, *Water Budgets: Foundations for Effective Water-resources and Environmental Management* (U.S. Geological Survey Circular 1308, Reston, Virginia, 2007)
38. https://en.wikipedia.org/wiki/Atmosphere-of-Earth
39. S. Bashkin, J. Stoner, *Atomic Energy Levels and Grotrian Diagrams*, vol. 1–4 (Amsterdam, North Holland, 1975–1982)
40. J.P. Peixoto, A.H. Oort, *Physics of Climate* (American Institute of Physics, Washington, 1992)
41. K.E. Trenberth, L. Smith, T. Qian et al. J. Hydrometeorol. **8**, 758 (2007)
42. https://en.wikipedia.org/wiki/water-circle
43. https://water.usgs.gov/edu/watercyrcleatmosphere.html
44. https://en.wikipedia.org/wiki/Properties-of-water
45. *Handbook of Chemistry and Physics*, 86 edn, ed. D.R. Lide (London, CRC Press, 2003–2004)
46. P.C. Reist, *Introduction to Aerosol Science* (Macmillan Publishing Company, New York, 1984)
47. W.C. Hinds, *Aerosol Technology: Properties, Behavior and Measurement of Airborne Particles* (Wiley, New York, 1999)
48. R.F. Alder et al. J. Hydrometeor**4**, 147 (2003)
49. G.J. Huffman, R.F. Alser, D.T. Bolvin, G. Gu, Geoph. Res. Lett. **36**, L17808 (2009)
50. https://en.wikipedia.org/wiki/Electromagnetic-absorption-by-water
51. C.M.R. Platt, Quart. J. Roy. Meteorolog. Soc. **102**, 553 (2006)
52. J. Warner, Tellus **7**, 450 (1955)
53. B.J. Mason, *The Physics of Clouds* (Claredon Press, Oxford, 1971)

54. W.R. Leaitch, G.A. Isaak, Atmosp. Environ. **25**, 601 (1991)
55. https://en.wikipedia.org/wiki/Liquid-water-content
56. H. Svensmark, E. Frils-Christensen, J. Atmos. Terr. Phys. **59**, 1225 (1997)
57. H. Svensmark et al. Proc. Roy. Soc. **A463**, 385 (2007)
58. H. Svensmark, T. Bondo, J. Svensmark. Geophys. Res. Lett. **36**, L151001 (2009)
59. H. Svensmark, M.B. Enghoff, J.O.P. Pedersen, Phys. Lett. A **377**, 2343 (2013)

Chapter 7
Greenhouse Effect in Atmospheres of Earth and Venus

Abstract The greenhouse phenomenon in the atmosphere that results from emission of its molecules and particles in the infrared spectrum range is determined by atmospheric water in the form of molecules and microdrops and by carbon dioxide molecules for the Earth atmosphere and by carbon dioxide molecules and dust for the Venus atmosphere. The line-by-line method used the frequency dependent radiative temperature for atmospheric air with a large optical thickness in the infrared spectral range, allows one to separate emission of various components in atmospheric emission. This method demonstrates that the removal of carbon dioxide from the Earth's atmosphere leads to a decrease of the average temperature of the Earth's surface by 4 K; however, doubling of the carbon dioxide amount causes an increase of the Earth's temperature by 0.4 K from the total 2 K at CO_2 doubling in the real atmosphere, as it follows from the NASA measurements. The contribution to this temperature change due to injections of carbon dioxide in the atmosphere due to combustion of fossil fuel, and it is 0.02 K. The infrared radiative flux to the Venus surface due to CO_2 is about 30% of the total flux, and the other part is determined by a dust.

7.1 General Principles of Atmospheric Greenhouse Effect

7.1.1 Nature of Atmospheric Greenhouse Effect

The nature of the greenhouse effect was understood almost two century ago [1, 2]; it is determined by a property of the matter over a surface. In a usual greenhouse, the Earth surface is heated as a result of absorption of solar radiation and it is cooled by emission of infrared radiation. This balance establishes the surface temperature. If place a partition over the surface which is transparent for visible radiation and returns to the surface partially infrared radiation, its heat balance changes and the surface temperature increases. In the Earth's atmosphere the role of this partition plays its atmosphere that must be transparent for visible solar radiation and it is not transparent for infrared radiation. As a result, the surface temperature increases compared with the case if the atmosphere is absent.

V. Krainov and B. M. Smirnov, *Atomic and Molecular Radiative Processes*, Springer Series on Atomic, Optical, and Plasma Physics 108,
https://doi.org/10.1007/978-3-030-21955-0_7

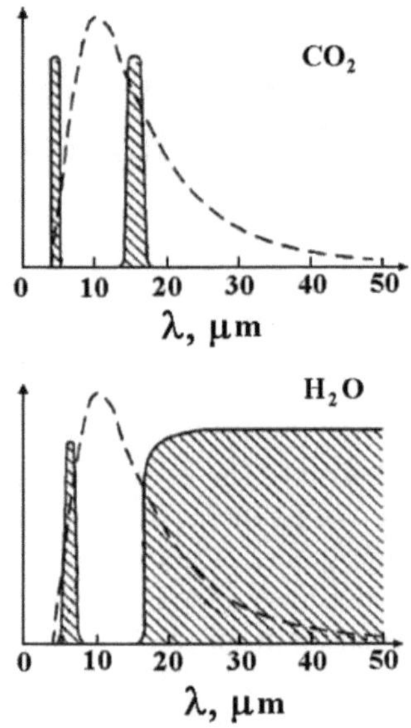

Fig. 7.1 Spectrum of emission of carbons dioxide molecules and water molecules located in atmospheric air and spectrum of equilibrium radiation at the temperature of 288 K [8, 9]

Infrared emission of the Earth's atmosphere is determined by vibration-rotation transitions of some atmospheric molecules, mostly H_2O and CO_2, and also by radiation of aerosols and atmospheric particles [3]. As it was indicated above, Fourier [1, 2] describes the nature of the greenhouse effect in the Earth's atmosphere. The next step in understanding the greenhouse effect of the Earth atmosphere was made by Tyndall [4–6] who shown on the basis of his experiments with molecular gases [7] that greenhouse atmospheric properties are determined by molecular gases, such as a water vapor, carbon dioxide and methane.

It should be noted the peculiarity of the greenhouse effect which is demonstrated by Fig. 7.1 [8, 9]. This figure exhibits emission of atmospheric molecules of carbon dioxide or water in a restricted spectrum range, and radiation of water molecules is more important than that due to CO_2 molecules for following reasons. First, water molecules have a dipole moment and hence they are more optically active objects. In atmospheric radiation it results in radiative rotation transitions without the change of a vibration molecular state. Rotation transitions determine a long-wave spectrum range. Second, the number density of water molecules in the atmosphere is higher than that for carbon dioxide molecules. In particular, the average number density of water molecules near the Earth's surface is $4 \cdot 10^{17}\,\mathrm{cm}^{-3}$ compared to $1 \cdot 10^{16}\,\mathrm{cm}^{-3}$ which we used above for the mean number density of atmospheric CO_2 molecules. This shows an important role of aerosols in atmospheric emission.

Because the greenhouse phenomenon is connected with the atmosphere energetics involving solar radiation in the visible spectrum range as well as infrared radiation of the atmosphere, it is of importance for the energetic balance of the Earth and its atmosphere. These data represented in Fig. 7.2 are taken from books [10–13] of one of authors which in turn are based on the NASA data, in particular, on [14]. These

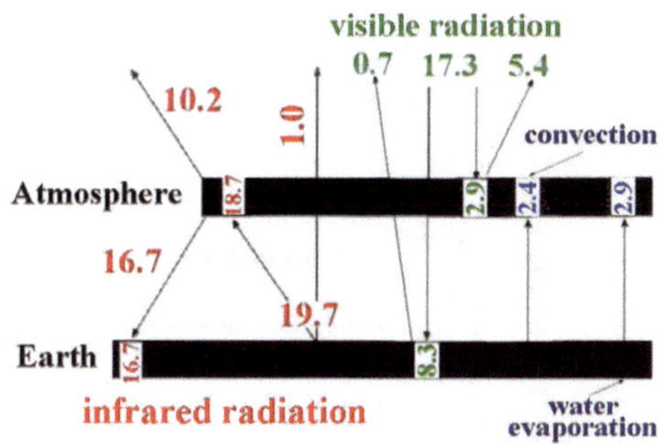

Fig. 7.2 Expressed in 10^{16} W the powers of indicated processes which lead to obtaining or loss of the energy by the Earth as a whole as well as by its atmosphere. Absorbed powers are given inside corresponding rectangulars, consumed powers are indicated near arrows

data are in accordance with those contained in books [15–20] and also with those of many papers, in particular, [21–26]. Evidently, the reason of this coincidence within the limits of a few percent for basic channels is that the data are taken from the same source—the data of NASA [14]. In addition, because these data relate to various times within one half century, one can conclude that the rates of these energetic processes have the natural character and vary weakly during this time range.

In evaluation some atmospheric parameters it is convenient to use the average energy fluxes. In consideration radiative processes in the atmosphere, it is convenient to operate with average radiative fluxes which result from dividing the total power of Fig. 7.2 to the area $S = 5.1 \cdot 10^{14}\,\mathrm{m}^2$, though the accuracy of these data is not enough. The average radiative flux of solar radiation in the visible spectrum range is $340\,\mathrm{W/m^2}$, the average radiative flux of from the Earth's surface in the infrared spectrum range is $386\,\mathrm{W/m^2}$, the average radiative fluxes of from the atmosphere in the infrared spectrum range are 327 and $200\,\mathrm{W/m^2}$ towards the Earth and outside correspondingly. Additionally, the average radiative flux from the Earth's surface that is not absorbed by the atmosphere and goes outside is $20\,\mathrm{W/m^2}$. Note also that the average energy flux from the Earth to its atmosphere in accordance with data of Fig. 7.2 is equal $57\,\mathrm{W/m^2}$ due to water evaporation from the earth's surface and $47\,\mathrm{W/m^2}$ due to convection.

From the energetic balance of the Earth and its atmosphere as a whole given in Fig. 7.2 it follows that the basis of this balance is solar radiation which is penetrated in the Earth's atmosphere and it is converted partially in infrared radiation through emission and its absorption by the Earth and atmosphere. Note that the Earth emits infrared radiation almost as a black body. Since the grey coefficient of various objects on the Earth's surface in the IR spectral range is close to one [27], the Earth's surface may be considered as a black body for emission, that leads to the average Earth's temperature to be $T = 287\,\mathrm{K}$, whereas the standard atmosphere model [28] gives for the global Earth's temperature $T = 288\,\mathrm{K}$. We note also that the atmosphere removal leads to the global temperature $T = 278\,\mathrm{K}$, if the Earth absorbs solar radiation completely and emits as a black body.

Let us represent the temperature T at a given point of the globe surface as $T = T_g + \delta T$, where T_g is the global temperature. Assuming that the Earth's surface emits

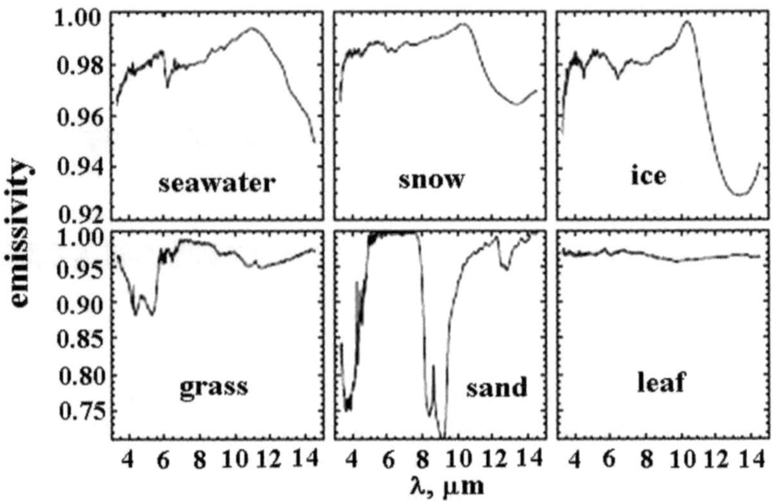

Fig. 7.3 Emissivity of various surfaces via the photon wavelength [27]

as a black body in the IR spectral range, one can obtain for the radiation power P_s of the Earth's surfa0 parameter $\delta T / T_g$

$$P_s = \sigma T_g^4 S \left[1 + 6 \left(\frac{\delta T}{T_g} \right)^2 \right], \tag{7.1.1}$$

where $S = 5.1 \cdot 10^{14}\,\text{m}^2$ is the area of the Earth's surface. Because a typical values the temperature fluctuation δT are several Kelvin, the second term of formula (7.1.1) is small. Therefore, one can use the average surface temperature for its emission.

One can justify the black body model for the Earth's surface by data of Fig. 7.3, where the grey coefficient or emissivity are given in the infrared spectrum range for objects which can determine the Earth's emission. The grey coefficient is the ratio of the photon flux from the surface of a given object with a certain temperature to that of the black body surface with this temperature. This allows one to consider the Earth's surface as a black body in evaluation of its emission, and the accuracy of this replace is about of several percents.

We now step aside from radiative processes and consider atmospheric phenomena which are important for the greenhouse effect in the atmosphere. Because this effect relates to the entire Earth, in its analysis it is convenient to deal with parameters averaged over the globe and time including a time of day and season. This leads to the model of standard atmosphere [28] which is characterized by average parameters and relates roughly to the USA atmosphere. Though atmospheric air consists of nitrogen (79%) and oxygen (20%), we below assume air molecules to be identical with the molecular weight $m = 29\,\text{a.u.m.}$ The dependence on the altitude h for the

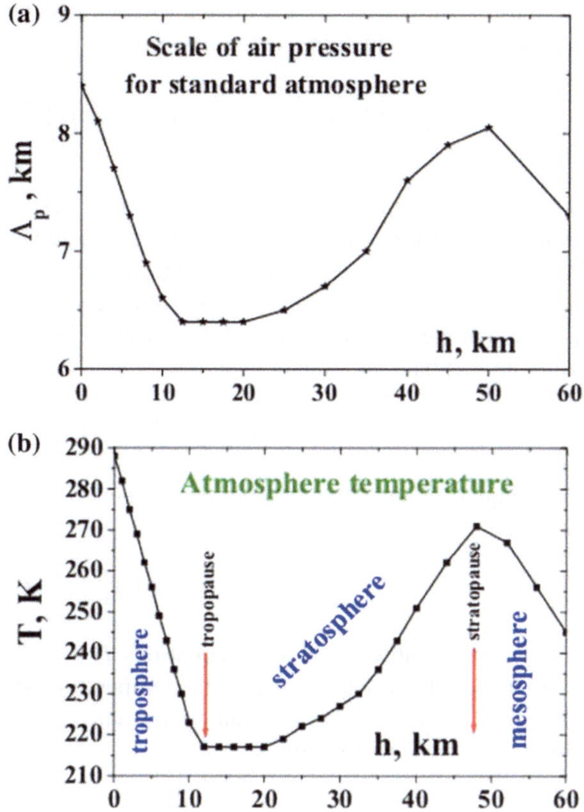

Fig. 7.4 Scaling parameter of formula (7.1.2) (**a**) and the atmosphere temperature (**b**) as an altitude function for standard atmosphere

number density $N_a(h)$ of air molecules, as well as other atmospheric molecules which are mixed with air ones, is given by formula

$$N(h) = N(0) \exp\left(-\frac{h}{\Lambda}\right) \tag{7.1.2}$$

The scaling parameter Λ follows from approximation the data of standard atmosphere by this formula, is represented in Fig. 7.4a. Figure 7.4b contains the altitude dependence of the atmosphere temperature and notations for atmospheric lower layers. Being guided by the greenhouse effect, we are restricted by the troposphere and lower stratosphere only, where the temperature gradient in the troposphere is close to $dT/dh = -6.5$ K/km, as it follows from Fig. 7.4b.

In order to understand the character of atmospheric emission, one can introduce the effective temperature of an uniform atmosphere for each frequency. In other words, in the framework of the line-by-line model, we introduce the radiative temperature T_ω which determines the emission flux at each frequency of atmospheric CO_2 molecules in the total emission of the Earth's atmosphere, we first use the radiative flux towards the Earth's surface as the parameter which characterizes atmospheric radiators in accordance with equation

$$J_{\downarrow} = \int_0^{\infty} \frac{\hbar\omega^3 d\omega}{4\pi^2 c^2}\left[\exp\left(\frac{\hbar\omega}{T_\omega}\right) - 1\right]^{-1} = 327\,\mathrm{W/m^2}, \; J_{\uparrow} = \int_0^{\infty} \frac{\hbar\omega^3 d\omega}{4\pi^2 c^2}\left[\exp\left(\frac{\hbar\omega}{T_\omega}\right) - 1\right]^{-1} = 200\,\mathrm{W/m^2}, \tag{7.1.3}$$

where $J_{\downarrow}$, $J_{\uparrow}$ are the radiative fluxes toward the Earth and outside it, and the radiative temperatures for radiation directed to the Earth and outgoing radiation are different. We below consider various models on the basis of formula (7.1.3). According to the simplest model, the atmospheric absorption coefficient is independent of frequency ω, and then the radiative temperature towards the Earth $T_{\downarrow}$ and outside it $T_{\uparrow}$ follow from the Stephan-Boltzmann equation

$$J_{\downarrow} = \sigma T_{\downarrow}^4, \; J_{\uparrow} = \sigma T_{\uparrow}^4,$$

These equations give for temperatures of layers which are responsible for atmospheric emission [19, 29]

$$T_{\downarrow} = 276\,\mathrm{K}, \; T_{\uparrow} = 244\,\mathrm{K} \tag{7.1.4}$$

These temperatures follow from equations $T_{\uparrow} = T([h_{\uparrow}])$ and radiation towards the Earth $T_{\downarrow} = T[h_{\downarrow}]$, where $T(h)$ is the temperature at an altitude h. One can use the following connection between an appropriate of the atmospheric temperature and altitude for these temperatures

$$T_{\downarrow} = T_E - h_{\downarrow}\frac{dT}{dh}, \; T_{\uparrow} = T_E - h_{\uparrow}\frac{dT}{dh}, \tag{7.1.5}$$

where within the framework the standard atmosphere model the average temperature gradient is $dT/dh = 6.5\,\mathrm{K/km}$, and $T_E = 288\,\mathrm{K}$ is the temperature of the Earth's surface. From formulas (7.1.4) and (7.1.5) it follows for these altitudes

$$h_{\downarrow} = 1.9\,\mathrm{km}, \; h_{\uparrow} = 6.8\,\mathrm{km} \tag{7.1.6}$$

It is of importance the local thermodynamic equilibrium for vibrationally excited molecules which accompanies the atmosphere greenhouse effect and requires large radiative times compared with collision times involving excited molecules in atmospheric air. Therefore radiative processes do not violate the Boltzmann distribution over excited vibrational and rotational states; the emission of this gas is determined by excited molecules which result from collisions of nonexcited molecules with gaseous ones, rather than from absorption of the radiation. From this consideration it follows that radiative transport does not give a contribution to the emission of this gas.

Indeed, the emission of photons created by vibrationally excited molecules does not violate this distribution if the rate of excitation of an excited state exceeds significantly the rate of radiative decay of this state. In particular, for the basic radiative transition $01^00 \rightarrow 00^00$ we have the rate constant of destruction of the upper state $k_{\rm rel} = 5 \cdot 10^{-16}\,{\rm cm}^3/{\rm s}$ at room temperature [70]. Then we obtain the condition for the number density of air molecules N_a of the Boltzmann distribution

$$N_a \gg \frac{1}{k_{\rm rel}\tau_r} \tag{7.1.7}$$

Taking the radiative lifetime $\tau_r = 0.33\,{\rm s}$ according to Fig. 4.3 data, we obtain the criterion (7.1.7) as $N_a \gg 3 \cdot 10^{15}\,{\rm cm}^{-3}$. This criterion is violated only outside the stratosphere.

We thus are based on the condition of local thermodynamic equilibrium in atmospheric air, and that the optical thickness of the atmosphere is large. Under these conditions, we reduce the atmosphere with am alternative temperature to that with a constant temperature. This is possible at a small temperature gradient that is a basis of expansion over a small parameter. Such an expansion allows us to find the radiative temperature T_ω that characterizes the radiative flux at this frequency.

7.1.2 Global Properties of the Earth's Atmosphere

We below consider the atmospheric greenhouse effect in detail. The greenhouse effect is created by three atmospheric components, namely, by water vapor, carbon dioxide and water aerosols; though other greenhouse components gives a small contribution to this effect, their influence on the greenhouse effect will be considered also. Note that molecular greenhouse components, molecules of water and carbon dioxide located in atmospheric air, create infrared radiation as a result of vibrational-rotational or rotational transitions of molecules. Rates of such radiative transitions are relatively low, and hence they do not influence on thermodynamic equilibrium in air between vibrational and rotational states of greenhouse molecules. Hence, radiation emitted by the atmosphere is characterized by the air temperature.

Let us introduce the global Earth's temperature as the temperature of the Earth's surface averaged over all the globe surface and time. The global temperature is a parameter of the Earth's energetic balance. Variation of the global temperature in time characterizes the climate change, and our task is to determine the rate of change of the global temperature dT/dt, and the range of temporal variation which is measured in years. The problem is that chaotic variations of the global temperature, as well as the temperature at a given geographical point, in the course of a day and season reach tens degrees, so that the global temperatures averaged over year are of the order of 1 K. But variation of the global temperature for a century does not exceed 1 K.

The method [30] for variation of global temperature allows one to overcome the indicated problem and to decrease the fluctuations in determination of the global temperature by one order of magnitude. Within this method we compare temperatures at the same geographical point and time of day and season and construct the difference

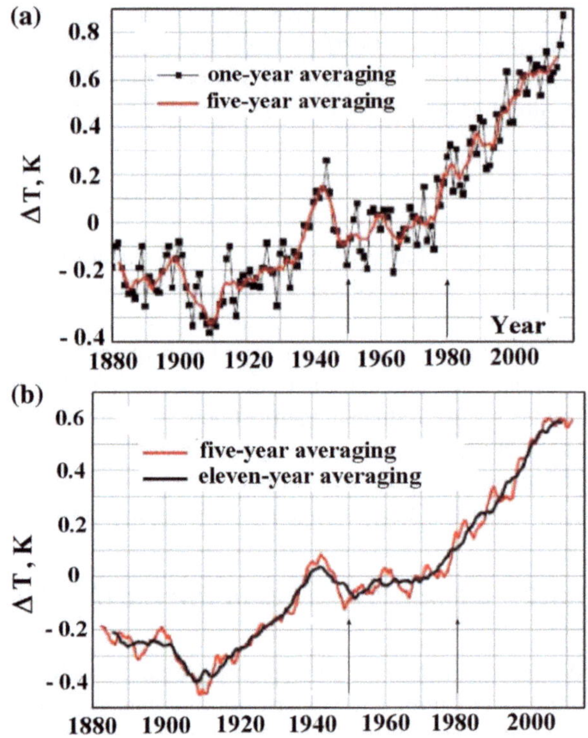

Fig. 7.5 Evolution of the global temperature with averaging for one and fife years [31] (**a**), and for five and fifteen years [32, 33] (**b**). Arrows indicate years when the global temperature does not vary in average and is taken as a basic one

of these temperatures ΔT at different years. Next this temperature difference is averaged over the Earth's surface and also over say and season time during the year. In this manner one can determine variation ΔT of the global temperature in time and the rate $\mathrm{d}\Delta T/\mathrm{d}t$ of its variation. Fluctuations of variation of the global temperature in this method are estimated as 0.1 K.

Realization of this method requires large information and a labor-intensive work. Nevertheless, this information follows from data of meteorological stations. A number of such meteorological stations was about 6 thousands in the second half of 19th century and now their number decreases in three times only, because the basic information follows from satellite measurements. Treatment of this information is made within the framework of NASA programs [Gottard Institute for Space Studies—GISS], and results of this treatment are given in [31–36]. Comparison of the global temperature change in summer and winter, as well as in daytime and night time or in North and South hemispheres gives that indicated values of ΔT during a year does not exceed 0.2 K [32, 33].

Figure 7.5 give evolution of the global temperature with averaging over year, five years and fifteen years. One can see that the larger time of averaging, the smother this dependence. From Fig. 7.5 it follows that year fluctuations of the global temperature are of the order of 0.1 K. One can see a non-monotonic evolution of the global temperature in time. Indeed, during 1880–1910 a weak cooling was observed that

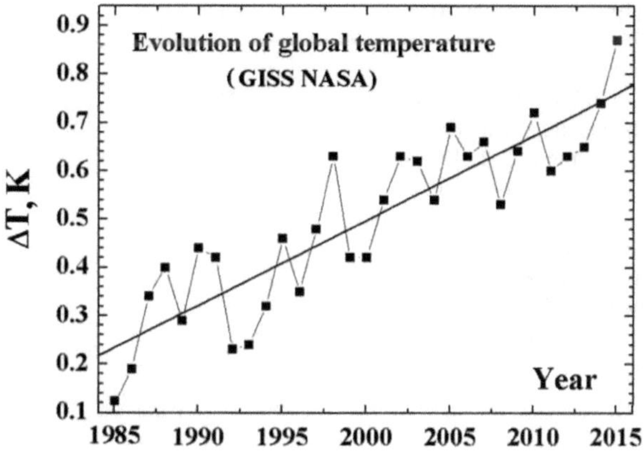

Fig. 7.6 Evolution of global temperature in last years [32]. The temperature variation is counted from the average global temperature in 1950–1980

was changed by a weak heating in 1910–1940 which was continued by cooling during 1940–1950. Next, during 1950–1980 the global temperature did not vary within the limit of its accuracy, and the Earth heating takes place after 1980. Basing on these data, we give in Fig. 7.6 evolution of the global temperature after 1985 when the global temperature was increased monotonically [37]. Approximation of these data by the linear time dependence gives for evolution of average change of the global temperature

$$\frac{\mathrm{d}\Delta T}{\mathrm{d}t} = 0.018\,\mathrm{K/yr}, \qquad (7.1.8)$$

and the fluctuation (standard deviation) for data of Fig. 7.6 is equal $\Delta = 0.09\,\mathrm{K}$. The latter means that a smooth (linear) time dependence for the global temperature change ΔT is valid at time intervals above 5 years.

One of a greenhouse atmospheric component is carbon dioxide. The equilibrium of atmospheric carbon dioxide results from its conversion in a hard carbon in the photosynthesis processes, and the inverse process by processes of putrefaction and breathing of plants. As it follows from the carbon balance in the atmosphere [38–41], the rate of each process is approximately $2 \cdot 10^9$ ton/yr, while the total carbon mass in atmospheric carbon dioxide is approximately $2 \cdot 10^9$ ton/yr. This means that the average residence time of an atmospheric CO_2 molecule is 4 years. In addition, the total carbon mass in fossil fuels, i.e. in coal, oil and methane, is $1 \cdot 10^8$ ton/yr. This means that only 5% of the rate of carbon dioxide injections in the atmosphere corresponds to combustion of fossil fuels, i.e. the latter is not important for the balance of atmospheric carbon dioxide.

Detailed information about atmospheric carbon dioxide follows from monitoring of atmospheric CO_2 that is made from 1959 in the Mauna Loa observatory (Hawaii, USA) [43–46]. This observatory is located at altitude 3400 m above the sea level $<$ that is far from sources or absorbers of carbon dioxide. Some results of this monitoring are given in Fig. 7.7. As is seen, the concentration of carbon dioxide

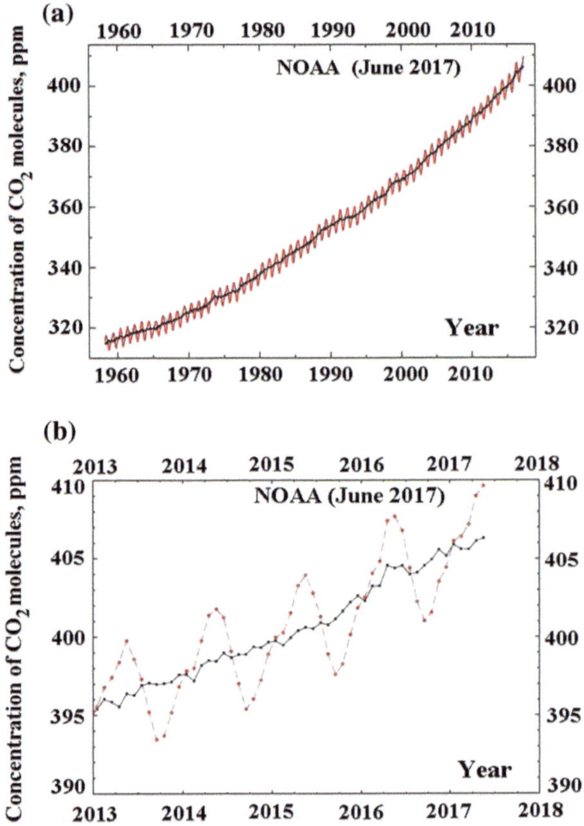

Fig. 7.7 Concentration of CO_2 molecules in atmospheric air during the last half century (**a**) and for the last five years (**b**) according to [42, 43]; open circles corresponds to an average during a month, and filled squares relate to averaged data for year (one half year before and after an indicated data)

molecules in atmospheric air increases from 316 ppm in 1959 up to 409 ppm in 2017. In addition, the rate of an increase of the carbon dioxide concentration grows in time from 0.7 ppm/yr in 1959 up to approximately 2.1 ppm in 2017. The reason of season oscillations of the CO_2 concentration is explained that the photosynthesis process is stronger in the northern hemisphere, where it proceeds mostly in the period from May to September, and the photosynthesis is absent practically in the period from October to March. The data of Fig. 7.7a give the contemporary rate of variation of CO_2 concentration as

$$\frac{d \ln c}{dt} = 0.006\,\mathrm{yr}^{-1}, \tag{7.1.9}$$

with an accuracy approximately 10%.

Within the framework of the Arrenius concept [47], one can introduce the equilibrium climate sensitivity (ECS) [48] as a change of the global Earth temperature at the doubling of the atmospheric carbon dioxide concentration. This value is given by

$$\mathrm{ECS} = \Delta T \frac{\ln 2}{\ln(c_2/c_1)}, \tag{7.1.10}$$

where c_1 and c_2 are concentration of CO_2 molecules at the initial and final time, and ΔT is the global temperature variation for this time interval. This value may be determine twofold. In the first case it follows from formulas (7.1.8) and (7.1.9). In the second case we take the concentration of CO_2 molecules to be $c_1 = 280\,\mathrm{ppm}$ in a pre-industrial period, and $c_2 = 410\,\mathrm{ppm}$ at the present time, and the temperature change is $\Delta = (0.8 \pm 0.1)\,\mathrm{K}$ [49] for this time. As a result, we obtain

$$\mathrm{ECS} = (2.0 \pm 0.3)\,^{\circ}\mathrm{C} \tag{7.1.11}$$

Note that this value follow from NASA measurements, rather than some evaluations.

The mean amount of atmospheric water is $1.3 \cdot 10^{19}\,\mathrm{g}$ [50–53] that corresponds to the mean water density in the atmosphere as $3\,\mathrm{g/m^3}$. One can compare the water mass in the atmosphere with the atmosphere one $5.1 \cdot 10^{21}\,\mathrm{g}$ which relates to nitrogen and oxygen. This corresponds to the average concentration of water molecules in atmospheric air as approximately 0.4%, whereas near the Earth's surface the average concentration of water molecules in air is equal 1.7%. The total rate of water evaporation from the Earth's surface is $3.9 \cdot 10^{20}\,\mathrm{g/yr}$ [54–57] (only $1.0 \cdot 10^{18}\,\mathrm{g}$ in the form of snow), and the same rate relates to water returning to the Earth's surface. As it follows from this consideration, the power of the evaporation process is equal $2.4 \cdot 10^{16}\,\mathrm{W}$ (see Fig. 7.2) which is returned as a result of water condensation in the atmosphere. In this manner, the above power is transferred from the Earth to atmosphere.

From the amount of atmospheric water and the rate of its formation it follows that an average time of residence of water molecules in the atmosphere is approximately 9 days [57]. In the water balance between the land, ocean and atmosphere, precipitation of atmospheric water on the Earth's surface and uniform distribution over it gives a layer of liquid water of a thickness 2.5 cm [58]. The height of a precipitated layer may be used as a unit for amount of a water vapor in the atmosphere [59–63]. As it follows from data [59, 60], approximately 80% of atmospheric water is located at altitudes below 3 km. According to data [61–63], the rate of water precipitation on the Earth's surface is higher for oceans, whereas season variations of this rate are more for land.

Atmospheric water contains a small part of Earth's water with the mass of $1.4 \cdot 10^{24}\,\mathrm{g}$. If this water would be distributed over the Earth's surface uniformly, the layer thickness will be 2.7 km. From this it follows that the most part of this water is located underground. Note that 96% of Earth's water is salty. In addition, open water located on the Earth surface is a source of atmospheric water and is found in equilibrium with it.

Near the Earth's surface the average partial pressure of water is about 2 Torr, whereas the saturated water pressure is 4.7 Torr at the temperature [64, 65]. Therefore atmospheric water is in the form of a vapor mostly, and a small part of water exist in the form of aerosols, at least, at altitudes below 3 km. Because of the condensation process, the number density of water molecules decreases with an altitude h sharper than that in the case of air molecules. By analogy with formula (7.1.2), the dependence of the number density of water molecules $N(H_2O)$ on the altitude h may be approximated by

$$N(H_2O) = N_o \exp(-h/\lambda), \tag{7.1.12}$$

and the parameters of formula (7.2.27) are equal according to measurements [61–63] of the global atmospheric moisture at various altitudes

$$N_o = (4.2 \pm 0.2) \cdot 10^{17}\,\mathrm{cm}^{-3}, \ \ \lambda = (2.0 \pm 0.2)\,\mathrm{km}, \tag{7.1.13}$$

Here the data of 1950–1960 and 2000–2010 are used, so that the values of these parameters vary during this time inside the indicated accuracy. In reality, the total amount of atmospheric water grows in time with the rate 0.07 g/kg per decade (1 g of water per 1 kg of air) [66–68], that corresponds to variation of the concentration of atmospheric water c as

$$\frac{d \ln c}{dt} = 7 \cdot 10^{-3}\,\mathrm{yr}^{-1} \tag{7.1.14}$$

From the end of 19th century an increase of the air moisture near the Earth's surface is equal approximately 4% [69].

7.1.3 Models of Emission from Optically Dense Gaseous Layer

We below represent the models for emission of the atmosphere considering it in average as a gaseous layer above the Earth's surface; its parameters depend on the altitude only because a typical thickness of the atmospheric layer which is responsible for its emission is small compared to the Earth's radius. But it is necessary to take into account a nonuniform distribution of the atmospheric temperature and number density of radiating and excited molecules. In this analysis it is of importance the local thermodynamic equilibrium for vibrationally excited atmospheric molecules because radiative times are large compared with collision times involving excited molecules in atmospheric air. Therefore radiative processes do not violate the Boltzmann distribution over excited vibrational and rotational states. From this it follows that radiation does not influences on the distribution, and radiative transport involving reabsorption processes is negligible. This fact decreases a number of models [70] which may be used in the analysis of atmospheric emission in the infrared spectrum range.

We now use a simple model on the basis of the assumption that the atmospheric absorption coefficient is independent of the frequency, and also it is a monotonous function of an altitude [19, 29]. Under these conditions one can determine the average optical thickness of the atmosphere. Let us approximate the altitude dependence of the absorption coefficient as

$$\kappa_\omega \equiv \frac{du_\omega}{dh} = A \exp\left(-\frac{h}{\lambda}\right), \tag{7.1.15}$$

In order to determine the parameters of this formula, we use formula (2.2.37) which allows one to find the altitude of the layers which are responsible for emission toward the Earth $h_\downarrow$ and outside the atmosphere $h_\uparrow$. These altitudes satisfy to equations (2.2.31)

$$\int_{h_\uparrow}^{\infty} k_\omega dh = 2/3, \quad \int_{0}^{h_\downarrow} k_\omega dh = 2/3 \tag{7.1.16}$$

These equations allow one to determine the parameters of formula (7.1.15) [19, 29]

$$A = 0.35\,\mathrm{km}^{-1}, \ \lambda = 6.0\,\mathrm{km}, \ u = A\lambda = 2.1, \tag{7.1.17}$$

where u is the total optical thickness of the atmosphere. Thus, this model [19, 29] with the average absorption coefficient over the infrared spectrum range proves the validity of the model of a weakly varied optical thickness as an altitude function for optically thick layers under consideration.

Note that the used model for emission of a nonuniform gaseous layer is used a small parameter (2.2.37) which in the limit $\hbar\omega > T$ has the form

$$\alpha = \frac{5}{18}\left(\frac{\hbar\omega}{T^2} \cdot \frac{dT}{du}\right)^2 \tag{7.1.18}$$

Let us determine altitudes which are responsible for emission toward the Earth and outside the atmosphere, the values of this small parameter are equal $\alpha_\downarrow = 0.03$ and $\alpha_\uparrow = 1/4$.

We now extract one component from the absorbed gas and present the absorption coefficient of the entire gas K_ω in the form

$$K_\omega = \kappa + k_\omega, \tag{7.1.19}$$

where k_ω is the absorption coefficient of an extracted gaseous component and κ is that for other absorbed component. For example, if the standard atmosphere contains three main absorbed components: water molecules, water microdrops and carbon dioxide molecules, one can refer κ to water molecules and water microdrops, while k_ω is the absorption coefficient of CO_2 atmospheric molecules, and then one can find the contribution of CO_2 molecules to atmospheric absorption. In another example k_ω refers to trace gases, and κ is the absorption coefficient of three basic absorbed components of the atmosphere. Below we focus on the first example because it is of the practical interest.

Let us formulate the algorithm to determine the contribution of a certain component of atmospheric air to its emission under given conditions. For obviousness, we will consider below as this example the emission of atmospheric carbon dioxide toward the Earth. One can use two models for determination the radiative flux j_ω due to carbon dioxide, as well as the total radiative flux J_ω to the Earth's surface due to all atmospheric components. According to formulas (7.1.15) and (7.1.16) one can characterize radiation at a frequency ω by the radiative temperature T_ω that is given by

$$T_\omega = T(h_\omega), \; h_\omega = \frac{2}{3K_\omega} = \frac{2}{3(\kappa + k_\omega)} \tag{7.1.20}$$

This formula describes the emission towards the Earth and is based on the assumption $h_\omega \ll \lambda$. Next, the equality of the radiative temperature T_ω for a given frequency and the temperature of an atmospheric layer located at an effective altitude h_ω means local thermodynamic equilibrium for radiating molecules.

The radiative temperature T_ω allows one to determine the radiative flux toward the Earth J_ω at a given frequency ω according to the Planck formula (2.2.37) [71, 72]

$$J_\omega = \frac{\omega^3}{4\pi^2 c^3 \left\{\exp\left[\hbar\omega / T_\omega\right] - 1\right\}}, \tag{7.1.21}$$

and this gives the total radiative flux as the integral of the partial ones (7.1.21). Using the partial radiative temperature T_ω corresponds to the line-by-line model that means determination of partial radiative fluxes at each frequency and subsequent integration of these fluxes in the total radiative flux.

Another model, the absorption band model [19, 70, 73], is more rough and assumes that the emission takes place in a certain frequency range which is called the absorption band. This model holds true at high pressures where frequency oscillations of the absorption coefficient are absent, i.e. according to formula (4.2.16) $k_\omega = \chi_\omega$. In this case we are based on a sharp dependence $\chi(\omega)$, and one can introduce boundary frequencies ω_1 and ω_2 of the absorption band on the basis of the relations

$$k_{\omega_1} = k_{\omega_2} = \kappa \tag{7.1.22}$$

It is clear that this model does not work in a transient range of frequencies, whereas oscillations expand the transient range and hence decrease of the accuracy of this model. We below analyze the above models of atmosphere emission for main greenhouse

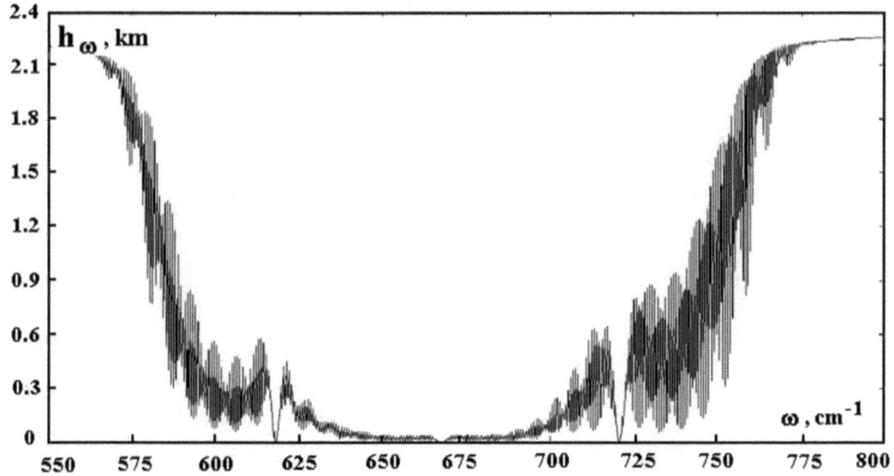

Fig. 7.8 Effective altitude h_ω which determines the radiative flux at a given frequency at the contemporary concentration of atmospheric carbon dioxide under the assumption that the absorption coefficient of atmospheric water is independent of the frequency [74]

components of the atmosphere, namely, carbon dioxide molecules, water molecules and water microdrops—aerosols.

7.2 Greenhouse Effect in Atmospheres

7.2.1 *Emission of Atmospheric CO_2 Molecules Towards the Earth*

We now evaluate the radiative flux toward the Earth due to atmospheric molecules of carbon dioxide. For this it is necessary to take into account vibrational transitions in accordance with formula (4.2.27) which create the radiative flux towards the Earth. We use formulas (4.2.17), (4.2.18), and (4.2.21) for the absorption coefficient due to each vibrational transition, and also formula (4.2.25) for the absorption coefficient of P and R branches of atmospheric CO_2 molecules, as well formulas (4.2.31) and (4.2.32) for Q-branch. All these evaluations relate for a high optical density of the atmosphere and the local thermodynamic equilibrium.

We first determine the radiative temperature T_ω for atmospheric emission in the frequency range where it is created by atmospheric CO_2 molecules. For the standard atmosphere model where the temperature gradient is $\mathrm{d}T/\mathrm{d}h = 6.5\,\mathrm{K/km}$, formulas (7.1.5) and (7.1.18) give for the radiative temperature T_ω

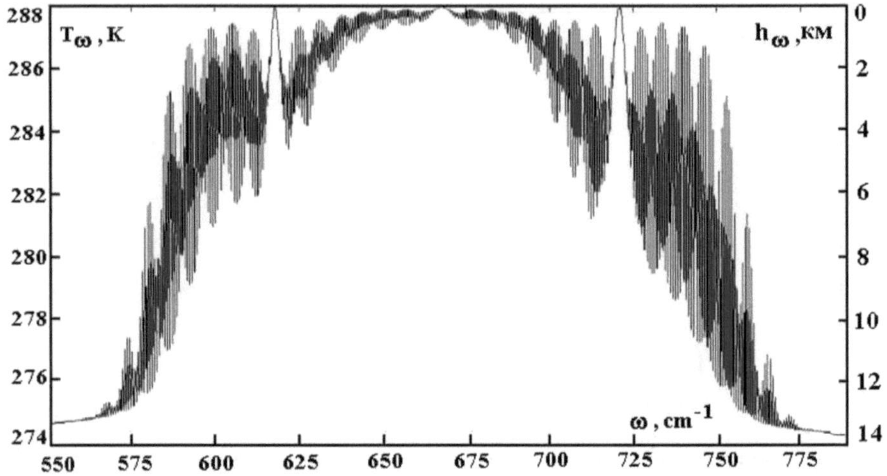

Fig. 7.9 Radiative temperature T_ω at the contemporary concentration of atmospheric carbon dioxide for the line-by-line method under assumption that the absorption coefficient of atmospheric water is independent of the frequency [74]

$$T_{\omega_1} = T_E - h_\omega \frac{dT}{dh} = T_E - \frac{\alpha}{\kappa + k_\omega}, \tag{7.2.1}$$

where $T_E = 288\,\mathrm{K}$ is the Earth's temperature, h_ω is the altitude which is responsible for radiation towards the Earth, $\alpha = 4.3\,\mathrm{K/km}$. Figure 7.8 contains the effective altitude h_ω which is responsible for radiation at a given frequency and is given by formula (7.1.20). Figure 7.9 gives the radiative temperature due to atmospheric carbon dioxide in accordance with formula (7.2.1). Roughly, this dependence may be approximated by a table-form that corresponds to the absorption band model. This means that inside the absorption band, the radiative temperature is equal to the Earth's temperature because of a high optical thickness of the atmosphere due to atmospheric CO_2 molecules. Outside the absorption band CO_2 molecules do not partake in origin of the radiative temperature, and within the framework of a constant temperature model the radiative temperature in this range is independent of the frequency.

Within the framework of the absorption model, we have on the basis of formula (7.1.22) for boundaries frequencies left ω_{1b} and right ω_{1b} from the absorption band at the contemporary amount of atmospheric carbon dioxide molecules

$$\omega_{1b} = 585\,\mathrm{cm}^{-1}, \ \omega_{2b} = 750\,\mathrm{cm}^{-1} \tag{7.2.2}$$

Within the framework of the absorption band model, this gives the radiation flux $J_\downarrow$ toward the Earth's surface due to atmospheric CO_2 molecules

$$J_{\downarrow} = \int_{\omega_{1b}}^{\omega_{2b}} j_{\omega} d\omega = \int_{\omega_{1b}}^{\omega_{2b}} \frac{\hbar\omega^3 d\omega}{4\pi^2 c^2} \left[\exp\left(\frac{\hbar\omega}{T_{\omega}}\right) - 1\right]^{-1}, \tag{7.2.3}$$

where T is the atmosphere temperature near the Earth's surface. Using parameters of this formula for the standard atmosphere model, we obtain the value of the radiative flux at the contemporary number density of atmospheric CO_2 molecules ($N(CO_2) = 1 \cdot 10^{16}\,\text{cm}^{-3}$) [74]

$$J_{\downarrow} = 67\,\text{W/m}^2 \tag{7.2.4}$$

One can fulfil the same operation at doubling number density of the amount of atmospheric carbon dioxide. Then instead of the values (7.2.2) we have the boundaries of the absorption band

$$\omega'_{1b} = 581\,\text{cm}^{-1}, \ \omega'_{2b} = 755\,\text{cm}^{-1}, \tag{7.2.5}$$

that corresponds to the following radiative flux toward the Earth's surface [74]

$$J'_{\downarrow} = 71\,\text{W/m}^2 \tag{7.2.6}$$

On the basis of the line-by-line model, the radiative flux toward the Earth is given by

$$J_{\downarrow} = \int_{\omega_{1b}}^{\omega_{2b}} j_{\omega} d\omega = \int_{\omega_{1b}}^{\omega_{2b}} \frac{\hbar\omega^3 d\omega}{4\pi^2 c^2} \left[\exp\left(\frac{\hbar\omega}{T_{\omega}}\right) - 1\right]^{-1} \tag{7.2.7}$$

The emission radiative flux $J_{\downarrow}$ toward the Earth's surface at the contemporary concentration of CO_2 molecules and that $J'_{\downarrow}$ at its doubled concentration are equal according to this formula [74]

$$J_{\downarrow} = 61\,\text{W/m}^2, \ J'_{\downarrow} = 68\,\text{W/m}^2 \tag{7.2.8}$$

As a result, we have for the average radiative flux toward the Earth under these conditions [74]

$$J_{\downarrow} = (64 \pm 3)\,\text{W/m}^2, \ J'_{\downarrow} = (70 \pm 2)\,\text{W/m}^2 \ \Delta J_{\downarrow}(CO_2) = (5 \pm 2)\,\text{W/m}^2, \tag{7.2.9}$$

where $\Delta J_{\downarrow}(CO_2)$ is the difference of radiative fluxes toward the Earth due to atmospheric CO_2 molecules as a result of doubling of the atmospheric concentration of CO_2 molecules.

Let us assume that inside the absorption bands for atmospheric molecules of carbon dioxide which boundaries are given by formulas (7.2.2) and (7.2.5) the radiative atmospheric temperature T_ω coincides with the global Earth's temperature $T = 288\,\mathrm{K}$. Then the atmospheric radiative flux is equal $J = 67\,\mathrm{W/m^2}$ at the contemporary concentration of carbon dioxide molecules and $J' = 71\,\mathrm{W/m^2}$ at the doubled concentration of carbon dioxide molecules. One can see that these values coincide within the accuracy limits with evaluations (7.2.9) on the basis of real radiative temperatures. The difference of these radiative fluxes coincides with the accepted flux difference $4\,\mathrm{W/m^2}$ [16, 75], but this is not the change of the radiative flux toward the Earth due to doubling of the amount of atmospheric carbon dioxide because of a different spectral width of the absorption band. We below determine the change of the radiative flux toward the Earth due to doubling of the amount of atmospheric carbon dioxide. Next, from this it follows that the contribution of emission of CO_2 molecules to the total radiative flux toward the Earth is approximately 20%.

We now improve the model of frequency independent absorption coefficient which allowed one to determine the altitude h_ω of an atmospheric layer which is responsible for emission toward the Earth's surface and the temperature of this layer $T_\downarrow$ according to formula (7.1.6) and (7.1.4) as $h_\downarrow = 1.9\,\mathrm{km}$, $T_\downarrow = 276\,\mathrm{K}$. We now account for a heightened radiative flux $J_\downarrow(CO_2)$ due to CO_2 molecules in their spectral range between ω_1 and ω_2. Then we have the following equation for the radiative flux

$$J_\downarrow = \int\limits_0^{\omega_1} \frac{\hbar\omega^3 \mathrm{d}\omega}{4\pi^2 c^2}\left[\exp\left(\frac{\hbar\omega}{T}\right) - 1\right]^{-1} + J_\downarrow(CO_2) + \int\limits_{\omega_2}^{\infty} \frac{\hbar\omega^3 \mathrm{d}\omega}{4\pi^2 c^2}\left[\exp\left(\frac{\hbar\omega}{T}\right) - 1\right]^{-1} = 327\,\mathrm{W/m^2}, \tag{7.2.10}$$

where T is the radiative temperature outside the absorption band by CO_2 molecules. As a result, we have for parameters of this range

$$T_\downarrow = 274\,\mathrm{K},\ h_\downarrow = 2.2\,\mathrm{km}\ \kappa = 0.30\,\mathrm{km}^{-1} \tag{7.2.11}$$

We now check the validity of the method which allowed one to reduce the atmosphere with a varied temperature to the one where the temperature is independent of the altitude. It is based on a smallness of the parameter α which is defined by formula (7.1.18). Outside the absorption band we have

$$\frac{\mathrm{d}T}{\mathrm{d}u} = \frac{\mathrm{d}T}{\kappa \mathrm{d}h} = 22\,\mathrm{K},$$

and formula (7.1.18) gives at boundaries (7.2.2) of the absorption band $\alpha(\omega_{1b}) = 0.016$, $\alpha(\omega_{2b}) = 0.026$ that confirms the validity of the method used.

Let us analyze the change of the radiative flux $\Delta J_\downarrow$ toward the earth's surface as a result of doubling of the atmospheric carbon dioxide amount in order to determine the ECS (equilibrium climate sensitivity) subsequently. This increase leads to an increase of the radiative temperature for radiation directed toward the Earth, and this change is represented in Fig. 7.10 [74]. This allows one to determine an increase $\Delta J_\downarrow$ of the radiative flux toward the Earth as a result of doubling of the concentration of atmospheric CO_2 molecules as

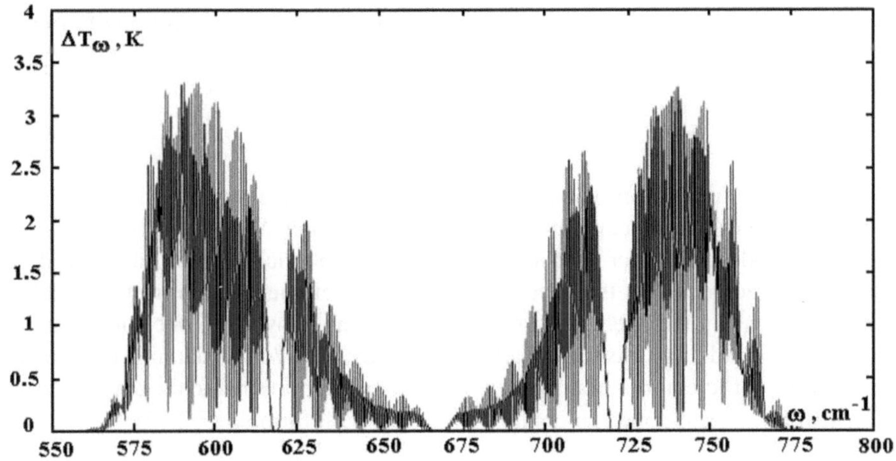

Fig. 7.10 Increase of the radiative temperature for the flux toward the Earth resulted from doubling of the concentration of carbon dioxide molecules within the framework of the line-by-line model [74]

$$\Delta J_{\downarrow} = \int_{\omega_1}^{\omega_2} \frac{\hbar\omega^3 d\omega}{4\pi^2 c^2} \left\{ \left[\exp\left(\frac{\hbar\omega}{T_\omega}\right) - 1 \right]^{-1} - \left[\exp\left(\frac{\hbar\omega}{T'_\omega}\right) - 1 \right]^{-1} \right\}, \quad (7.2.12)$$

where T_ω and T'_ω are the radiative temperatures at the contemporary and doubled concentrations of atmospheric carbon dioxide molecules respectively. Note that this difference determines an additional radiative flux toward the Earth due to doubling of the concentration of atmospheric CO_2 molecules, rather the value $\Delta J_{\downarrow}(CO_2)$ in formula (7.2.9).

It is clear that the used model in which the frequency is independent of the absorption coefficient, is rough because the absorption coefficient as a frequency function oscillates strongly in reality. Therefore we take it in a more general form

$$\kappa = \kappa_o + a \cos[b(\omega - \omega_o)], ; \quad (7.2.13)$$

where $a < \kappa_o$, $b \sim 1\,\text{cm}^{-1}$, and $\kappa_o = 0.3\,\text{km}^{-1}$, i.e. the average absorption coefficient in the range outside of the absorption band due to CO_2 molecules is κ_o and (7.2.10) is fulfilled. Then formula (7.2.12) gives after averaging over the parameter b.

$$\Delta J_{\downarrow} = (1.0 \pm 0.2)\,\text{W/m}^2 \quad (7.2.14)$$

As is seen, this difference $\Delta J_{\downarrow} = J'_{\downarrow} - J_{\downarrow}$ is several times less than the that $\Delta J_{\downarrow}(CO_2)$ defined by formula (7.2.9) as the difference of radiative fluxes toward the Earth which are created by CO_2 molecules. This means that increase of the flux due to CO_2 molecules is accompanied by a more strong absorption of emitted radiation. It is convenient to represent formula (7.2.14) in the form

$$\frac{\Delta J_{\downarrow}(CO_2)}{d \ln c} = (1.4 \pm 0.3)\,\mathrm{W/m^2} \tag{7.2.15}$$

In the same manner, one can determine the change of the radiative flux as a result of the change of the concentration of atmospheric water. Within the framework of the above model, the water contribution to the atmospheric radiative flux toward the Earth is characterized by the average absorption coefficient κ in formula (7.1.19). Indeed, increasing the average absorption coefficient κ by 10%, one can obtain an increase in the radiative flux to the Earth's surface by 4.7 W/m^2. Assuming the average absorption coefficient κ of atmospheric water to be proportional to the concentration of atmospheric water, one can obtain by analogy with formula (7.2.15)

$$\frac{\Delta J_{\downarrow}(H_2O)}{dc(H_2O)} \approx 47\,\mathrm{W/m^2} \tag{7.2.16}$$

7.2.2 Water as Atmospheric Radiator

As we indicate above, basic atmospheric radiators are atmospheric carbon dioxide and water. From the above evaluations based on spectroscopic parameters of CO_2 molecules it follows that carbon dioxide creates approximately 20% of the radiative flux in the infrared spectrum range that falls on the Earth. Hence atmospheric water in the form of H_2O molecules and water microdrops is responsible for approximately 80% of the radiative flux which is absorbed by the Earth in the infrared spectrum range.

Atmospheric water includes water molecules and water microdrops including those of clouds. One can evaluate the radiative flux due to atmospheric water molecules in the same matter as we done above in the case of CO_2 molecules on the basis of spectroscopic parameters these molecules taken from the HITRAN data bank. But because of the complex character of the spectroscopy of water molecules, the visualization will be lost in this way; since our goal is to represent the physical picture of the greenhouse phenomenon and to show the role of spectroscopy in creation of this phenomenon, we use a more rough method for estimation of the role of water molecules.

In this consideration we return to the concept which is presented in Fig. 7.1. In spite of a low accuracy of spectra of this Figure compared to contemporary data, it demonstrates an important concept. Namely, molecules are optically active in a certain spectral range; continuous radiators, as water aerosols, are necessary to cover other spectral range. Basing of this concept, we divide the infrared spectrum which provides the atmospheric greenhouse effect in parts related to certain molecules; the rest spectrum part is covered by aerosols. Roughly, the absorption band due to carbon

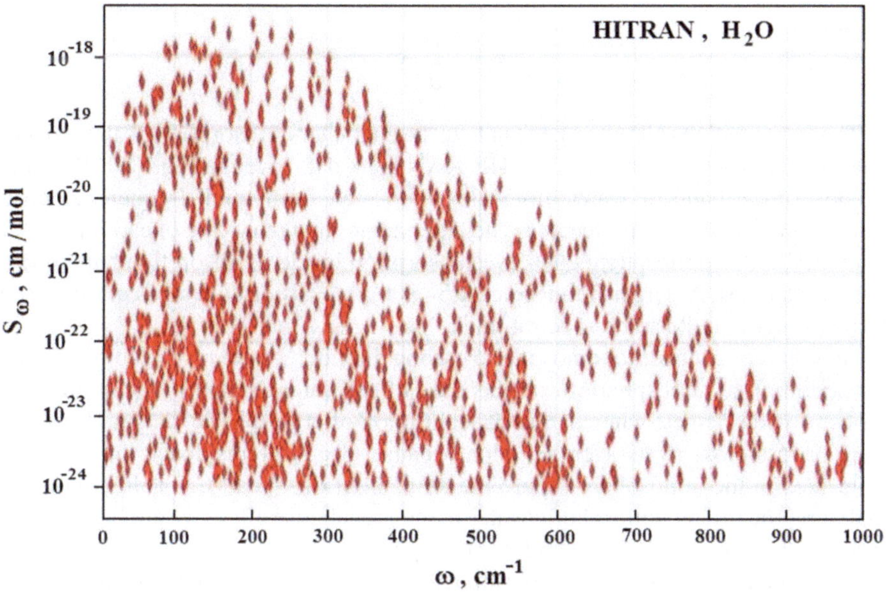

Fig. 7.11 Spectral intensity of atmospheric water molecules according to the HITRAN data bank at the temperature $T = 296\,\mathrm{K}$ [76]

dioxide molecules is restricted by the spectral range (7.2.2), and we below estimate boundaries of the absorption band for atmospheric water molecules..

Let us assume that the boundary frequencies of the absorption band for water molecules is created at altitudes $h \sim 2\,\mathrm{km}$, as it takes place for CO_2 molecules. According to formula (2.2.38) this corresponds to the absorption coefficient $k_\omega \sim 3 \cdot 10^{-6}\,\mathrm{cm}^{-1}$. In order to reduce this estimation to data of Fig. 7.11, we express the absorption coefficient k_ω averaged over oscillations through the spectral intensity S_ω according to formula (4.2.16)

$$k_\omega = N(\mathrm{H_2O})S_\omega \lambda_d,$$

where $N(\mathrm{H_2O}) \sim 10^{17}\,\mathrm{cm}^{-3}$ is the number density of water molecules; taking an average energy difference $d \sim 0.1\,\mathrm{cm}^{-1}$ for neighboring levels which is comparable with the line width, one can obtain $\lambda_d \sim 10\,\mathrm{cm}$ for the wavelength of photons which determine the radiative flux emitted by the atmosphere due to H_2O molecules. Though the used formula (4.2.16) relates to molecules with a regular spectrum structure, it may be taken as an estimation.

From this one can obtain for boundary values of the spectral intensity $S_\omega^* \sim 3 \cdot 10^{-24}\,\mathrm{cm/mol}$. In addition it would be required the difference of energies for neighboring resonances d must be less than $\nu \sim 0.1\,\mathrm{cm}$. If this relation is not fulfilled, a dip occurs in a dependence k_ω between neighboring energy levels, and the absorption at indicated frequencies takes place if

$$S_\omega > \frac{\nu^2}{d^2} S^*_\omega \tag{7.2.17}$$

These estimations allows one to analyze the character of absorption of atmospheric water molecules on the basis of the HITRAN bank data given in Fig. 7.11. Roughly, the absorption band of water molecules is located from small frequencies up to absorption band of CO_2 molecules. The latter means that removing of carbon dioxide from the Earth's atmosphere conserves absorption inside its absorption band due to water molecules. At larger frequencies $\omega > 800\,\mathrm{cm}^{-1}$ absorption takes place only in narrow ranges due to some spectral lines.

Note that in contrast to atmospheric carbon dioxide, we are restricted by a rough consideration of atmospheric water because a real water amount in the atmosphere varies in wide ranges, whereas the concentration of atmospheric CO_2 molecules is more or less close to its average value because of a large residence time of these molecules in the atmosphere (approximately about of 4 years). For this reason, in analyzing the role of atmospheric water in the greenhouse effect we are based on the energetic balance of the Earth and its atmosphere.

Let us analyze the character of the atmospheric emission directed toward the Earth. We assume on the basis of Fig. 7.11 that atmospheric water molecules provides emission of the atmosphere at frequencies below $800\,\mathrm{cm}^{-1}$. This range includes also the absorption band due to CO_2 molecules according to (7.2.2), i.e. the absorption bands due to atmospheric carbon dioxide and atmospheric water molecules are overlapped. Next, one can take the radiative temperature for atmospheric emission toward the Earth to be equal to the temperature of the Earth's surface, as we obtain above inside the absorption band due to CO_2 molecules. This gives the radiative flux toward the Earth $J_\downarrow(\omega < 800\,\mathrm{cm}^{-1}) = 233\,\mathrm{W/m}^2$, so that according to Fig. 7.2 data the total flux is $J_\downarrow = 327\,\mathrm{W/m}^2$; the other part of the flux is $J_\downarrow(\omega > 800\,\mathrm{cm}^{-1}) = 94\,\mathrm{W/m}^2$. Evidently, the latter is created by water microdrops since absorption by CO_2 and H_2O molecules in this spectral range is weak. Accounting for also the radiative flux (7.2.9) that is determined by atmospheric carbon dioxide, one can obtain that the atmospheric radiative flux toward the Earth is created roughly in 20% by CO_2 molecules, in 50% by water molecules, and in 30% by water microdrops.

Water microdrops are of importance for atmospheric emission along with atmospheric water molecules and CO_2 molecules, in spite of a small amount of atmospheric water mass in the form of microdrops compared with that in the form of molecules. Because of a high optical thickness of the atmosphere at frequencies inside the absorption bands due to water and carbon dioxide molecules, the outgoing atmospheric radiation emitted by the Earth surface, is located in the spectrum range $\omega > 800\,\mathrm{cm}^{-1}$). It is approximately $J_p = 20\,\mathrm{W/m}^2$ according to data of Fig. 7.2 or 5% of total emitted power from the Earth's surface.

Let us estimate the radiative temperature T_ω and an effective altitude h_ω for emission of aerosols in the range $\omega > 800\,\mathrm{cm}^{-1}$). Above we found the radiative flux in this frequency range to be $J_\downarrow(\omega > 800\,\mathrm{cm}^{-1}) = 94\,\mathrm{W/m}^2$; according to the Planck formula this corresponds to the radiative temperature $T_\omega = 265\,\mathrm{K}$, that is realized at the altitude $h_\omega = 3.6\,\text{km}$. The optical thickness u of the atmosphere in the spectral range where absorbers are water microdrops in accordance with the character of radiation propagation, follows from the equation

$$J_p = J'_E \int_0^\infty f(u)\mathrm{d}u \int_0^1 \exp\left(-\frac{u}{\cos\theta}\right) d\cos\theta, \tag{7.2.18}$$

where J'_E is the radiative flux from the Earth in this spectral range, $f(u)$ is the distribution function over optical thicknesses, θ is the angle between the normal to the Earth's surface and direction of photon motion. As a blackbody, the the Earth emits the radiative flux $J = 153\,\text{W/m}^2$ at frequencies above $800\,\text{cm}^{-1}$. Evidently, the flux $J_t = 20\,\text{W/m}^2$ passes through the atmosphere in this spectral range, and the probability for a photon to pass through the atmosphere in the indicated spectral range is $P = 0.13$. On the other hand, taking for simplicity the distribution function in formula (6.3.2) as $f(u) = \exp(-u/u_o)$, we take for the probability of surviving of an emitted photon by the Earth

$$P = \int_0^1 \frac{\cos\theta}{u_o + \cos\theta} d\cos\theta = 0.13$$

Solution of this equation gives $u_o = 3.2$, i.e. microdrops form clouds with a large average optical thickness. It is clear that in reality the optical thickness of the atmosphere varies in time.

In addition, one can estimate the amount of the atmospheric water in microdrops, being guided by experiment [77]. According to this experiment, the average absorption cross section σ_{abs} of infrared radiation at the wavelength $\lambda = (10 - 12)\,\mu\text{m}$ by water microdrops is $\sigma_{\text{abs}} = (1.5 - 2.1) \cdot 10^{-6}\,\text{cm}^2$. The cross section of a typical microdrop which radius is $r_o = 8\,\mu\text{m}$ is $\sigma = \pi r_o^2 = 2.0 \cdot 10^{-6}\,\text{cm}^2$ corresponds to the measured value. On the other hand, the absorption cross section for a ball of a radius r_o with blackbody properties of its surface is given by [78]

$$\sigma_{\text{abs}} = \frac{\pi r_o^3}{r_o + C\lambda},$$

where λ is the wavelength, and the numerical coefficient $C \sim 1$. Let us denote the drop number density in clouds as N, and the average thickness of the layer as L. Then the optical thickness of this layer is

$$u = NL\sigma_{\text{abs}}$$

On the other hand, the water mass in microdrops per unit area of the Earth's surface under these conditions is

$$M = \frac{4}{3}\pi r_o^3 \rho N L,$$

where $\rho = 1\,\mathrm{g/cm^3}$ is the density of liquid water. From this we have

$$M = \frac{4\pi r_o^3}{3\sigma_{\mathrm{abs}}}\rho u = \frac{4(r_o + C\lambda)}{3}\rho u \tag{7.2.19}$$

Under these conditions we obtain $M \sim 10^{-2}\,\mathrm{g/cm^2}$, i.e. of the order of 0.3% of the atmospheric water mass is contained in microdrops, i.e. this atmospheric water is found in clouds.

Basing on the observed data for evolution of global atmospheric parameters and the above spectroscopic analysis of the atmospheric greenhouse phenomenon, we below consider evolution of the global temperature as a result of an increase of the amount of atmospheric water and carbon dioxide separately. Let us represent the change of the global temperature T_g, if it is determined by variation of the concentrations $c(CO_2)$ of atmospheric CO_2 molecules and that $c(H_2O)$ of water molecules, in the form

$$\mathrm{d}T_g = a\frac{\mathrm{d}c(CO_2)}{c(CO_2)} + b\frac{\mathrm{d}c(H_2O)}{c(H_2O)}, \tag{7.2.20}$$

where $a = 0.7\,\mathrm{K}$ according to (7.1.18). In order to determine the parameter b, we assume that the change (7.2.20) of the global temperature in a real atmosphere as a result of doubling of the concentration of CO_2 molecules is determined by atmospheric carbon dioxide and water, and formula (7.1.18) gives that for CO_2 molecules. In addition, we have that doubling of the concentration of CO_2 molecules is caused by an increase of the radiative flux $\Delta J = 1\,\mathrm{W/m^2}$ due to these molecules and $\Delta J_\downarrow = 4\,\mathrm{W/m^2}$ due to other atmospheric radiators assuming those to be water molecules.

We above determine the average absorption coefficient $\kappa_\omega = 0.33\,\mathrm{km^{-1}}$ for the contemporary concentration of CO_2 molecules from equation $J_\downarrow = 327\,\mathrm{W/m^2}$. Correspondingly, doubling of the carbon dioxide concentration leads to equation $J_\downarrow = 332\,\mathrm{W/m^2}$ that gives $\kappa_\omega = 0.35\,\mathrm{km^{-1}}$. Since the global temperature change due to atmospheric radiators other than CO_2 molecules, is 1.6 K according to formulas (7.2.27) and (7.2.29), one can obtain from this that $b = 25\,\mathrm{K}$. As it is seen, $b \gg a$, because on the one hand, the concentration of water molecules near the Earth surface is 40 times more than that of carbon dioxide molecules, and, on the other hand, a water molecule due to its dipole moment has a higher optical activity compared to the CO_2 molecule. Note also that the correlation between evolution of the CO_2 concentration and global temperature was used in this operation that exists only last 40 years.

In considering emission of atmospheric water in IR spectral range, we were based on simple models and measurements related to atmospheric water. One can expect a more fruitful method is to evaluate the absorption coefficient of atmospheric water on the basis of high-resolved spectroscopy of water molecules [79] in the same manner as it is made for atmospheric carbon dioxide. But it is not reliable because of strong fluctuations in the concentration of water molecules and an important contribution of water microdrops which in spite a relatively small mass close the windows of transparency of water and carbon dioxide atmospheric molecules. Additional data for atmospheric carbon dioxide and water allow one to increase the reliability of parameters described the greenhouse atmospheric phenomenon.

In conclusion of the analysis of the greenhouse phenomenon in the Earth's atmosphere we note that atmospheric CO_2 molecules give the contribution of approximately 20% both in the radiative flux of infrared radiation toward the Earth and the change of the global temperature due to growth of the amount of atmospheric carbon dioxide. The other part of these values relate to atmospheric water in the form of H_2O molecules and water microdrops. We determine the contribution from atmospheric carbon dioxide on the basis of the spectroscopy analysis because of a large residence time of CO_2 molecules in the atmosphere that is approximately 4 years. The amount of atmospheric water varies in irregular manner depending on time and geographic point. Therefore we are based on the energetic balance of the Earth's atmosphere that is reliable. As a result, we obtain roughly that the contribution to the greenhouse effect due to CO_2 molecules is 20%, due to water molecules is 50% and due to water microdrops is approximately 30%.

7.2.3 Climate Sensitivity

We above have determined the variation of the radiative flux toward the Earth due to a change of the concentration of atmospheric CO_2 molecules. This causes a change of the Earth's temperature, and our task is to determine the connection between these values. The characteristic of this change is according to Arrenius [47] so called the equilibrium climate sensitivity [48] which is the change of the global temperature at doubling of the atmospheric concentration of CO_2 molecules. But this problem may be formulated in a general form, as the global temperature change as a result of variation of the radiative flux $\Delta J_\downarrow$ toward the Earth/s surface. Let us define the climate sensitivity [16, 80] S as the ratio of the global temperature variation ΔT to the change $\Delta J_\downarrow$ of the radiative flux toward the Earth, and this change causes the change of the global temperature ΔT. We have

$$S = \frac{\Delta T}{\Delta J_\downarrow} \tag{7.2.21}$$

It is convenient to operate with a reciprocal value $F = 1/S$, which is called the radiative forcing, since the total radiative forcing is the sum of partial ones. In this consideration we assume the interaction of solar radiation with the atmosphere and Earth is not varied at a small Earth temperature change. This also takes into account that optical parameters of the atmosphere in the visible spectral range are not varied if these parameters in the infrared spectral range vary slightly, including the conservation of the Earth albedo. In this consideration, we are based on the standard atmosphere model, so that the Earth's temperature is $T_E = 288\,\mathrm{K}$, the temperature gradient $\mathrm{d}T/\mathrm{d}h = -6.5\,\mathrm{K/km}$ is constant up to the tropopause at $h_o = 11$ km, where the atmospheric temperature equals $T_{\min} = 217\,\mathrm{K}$, and the temperature difference between the Earth and tropopause is $\delta T = T_e - T_{\min} = 61\,\mathrm{K}$. Let the temperature change at the Earth be ΔT, while the tropopause temperature is unvaried. We then determine the change of energy fluxes under the above conditions, where the water concentration is supported to be a constant.

One can obtain the temperature T_h at a given altitude h and its change ΔT_h under given conditions

$$T_h = T_E\left(1 - \frac{\delta h}{h_o}\right), \quad \Delta T_h = \Delta T\left(1 - \frac{h}{h_o}\right) \tag{7.2.22}$$

Correspondingly, a change ΔJ_h of the radiative flux, if it is created at the altitude h, is

$$\Delta J_h = 4\sigma T_h^3 \Delta T_h = \frac{4J_h \Delta T_h}{T_h}, \tag{7.2.23}$$

where σ is the Stephan-Boltzmann constant. In determination of the convection flux change ΔJ_c, we assume the energy flux due to convection to be proportional to the temperature gradient, that gives

$$\Delta J_c = J_c \frac{\Delta T}{\delta T} \tag{7.2.24}$$

We use these relations below in determination the

The radiative flux from the Earth's surface $J_E = 386\,\mathrm{W/m^2}$ is compensated partially by atmospheric emission through carbon dioxide and water molecules which equals approximately $\Delta J_E = 216\,\mathrm{W/m^2}$ in the range $\omega < 800\,\mathrm{cm^{-1}}$. Assuming the radiative temperature for emission of atmospheric molecules to be equal to the Earth's temperature, one can find the radiative forcing F_E due to emission of the Earth and atmospheric molecules

$$F_E = \frac{4\Delta J_E}{T_E} = 2.4\,\mathrm{W/(m^2\,K)} \tag{7.2.25}$$

In accordance with the energetic balance of the Earth given in Fig. 7.2, we take the average energy flux due to water evaporation from Earth's surface to be $J_{ev} = 57\,\mathrm{W/m^2}$ and due to atmospheric convection $J_c = 47\,\mathrm{W/m^2}$. The temperature dependence of the flux due to water evaporation is $\sim \exp(-\Delta\varepsilon/T)$, where $\Delta\varepsilon = 0.43\,\mathrm{eV}$ is the binding of the water molecule for the liquid phase. The flux change ΔJ_c due to convection we assume to be proportional to the temperature change. As a result, we obtain for the change of the energy flux for these channels as

$$\Delta J_{\text{ev}} = J_{\text{ev}} \frac{\Delta \varepsilon \Delta T}{T_E^2}, \ \Delta J_c = \frac{\Delta T}{T_E} J_c, \tag{7.2.26}$$

and these channels give $F_{\text{ev}} = 3.4\,\text{W}/(\text{m}^2\,\text{K})$, $F_c = 0.16\,\text{W}/(\text{m}^2\,\text{K})$, $F_{\text{ev}} + F_c = 3.6\,\text{W}/(\text{m}^2\,\text{K})$.

We determine also the feedback which takes into account the connection between an increase of the global temperature ΔT and the concentration of optically active components. Let us use the Pauling concept [81, 82] according to which an increase of the carbon dioxide concentration in the atmosphere results from an increase of the global temperature, but not vise versa, as it is assumed in the standard consideration. In this case an atmospheric amount of carbon dioxide follows from equilibrium with ocean carbon dioxide which is dissolved in oceans and is located there in the form of HCO_3^- and other compounds. This equilibrium proceeds through the formation of

$$CO_2 + H_2O(\text{liq}) \longleftrightarrow H_2CO_3, \tag{7.2.27}$$

that means an equilibrium between free atmospheric CO_2 molecules and bound ones in an indicated compound with liquid water, where the enthalpy of this transition is $\Delta H = 178\,\text{kc J/mol}$ or $1.8\,\text{eV}$ [83]. Because an amount of atmospheric carbon dioxide is less compared with bound carbon dioxide in this chemical compound, the concentration $c(CO_2)$ depends on the temperature as

$$c(CO_2) \sim \exp\left(-\frac{\Delta H}{T}\right), \tag{7.2.28}$$

that leads to the following relation of a concentration change for atmospheric carbon dioxide and a temperature change ΔT of the Earth's surface

$$\frac{Delta \ln c(CO_2)}{d\Delta T} = \frac{\Delta H}{T^2} = 0.26\,\text{K}^{-1} \tag{7.2.29}$$

This gives for the radiative forcing $F_f(CO_2) = 0.4\,\text{W}/(\text{m}^2\,\text{K})$ due to the feedback between the change of the atmospheric concentration of carbon dioxide and the temperature change ΔT if we use formula (2.2.15).

By analogy with formula (7.2.22), we have for concentration of atmospheric water molecules

$$c(H_2O) \sim \exp\left(-\frac{\Delta \varepsilon_b}{T}\right), \tag{7.2.30}$$

where $\varepsilon_b \approx 0.43\,\text{eV}$ is the binding energy of the water molecule on the surface of liquid water. This gives by analogy with formula (7.2.23)

$$\frac{\text{d} \ln c(H_2O)}{\text{d}\Delta T} = \frac{\varepsilon_b}{T^2} = 0.06\,\text{K}^{-1} \tag{7.2.31}$$

From this on the basis of formula (2.2.15) we have

$$F_f(H_2O) = \frac{\Delta J_\downarrow(H_2O)}{d\ln c}\frac{\Delta c(H_2O)}{d\Delta T} = 2.8\,W/(m^2\,K),\;\; F_f = F_f(H_2O) + F_f(CO_2) = 3.2\,W/(m^2\,K) \tag{7.2.32}$$

Because the radiative forcing due to an increase of the atmospheric concentration of water molecules due to an increase of the global temperature, we assume for simplicity, that the the radiative forcing as a result of the feedback due to the change of the atmospheric concentrations of carbon dioxide and water molecules to be equal. This gives for the radiative forcing due to this feedback $F_f = 0.3\,W/(m^2\,K)$.

Summarizing the above values, one can obtain the total radiative forcing

$$F_t = F_E + F_{ev} + F_c - F_f \approx 2.8\,W/(m^2\,K),\;\; S \approx 0.36\,m^2\,K/W \tag{7.2.33}$$

Note a low accuracy of the climate sensitivity since the radiative forcing is the difference of large values. The error increases also because atmospheric water molecules and water microdrops are not separated in this evaluation. In addition, according to evaluations [74] this value equals to $S = 0.42\,m^2\,K/W$.

It follows for the change of the global temperature as a result at doubling of the concentration of atmospheric CO_2 molecules.

$$\Delta T = (0.4 \pm 0.2)\,K, \tag{7.2.34}$$

where the error accounts for the accuracy of used values, whereas the result depends on processes included in the above scheme. Indeed, we assume the atmospheric and Earth's albedo, as well as another interaction of solar radiation with the atmosphere and Earth, to be unvaried in the course of the change of the concentration of CO_2 molecules, and also the content of atmospheric water is conserved. Because anthropogenic fluxes of carbon dioxide in the atmosphere resulted from combustion of fossil fuels is about 5%, the contribution of the human activity to ECS (the temperature change as a result of doubling of the atmospheric carbon dioxide amount) is

$$\Delta T = 0.02\,K, \tag{7.2.35}$$

i.e. injections of carbon dioxide in the atmosphere as a result of combustion of fossil fuels is not important for the greenhouse effect. In addition, total removal of CO_2 from the atmosphere causes a decrease of the radiative flux toward the Earth by approximately $9\,W/m^2$. This corresponds to a decrease of the global temperature by approximately 4 K.

The value (7.2.27) may be compared with the ECS for a real atmosphere on the basis of data for evolution of the global temperature during past 150 years [32, 33] and the monitoring of atmospheric carbon dioxide [43, 46]. This is given by formula (7.1.11)

$$\Delta T = (2.0 \pm 0.3)\,\text{K},$$

so that atmospheric carbon dioxide provides approximately 20% both the radiative flux toward the Earth and its change resulted from the change of the concentration of CO_2 molecules. The ECS which follows from treatment of data over past 65 million years is [80, 84, 85]

$$\Delta T = (3.5 \pm 1.3)\,\text{K}, \tag{7.2.36}$$

Evidently, a large error is determined by the change of conditions during different epochs.

For determination of the absorption coefficient (7.1.15), the spectroscopy information about these molecules is required, and it is the content of the data bank [86] that may be used for such evaluations. In the case of CO_2 molecules an additional simplification of a simple structure of this molecule allows one to use the regular or Elsasser model [87] for its vibrational-rotational spectrum with spectroscopy parameters according to [88]. As a result, this allows one to determine ECS under conditions, that the concentrations of all the atmospheric radiators except CO_2 molecules are constant at doubling of the concentration of carbon dioxide molecules. As a result, under these conditions, the change of the global temperature within the framework of the absorption band and line-by line models is equal [19, 29, 74]

$$\text{ECS} = (0.4 \pm 0.1)\,\text{K} \tag{7.2.37}$$

Comparing formula (7.2.37) with (7.1.11), one can conclude that approximately 20% of the change of the global temperature for a real atmosphere gives carbon dioxide, and then the contribution to the global temperature change from combustion of fossil fuels is approximately 1%.

In this consideration, we account for the total radiative flux toward the Earth equals $J_\downarrow = 327\,\text{W/m}^2$ according to (7.1.12). Considering this as equation for the absorption coefficient of other atmospheric radiators than atmospheric carbon dioxide, one can obtain for the contemporary concentration of atmospheric CO_2 molecules $\kappa_\omega = 0.33\,\text{km}^{-1}$. This corresponds to the altitude $h_\downarrow = 2\,\text{km}$ of a layer which is responsible for atmospheric emission, and its temperature $T_\downarrow = 275\,\text{K}$ is the radiative temperature in spectral ranges outside absorption of CO_2 molecules. An indicated temperature differs slightly from (7.1.4) where the absorption coefficient for atmospheric carbon dioxide is averaged over frequencies. At the doubled concentration of CO_2 molecules this equation gives $\kappa_\omega = 0.35\,\text{km}^{-1}$.

According to the above mentioned NASA investigations, the concentration of atmospheric CO_2 molecules increases in time from 0.028% at the preindustrial period which finishes in 19th century, up to 0.041% now. The amount of atmospheric carbon dioxide results from the equilibrium between atmospheric and surface carbon [38–41] and consists mostly of the transition between atmospheric carbon dioxide and solid carbon in composition of plants. The carbon transition from the atmosphere to the Earth's surface proceeds in the photosynthesis process, whereas the opposite transition results from breathing and degradation of plants, as well as from combustion of fossil fuels. Roughly, the Earth atmosphere includes 800 billion tons of carbon in atmospheric carbon dioxide, and the residence time of an individual CO_2 molecule in the atmosphere is approximately 4 years, i.e. an exchange between atmospheric and surface carbon is approximately $2 \cdot 10^8$ ton/yr. The total carbon amount in coal, oil and methane which is extracted from the Earth's interior annually, is approximately 10 billion tons, i.e. the human activity gives the contribution of 5% to the carbon circulation. Hence, the contribution of the man activity in the equilibrium climate sensitivity (7.2.37) is

$$\Delta T = 0.02\,\mathrm{K} \tag{7.2.38}$$

But a small contribution of injections of carbon dioxide in the atmosphere as a result of the combustion process does not mean that the influence of the man activity on global atmospheric parameters ia absent practically at the contemporary level of the human activity. Indeed, from the preindustrial period up to now the concentration of atmospheric CO_2 molecules increases by (40–50)%. This results from deforestation under the human activity. One can conclude from this analysis that contemporary using of fossil fuels which finally are injected in the atmosphere in the form of carbon dioxide and hence influence on the atmospheric spectroscopic properties; it gives the contribution to the global temperature change of approximately 1%.

7.2.4 Energy Balance of Venus and Its Atmosphere

We also apply the results of emission of a gas layer contained CO_2 molecules for the analysis of the energetic balance of the Venus and its atmosphere. Figure 4.10 gives the frequency dependence of the Venus atmosphere near its surface due to atmospheric CO_2 molecules, and below we use this to construct the energetic balance of the Venus and its atmosphere. Let us represent first parameters of the Venus atmosphere. The Venus atmosphere consists of carbon dioxide (96.5%) and nitrogen (3.5%) [89]. On the basis of this, for simplicity, we shall further assume that carbon dioxide is the only component of the Venusian atmosphere. The gas pressure at the surface of Venus is 92 atm, the temperature is 735 K [89, 90]. The temperature gradient is about -8 K/km in the altitude range from 0 to 60 km and then decreases monotonically, as the altitude increases, to almost zero at an altitude of 100 km [90]. On the basis of measurements [90], the scale of the change in the number density of carbon dioxide molecules Λ in the Venus atmosphere, which is introduced on the basis of formula (7.1.2), varies from $\Lambda = 19$ km at the surface Venus up to about 6 km at an altitude of 60 km, and 4 km at an altitude of 100 km. The number density of carbon dioxide molecules at the Venus surface is $N = 9.2 \cdot 10^{20}\,\mathrm{cm}^{-3}$, and also $N = 6.6 \cdot 10^{18}\,\mathrm{cm}^{-3}$ and $N = 1.2 \cdot 10^{15}\,\mathrm{cm}^{-3}$ at altitudes of 60 km and 100 km respectively.

Our task is to determine the contribution of carbon dioxide, located in the Venus atmosphere, the greenhouse effect. Indeed, the average flux of solar radiation per unit surface area of Venus is $(2622 \pm 6)\,\mathrm{W/cm^2}$ [91] (for the Earth's surface this value is $1365\mathrm{W/cm^2}$). For the observed surface temperature 735 K, the flux of infrared radiation under these conditions is $J_V = 1.7\,\mathrm{W/cm^2}$, i.e. heat fluxes are increased at approach to the planet surface. If the Venus absorbs all the solar radiation incident on it and then radiates as a black body, then its surface temperature will be 463 K.

In reality, most of the radiation penetrating into the Venus atmosphere is reflected, and the absorbed energy is consumed partially for creation of convective gas flow, while its main part is spent on atmosphere radiation. The Venus albedo is of 0.80 ± 0.02 according to [92] and 0.76 ± 0.01 according to [91]. As a result, the average radiative flux from the Venus atmosphere is $(157 \pm 6)\,\mathrm{W/m^2}$ [93]. This corresponds to an average radiation temperature of about 230 K, which agrees with measurements [94, 95]. This temperature is realized at an altitude of 70 km, where the carbon dioxide pressure is 28 Torr.

The analysis of the Venus energetic balance is beyond this work. Our task is to determine the role of carbon dioxide emission in the heat balance of the Venus. We will assume that the Venus atmosphere consists entirely of carbon dioxide and determine the radiative temperature in the spectrum range where carbon dioxide molecules emit. In the subsequent analysis, we use the representations of molecular spectroscopy (for example, [96–101]). Moreover, the CO_2 molecule is a linear symmetric molecule, that simplifies the analysis of its oscillations.

We use formula (7.1.20) to find the radiative temperature $T(\omega)$ of radiation leaving the Venus atmosphere in the range of the absorption spectrum of carbon dioxide molecules. Let us introduce the absorption band by molecules of carbon dioxide such, that the boundary frequencies for this band ω_{1b} and ω_{2b} in accordance with formula (7.1.22) are given by relations

$$k_\omega(\omega_{1b})\Lambda = k_\omega(\omega_{2b})\Lambda = \frac{2}{3}, \tag{7.2.39}$$

where the scale for the change of the number density of radiating molecules is determined by formula (7.1.2).

If the width of the absorption band is determined by wings of spectral lines for individual vibrational-rotational transitions, it is convenient to average the absorption coefficient (4.2.25) for P and R branches over oscillations within one period, i.e. over the frequency range for transitions between neighboring rotational momenta. Next, on the basis of formula (7.1.20), we determine the altitude of an atmosphere layer h_ω which is responsible for emission of photons at a given frequency, and the radiation temperature at a given frequency coincides with the temperature of this layer. Figure 7.12 contains the frequency dependence for this altitude. Correspondingly, the radiation temperature $T(\omega)$ of photons, which are produced by emission of carbon dioxide molecules, is given in an indicated range of photon frequencies in Fig. 7.13 in accordance with formula (7.1.20).

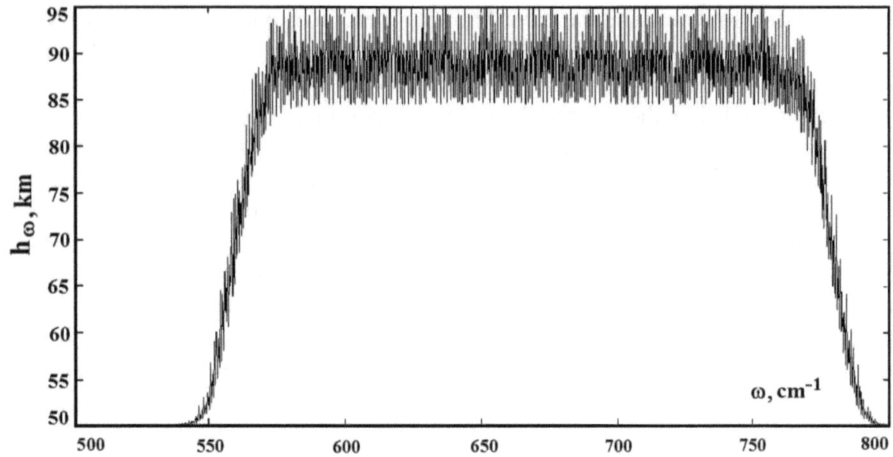

Fig. 7.12 Effective altitude of the Venus atmosphere which is responsible for creation of the radiative flux at a given frequency [102]

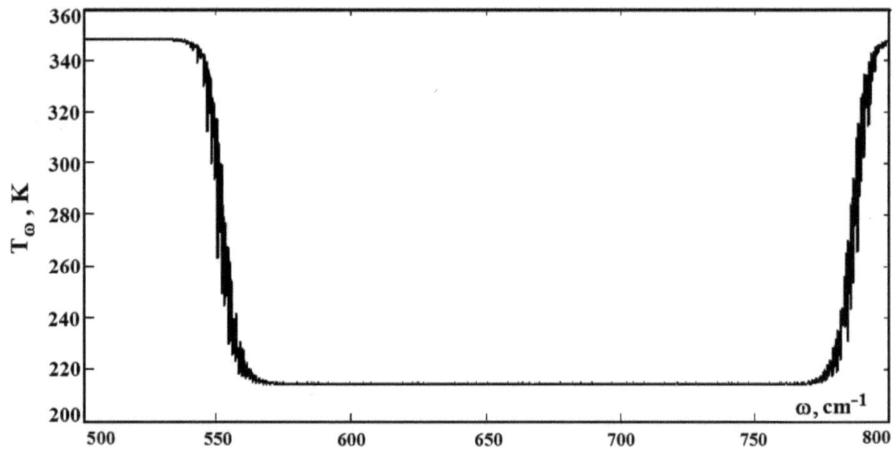

Fig. 7.13 Effective temperature of the Venus atmosphere which is responsible for creation of the radiative flux due to emission of atmospheric CO_2 molecules as a frequency function [102]

Let us determine the total outgoing radiation flux from the Venus atmosphere on the basis of formula (7.2.39) assuming that the radiative temperature for a part of the spectrum, which is not connected with carbon dioxide, is independent of the frequency. We have for the energy flux of radiation

$$J_{\downarrow} = \int_0^{\omega_1} \frac{\hbar\omega^3 d\omega}{4\pi^2 c^2 \left[\exp\left(\frac{\hbar\omega}{T_o}\right) - 1\right]} + \int_{\omega_1}^{\omega_2} \frac{\hbar\omega^3 d\omega}{4\pi^2 c^2 \left[\exp\left(\frac{\hbar\omega}{T(\omega)}\right) - 1\right]} + \int_{\omega_2}^{\infty} \frac{\hbar\omega^3 d\omega}{4\pi^2 c^2 \left[\exp\left(\frac{\hbar\omega}{T_o}\right) - 1\right]}, \tag{7.2.40}$$

and according to the measurements [93] this flux is $J = (157 \pm 6)\,\mathrm{W/m^2}$. Solving this equation, we find $T_o = (249 \pm 3)\,\mathrm{K}$, which corresponds to altitudes of 63–64 km. Note that clouds of the Venusian atmosphere, consisting of droplets of sulfuric acid, are concentrated in the altitude range of 60–70 km. Perhaps, these clouds are responsible for emission of thermal radiation in the range of the spectrum, not connected with molecules of carbon dioxide. We also note that at an altitude of 60 km the atmosphere temperature is 263 K, and the temperature gradient is $\mathrm{d}T/\mathrm{d}h = 4\,\mathrm{K/km}$, and the scale of the change in the number density of CO_2 molecules is $\Lambda = 6\,\mathrm{km}$. This gives a small parameter (7.1.18) $\alpha = 0.05$, which characterizes the validity of the assumptions used.

Let us give the parameters of Venus related to the problem under consideration. The atmosphere of Venus consists of carbon dioxide (96.5%) and nitrogen (3.5%) [89]. For simplicity, we assume carbon dioxide to be the only component of the Venus atmosphere. The gas pressure at the Venus surface is 92 atm, the temperature is 737 K [89, 90], that corresponds to the number density of molecules of carbon dioxide near the Venus surface $N = 9.2 \cdot 10^{20}\,\mathrm{cm^{-3}}$. The temperature gradient is about $-8\,\mathrm{K/km}$ in the altitude range from 0 to 60 km [90], and the scale of the change in the density of carbon dioxide molecules Λ in the Venus atmosphere, which is introduced on the basis of formula $\Lambda = \mathrm{d}\ln N/\mathrm{d}(1/h)$, where N is the number density of molecules in the atmosphere, h is the altitude above the Venus surface, and $\Lambda = 19\,\mathrm{km}$.

In constructing the energy balance of the surface of Venus and its atmosphere, we will model the Venus surface as a blackbody for IR radiation. We have that an absolutely black body with a temperature of 737 K creates an IR radiative flux $J_o = 16.7\,\mathrm{kW/m^2}$. On the other hand, the average flux of solar radiation per unit surface area of Venus $2.6\,\mathrm{kW/m^2}$ [91]. The Venus albedo is 0.80 ± 0.02 according to [92] and 0.76 ± 0.01 according to [91]; we take it to be 0.78. It follows that the average flux of solar radiation absorbed by the atmosphere and surface of Venus is $0.5\,\mathrm{kW/m^2}$. An additional contribution to the power absorbed by the surface of Venus is due to IR radiation produced by the molecules of carbon dioxide. To determine the radiative flux to the Venus surface, which contributes to the Venus energy balance, it is necessary to analyze the emission spectrum of carbon dioxide molecules in the infrared spectral range, which is presented in Fig. 4.3, so that the frequency dependence for the absorption coefficient is represented in Fig. 4.10.

The boundaries of absorption bands for the Venus atmosphere due to carbon dioxide molecules follow from the relation [13, 103]

$$k_\omega \Lambda = 2/3, \tag{7.2.41}$$

where $\Lambda = 19$ km is the scale of the change in the number density of molecules of carbon dioxide in the Venus atmosphere near its surface in accordance with formula (7.1.2). On the basis of this formula, we have for the boundaries of the first two absorption bands $\omega_1 = 493\,\mathrm{cm^{-1}}$ and $\omega_2 = 1174\,\mathrm{cm^{-1}}$, which gives IR radiative flux to the Venus surface $J_1 = 3.5\,\mathrm{kW/m^2}$. The third absorption band, which is created by the resonant radiation of the carbon dioxide molecule between the lower excited antisymmetric state and the ground vibrational state, has the boundaries

$\omega_1 = 2258\,\mathrm{cm}^{-1}$ and $\omega_2 = 2439\,\mathrm{cm}^{-1}$ that leads to IR radiative flux to the Venus surface $J_3 = 0.9\,\mathrm{kW/m}^2$. From this one can obtain the total radiative flux to the Venus surface, which consists of the absorbed flux of solar radiation and IR radiation generated by carbon dioxide molecules in the Venus atmosphere is $J = 4.9\,\mathrm{kW/m}^2$.

As it follows from the data presented, solar radiation absorbed by the Venus surface contributes 3% to the total energy flux absorbed by the Venus surface. In this case, the contribution of IR radiation of the Venus atmosphere absorbed by its surface is 26% from the total flux of radiation absorbed or emitted by the Venus surface. Note that in the case of the Earth's energy balance, the last channel gives 20% [16, 19, 75]. Thus, we arrive at a contradiction according to which the power of radiation absorbed by the surface of Venus, is small extent (about one-third) compared to the power of IR radiation from its surface at the observed temperature. In searching the cause of the discrepancy between the indicated powers, we first analyze the accuracy of finding the above powers. For this purpose, we use the absorption band model [19] for carbon dioxide molecules, according to which radiation at frequencies inside this band is created by carbon dioxide molecules, whereas outside the absorption band, carbon dioxide molecules do not contribute to atmospheric emission. This requires a sharp change of the absorption coefficient k_ω near the boundary of the absorption band which is given in Fig. 4.9. The accuracy of calculation the above radiation powers within the framework of the model used is better than 20%; the error in calculations also includes the fact that the radiative temperature of the Venus atmosphere near boundaries of the absorption bands is determined by the layers of the Venus atmosphere whose temperature differs from the temperature of its surface.

Let us ascertain which additional channels in the Venus energy balance can remove the contradiction obtained. Above we did not take into account the convective heat transfer from the Venus surface, associated with the vortex movement of carbon dioxide near its surface under the influence of the atmospheric temperature gradient. In the case of the Earth, the convective energy flux is about 10% of the total flux of solar radiation entering the Earth's atmosphere [16, 19, 75]. It can be expected that the relative contribution of convective transport in the energy balance of Venus does not exceed this value due to a high gas pressure in the atmosphere of Venus. Thus, convective transport in the atmosphere of Venus does not eliminate the above contradiction, especially since the convective transfer only increases the missing energy flow to the Venus surface, which is necessary for the fulfillment of the Venus energy balance.

However, convective heat transfer in the Venus atmosphere causes the dust to move from the Venus surface to its atmosphere. In the case of micron-size dust particles, the action of gravity leads to their return to the Venus surface through years. The weighted dust is optically thick in the IR spectral range and even at a low dust concentration the IR radiative flux may be provided by dust particles. Note that in the case of the Earth's atmosphere, the role of dust does not be important, since atmospheric dust is washed out by atmospheric water for 8–9 days. At the same time, the energy balance of Venus and its atmosphere begins with absorption of solar radiation by the atmosphere [104], and then the absorbed energy is transferred to the Venus surface as a result of convection and radiative transfer. Here we do not consider the radiative transfer from upper layers of the Venus atmosphere to its surface, but we solve a simpler problem.

One can estimate a dust amount of the atmosphere from the condition that the dust optical thickness u with respect to IR radiation is of the order of one. This condition has the form

$$u_\omega = k_\omega L = \sigma_\omega N L \sim 1, \tag{7.2.42}$$

where σ_ω is the photon absorption cross section of a dust particle, N is the number density of particles in the atmosphere, L is the thickness of the atmosphere layer in which the dust is located. Assuming dust particles to be spherical, we assume a typical dust radius r to be small compared to the wavelength $\lambda \sim 10\,\mu\text{m}$, which leads to the following estimate for the photon absorption cross section of the dust particle [78]

$$\sigma_\omega \sim \pi r^2 \cdot \frac{r}{\lambda} \sim \frac{V}{\lambda}, \tag{7.2.43}$$

where V is the average volume of dust particles. From this we obtain on the basis of formula (7.2.42)

$$\xi = VNL \sim \lambda u_\omega, \tag{7.2.44}$$

where ξ is a typical thickness of the dust layer, which ensures the absorption of infrared radiation of the planet, if atmospheric dust is collected near the Venus surface.

Comparing this amount of dust with the amount of carbon dioxide in the atmosphere of Venus, we obtain that the concentration of dust molecules is $\sim 10^{-7}$ (the number of molecules of atmospheric dust to the number of molecules of carbon dioxide in the atmosphere) provides the effect under consideration when this dust consisting from solid particles of micron and submicron sizes, creates the IR radiative flux absorbed subsequently by the Venus surface. If this dust is collected at the Venus surface, it forms a layer of thickness of several tens of microns (several IR wavelengths). Note that the microscopic dust of the Venusian atmosphere is near its surface, in contrast to the clouds in the Venus atmosphere [95, 105, 106], which are located at an altitude of 60–70 km and provide IR radiation which goes outside this planet.

Let us note one more feature of the conducted research. The performed calculations use information about spectroscopic parameters of carbon dioxide molecules taken from the HITRAN data bank [79] and include data for several hundred vibrational-rotational transitions of the carbon dioxide molecule. At present, this bank has information pertaining to hundreds of thousands of transitions involving carbon dioxide molecules [79] , i.e. a small part of existed information related to this problem is sufficient for the calculations of the above parameters.

As is seen, the disperse phase is the basic radiating component for outgoing radiation as well as for that directed to the Venus surface. In considering the radiative

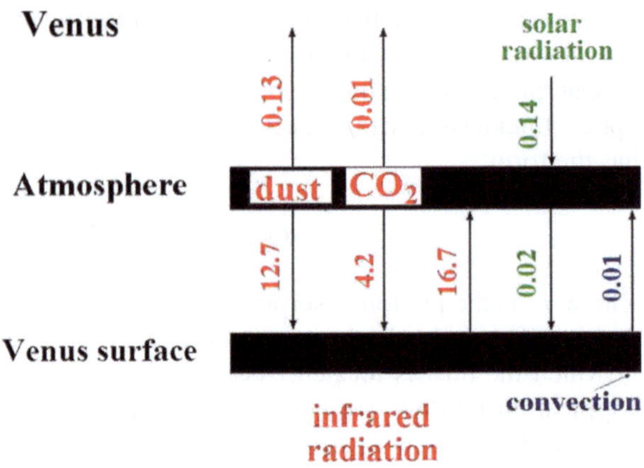

Fig. 7.14 Expressed in kW/m^2 average energy fluxes of the Venus and its atmosphere [102]

flux toward the Venus surface, we assume that this radiation is created mostly by dust particles which arose in the atmosphere from the surface under the action of convective atmospheric motion. We considered sand particles as a more probable candidate for these particles which are a source of infrared radiation directed toward the Venus surface. In order to satisfy the energetic balance between the Venus surface and its atmosphere, it was required that the optical thickness due to sand particles corresponds to formula (4.1.17) for an atmospheric layer starting from the surface up to altitude $h = 8\,\mathrm{km}$. From this one can find the total optical thickness of the Venus atmosphere due to sand particles as $u = 1.8$ under the assumption that the concentration of sand particles is independent of the altitude up to high altitudes. Hence, the optical density of sand particles starting from the altitude $h = 70\,\mathrm{km}$ up to infinity is 0.002. Hence, sand particles cannot be responsible for outside radiation.

One can expect that radiators are aerosols which constitute clouds of the Venus atmosphere [95, 105, 106], are located at altitudes of 60–70 km and provide infrared outgoing radiation of this planet. The Venus is covered by a thick layer of clouds that extends between 55 and 70 km above the surface. These rapidly-moving clouds are mainly composed of micron-sized droplets of sulphuric acid and other aerosols. One can estimate a specific mass of aerosols $\rho\xi$ which are responsible for emission of outgoing radiation of the Venus, where ρ is the mass density of aerosol material $<$ and ξ is given by formula (7.2.44). One can find a specific mass m of these aerosols in analogy with that for dust particles created the radiation toward the Venus surface. Taking an identical mass density for dust particles near the Venus surface and aerosol particles in upper layers ($\rho = 1.8\,\mathrm{g/cm^3}$ for sulfur acid), one can obtain the same specific mass of particles $M \approx 2\,\mathrm{mg/cm^2}$ because of the identical optical thickness of comparable layers. The specific mass of carbon dioxide is $50\,\mathrm{g/cm^2}$ for the Venus atmosphere above 70 km, that gives for the aerosol mass concentration $M/m \sim 4 \cdot 10^{-5}$. As is seen, the mass concentration of aerosols which create the outgoing radiation of the Venus atmosphere is two orders of magnitude higher than that for dust particles which determine radiation toward the Venus surface.

Summarized the above data, one can construct the energetic balance of the Venus and its atmosphere which is presented in Fig. 7.14. Though the convection flux is included in this Figure, its value is less that the error for radiative fluxes, i.e. the contribution of the convection flux into the the total one is negligible. Nevertheless, convection is of importance for Venus energetics because it lifts up dust particles from the Venus surface which provide its high temperature.

We above use the results for emission of carbon dioxide molecules in the infrared spectrum range for radiation of the Venus atmosphere, as a demonstration of the above methods for spectroscopy of molecules in gases. Note two peculiarities of this process. First, the temperature of the Venus surface is enough high, and the power of infrared radiation of the Venus atmosphere is two orders of magnitude larger than the power of solar radiation penetrated in the Venus atmosphere, but solar radiation is the basis of the energetic balance of the Venus. Second, under known parameters of the Venus atmosphere, one can evaluate precisely (with the accuracy better 20%) the power of emission due to atmospheric CO_2 molecules, and it is only one third from the power radiating by the Venus surface. This proves the presence of an atmospheric dust which amount is seven orders of magnitude less than the amount of atmospheric carbon dioxide, but this dust provides the atmosphere balance.

References

1. J.B.J. Fourier, Annal. Chem. Phys. **27**, 136 (1824)
2. J.B.J. Fourier, Mem. Acad. Roy. Sci. **7**, 569 (1827)
3. https://en.wikipedia.org/wiki/Greenhouse-gas
4. J. Tyndall, Proc. Roy. Inst. **3**, 155 (1959)
5. J. Tyndall, Philos. Mag. **22**, 169–273(1861)
6. J. Tyndall, Philos. Mag. **26**, 44 (1863)
7. https://en.wikipedia.org/wiki/John-Tyndall
8. Y.G. Houghton, *Physical Meteorology* (MIT Press, Cambridge, 1985)
9. S. Twomey, Atmos. Environ. **25A**, 2435 (1991)
10. B.M. Smirnov, *Introduction to Plasma Physics* (Moscow, Nauka, 1975; in Russian)
11. B.M. Smirnov, *Introduction to Plasma Physics* (Mir, Moscow, 1977)
12. B.M. Smirnov, *Physics of Weakly Ionized Gas* (Moscow, Nauka 1979; in Russian)
13. B.M. Smirnov, *Physics of Weakly Lonized Gases* (Mir, Moscow, 1980)
14. *Understanding Climate Change* (Washington, National Academy of Sciences, 1975)
15. J.H. Seinfeld, S.N. Pandis, *Atmospheric Chemistry and Physics* (Wiley, Hoboken, 2006)
16. M.L. Salby, *Physics of the Atmosphere and Climate* (Cambridge University Press, Cambridge, 2012)
17. H.N. Pollack, S.J. Hunter, R. Johnson, Rev. Geophys. **30**, 267 (1997)
18. O. Boucher, *Atmospheric Aerosols. Properties and Climate Impacts* (Dordrecht, Springer, 2015)
19. B.M. Smirnov, *Microphysics of Atmospheric Phenomena* (Springer, Switzerland, 2017)
20. G.R. North, K.-Y. Kim, *Energy Balance Climate Models* (Weinheim, Wiley, 2017)
21. J.T. Kiehl, K.E. Trenberth, Bull. Am. Meteorol. Soc. **78**, 197 (1997)
22. K.E. Trenberth, J.T. Fasullo, J.T. Kiehl, Bull. Am. Meteorol. Soc. **90**, 311 (2009)
23. K.E. Trenberth, J.T. Fasullo, Surf. Geophys. **33**, 413 (2012)
24. https://www.ssmi.com/msu

25. https://en.wikipedia.org/wiki/Earth's-energy-budget
26. https://en.wikipedia.org/wiki/Greenhouse-effect
27. M. Wendisch, P. Yang, *Theory of Atmospheric Radiative Transfer* (Wiley, Singapore, 2012)
28. *U.S. Standard Atmosphere* (U.S. Government Printing Office, Washington, 1976)
29. B.M. Smirnov, EPL **114**, 24005 (2016)
30. J.E. Hansen, D. Johnson, A. Lacis et al., *Science* **213**, 957 (1981)
31. J.E. Hansen, R. Ruedy, M. Sato, K. Lo, https://www.data.giss.nasa.gov/gstemp/2011
32. J. Hansen, M. Sato, R. Ruedy et al., https://www.columbia.edu/~jeh1/mailing/2016/20160120-Temperature2015
33. J. Hansen, M. Sato, R. Ruedy, https://www.columbia.edu/~jeh1/mailing/2014/20140121-Temperature2013
34. J.E. Hansen, R. Ruedy, J. Glascoe, M. Sato, J. Geophys. Res. **104**, 997 (1997)
35. J.E. Hansen, R. Ruedy, M. Sato et al., J. Geophys. Res. **106**, 947 (2001)
36. J.E. Hansen, R. Ruedy, M. Sato, K. Lo, Rev. Geophys. **48**, RG4004 (2010)
37. B.M. Smirnov, *Physics of Global Atmosphere* (Dolgoprudny, Intellect, 2017; in Russian)
38. T. Pedersen, Y. Rosental, S. Seitzinger et al., Science **290**, 291 (2000)
39. T.M. Wigley, D.S. Schimmel, *The Carbon Cycle* (Cambridge, Cambridge University Press, 2000)
40. D. Archer, *The Global Carbon Cycle* (Princeton University, Press, Princeton, 2010)
41. https://en.wikipedia.org/wiki/Carbon-cycle
42. https://www.esrl.noaa.gov/gmd/ccgg/trends
43. https://www.ersl.noaa.gov/gmd/cggg/trends
44. Ch.D. Keeling et al., Tellus **12**, 200 (1960)
45. Ch.D. Keeling et al., Tellus **28**, 538 (1976)
46. https://en.wikipedia.org/wiki/Mauna-Loa-Observatory
47. S. Arrhenius, Phil. Mag. **41**, 237 (1896)
48. https://en.wikipedia.org/wiki/Climate-sensitivity
49. https://climate.nasa.gov
50. I.A. Shiklomanov, *Water in Crisis: A Guide to the World's Fresh Water Resources* In: P.H. Gleick (Eds.) (Oxford, Oxford University Press, 1993; p. 13–24)
51. I.A. Shiklomanov, J.C. Rodda (eds.), *World Water Resources at the Beginning of the Twenty-First Century* (Cambridge University Press, Cambridge, 2003)
52. R.W.Healy, T.C. Winter, J.W. Labaugh, O.L. Franke, *Water Budgets: Foundations for Effective Water-Resources and Environmental Management* (Reston, Virginia, U.S. Geological Survey Circular 1308, 2007)
53. https://en.wikipedia.org/wiki/Atmosphere-of-Earth
54. S. Bashkin, J. Stoner, *Atomic Energy Levels and Grotrian Diagrams*, Vol. 1–4 (Amsterdam, North Holland, 1975–1982)
55. J.P. Peixoto, A.H. Oort, *Physics of Climate* (Washington, Amer. Inst. Phys., 1992)
56. K.E. Trenberth, L. Smith, T. Qian et al., J. Hydrometeorol. **8**, 758 (2007)
57. https://en.wikipedia.org/wiki/water-circle
58. https://water.usgs.gov/edu/watercycleatmosphere.html
59. https://www.climate4you.com/GreenhouseGasses.htm
60. https://atmospheres.gsfc.nasa.gov/meso/index.php
61. https://tamino.wordpress.com/2010/08/08/urban-wet-island/
62. https://www.c3headlines.com/greehouse-gases-atmosphereco2methanewater-vapor
63. https://www.c3headlines.com/natural-negativepositive-feedback
64. http://en.wikipedia.org/wiki/Properties-of-water
65. D.R. Lide, *Handbook of Chemistry and Physics*, Edn. 86 (CRC Press, London, 2003–2004)
66. A. Dai, J. Climate **19**, 3589 (2006)
67. K.M. Willett, P.D. Jones, N.P. Gillett, P.W. Thorne, J. Climate **21**, 5364 (2008)
68. D.I. Berry, E.C. Kent, Bul. Amer. Meteorol. Soc. **90**, 645 (2009)
69. https://tamino.wordpress.com/2011/05/17/hot-and-wet/

70. R.M. Goody, *Atmospheric radiation. Theoretical Basis* (Oxford, Oxford University Press, 1964)
71. F. Reif, *Statistical and Thermal Physics* (McGrow Hill, Boston, 1965)
72. L.D. Landau, E.M. Lifshitz, *Statistical Physics*, vol. 1 (Pergamon Press, Oxford, 1980)
73. R.M. Goody, Y.L. Yung, *Atmospheric Radiation* (Oxford University Press, New York, 1995)
74. B.M. Smirnov, JETP **126**, 446 (2018) §3
75. W.M. Haynes, *CRC Handbook of Chemistry and Physics*, Edn. 97 (London, CRC Press, 2016–2017)
76. https://www.hitran.iao.ru/home
77. C.M.R. Platt, Quart. J. Roy. Meteorolog. Soc. **102**, 553 (2006)
78. L.D. Landau, E.M. Lifshitz, *Electrodynamics of Continuous Media* (Pergamon Press, Oxford, 1984)
79. https://www.hitran.org/links/
80. Paleosens Project Members, Nature **491**, 683 (2012)
81. L. Pauling, *General Chemistry* (Freeman, San Francisco, 1970)
82. J.M. Kauffman, J. Sci. Explor. **4**, 723 (2007)
83. https://en.wikipedia.org/wiki/Calcium-carbonate
84. T.M. Cronin, *Paleoclimates: Understanding Climate Change Past and Present* (Columbia University Press, New York, 2010)
85. M.L. Bender, *Paleoclimate* (Princeton University Press, Princeton, 2013)
86. I.E. Gordon, L.S. Rothman, C. Hill, R.V. Kochanov, Y. Tan et al., J. Quant. Spectr. Rad. Transf. **203**, 3–69(2017)
87. W.M. Elsasser, Phys. Rev. **54**, 126 (1938)
88. L.S. Rothhman, W.S. Benedict, Appl. Opt. **17**, 2605 (1978)
89. https://en.wikipedia.org/wiki/Venus
90. https://en.wikipedia.org/wiki/Atmosphere-of-Venus
91. V.I. Moroz et al., Adv. Space Res. **5**, 197 (1985)
92. M.G. Tomasko et al., J. Geophys. Res. **85**, 8187 (1980)
93. https://asp.colorado.edu-espoclass/ASTR-5835-2015...Notes/Titov
94. J.T. Schofield, F.W. Taylor, Icarus **52**, 245 (1982)
95. L.V. Zasova et al., Planet Space Sci. **55**, 1712 (2007)
96. G. Herzberg, *Molecular Spectra and Molecular Structure* (Van Nostrand Reinhold, Princeton, 1945)
97. M.A. El'yashevich, *Molecular Spectroscopy* (Fizmatgiz, Moscow, 1963; in Russian)
98. H.C. Allen, P.C. Cross, *Molecular Vib-rotors; the Theory and Interpretation of High Resolution Infra-red Spectra* (Wiley, New York, 1963)
99. G. Herzberg, *Molecular Spectra and Molecular Structure: Electronic Spectra and Electronic Structure of Polyatomic Molecules* (Van Nostrand, New York, 1966)
100. C.N. Banwell, E.M. McCash, *Fundamentals of Molecular Spectroscopy* (McGraw-Hill, London, 1994)
101. https://en.wikipedia.org/wiki/Infrared-spectroscopy
102. B.M. Smirnov, JETP **127**, 48 (2018) §5
103. B.M. Smirnov, *Physics of Ionized Gases* (Wiley, New York, 2001)
104. D.V. Titov et al., https://asp.colorado.edu-espoclass/ASTR-5835-2015...Notes/Titov
105. L.V. Zasova, V.I. Moroz, L.V. Esposito, C.Y. Na, Icarus **105**, 92 (1993)
106. Y.J. Lee, D.V. Titov, S. Tellmann et al., Icarus **217**, 599 (2012)

Conclusion

The content of this book includes radiative processes involving atomic particles and atomic systems. Our analysis starts from general principles of interaction of electromagnetic waves with atomic particles or systems and uses as a basis theoretical quantum mechanics and statistic physics. This approach was elaborated a century ago, and the subsequent development of this area was based on using the new instruments which are radiators and detectors of radiation in a narrow spectral range, as well as new information about radiative properties of atomic particles as well as radiative processes with their participation. In this consideration we are restricted by the strongest radiative transitions as a result of dipole interaction between an electromagnetic wave and nonrelativistic atomic particles and systems that simplify the analysis.

The apparatus of quantum mechanics as the basis of the above analysis exists during a hundred of years, and its development consists in its application to some physical objects and processes. We above considered various elementary radiative processes which include radiative transitions between discrete atomic states, photoionization of atoms, photorecombination of electrons and ions, photoattachment of electrons to atoms, and also bremsstrahlung as a result of electron scattering on atoms or ions. In addition, radiative processes of interaction of an electromagnetic wave with a small particles are considered and include scattering and absorption of radiation by dielectric microparticles, metal nanoclusters and aerosols. On the basis of these processes, one can analyze two-photon processes including the Rayleigh and Raman scattering.

A list of applications of this area, in the first turn, includes processes of optical pumping by a resonant radiation interacting with a gas. These processes are the cooling of a gas by laser resonance radiation, light-induced drift of gas atoms which results from radiation tuned in the wing of the resonance spectral line, and the photoresonant plasma which is formed from irradiation of an atomic gas or vapor by resonance radiation. Another group of applications includes two-step ionization

V. Krainov and B. M. Smirnov, *Atomic and Molecular Radiative Processes*,
Springer Series on Atomic, Optical, and Plasma Physics 108,
https://doi.org/10.1007/978-3-030-21955-0

of atoms in a gas, interaction and reflection of radio-waves from the ionosphere, detection of submillimeter radiation on the basis of radiative transitions for Rydberg atoms.

Processes of reabsorption of resonance radiation are of importance in propagation of this radiation through a gas. In this case the main contribution to the outgoing radiation flux follows from wings of the resonance spectral line, and transport of resonance radiation through a radiating excited atomic gas is determined by this character of the radiative process. A general principle presented for emission of an optically dense gas is such, that the main contribution to emission of a gaseous system follows from layers with the optical thickness of the order of one from the boundary. This principle is applied to the solar photosphere and atmospheres of the Earth and Venus.

The experience of this book shows the importance of precise calculations for parameters of some processes. In this case, we above have presented the explicit evaluation of the radiation flux which goes through a boundary of the flat layer of a gas containing carbon dioxide, due to the thermal emission of carbon dioxide molecules. We have shown that a strict evaluation of the emission power for the Earth's and Venusian atmospheres leads to a certain position with respect to the greenhouse phenomenon in the atmosphere of these planets.

Bibliography

1. E.U. Condon, G.H. Shortley, *Theory of Atomic Spectra* (Cambridge University Press, Cambridge, 1949)
2. G. Herzberg, *Atomic Spectra and Atomic Structure* (Dover, New York, 1944)
3. H.G. Kuhn, *Atomic Spectra* (Longmans, London, 1969)
4. C. Herzberg, *Molecular Spectra and Molecular Structure*, vol. 1–4 (New York, van Nostrand, 1964–1966)
5. K.P. Huber, G. Herzberg, *Molecular Spectra and Molecular Structure* (Van Nostrand, Constants of Diatomic Molecules (New York, 1979)
6. J.D. Graybell, *Molecular Spectroscopy* (McGrawHill, New York, 1988)
7. S.V. Khristenko, A.I. Maslov, V.P. Shevelko, *Molecules and Their Spectroscopic Properties* (Springer, Berlin, 1998)
8. J.M. Brown, *Molecular Spectroscopy* (Oxford University Press, Oxford, 1998)
9. J.L. McHale, *Molecular Spectroscopy* (Prentice Hall, New Jersey, 1999)
10. P.F. Bernath, *Spectroscopy of Atoms and Molecules* (Oxford University Press, Oxford, 2005)
11. S. Svanberg, *Atomic and Molecular Spectroscopy* (Springer, Berlin, 2012)
12. R. Kakkar, *Atomic and Molecular Spectroscopy: Basic Concepts and Applications* (Cambridge University Press, Delhi, 2015)
13. A.S. Yatzenko, *Grotrian Diagrams of Neutral Atoms* (Novosibirsk, Nauka, 1993, in Russian)
14. T. Shirai, J. Sugar, W. Wiese et al., *Spectral Data and Grotrian Diagrams for Highly Ionized Atoms* (AIP, New York, 2000)
15. G.G. Telegin, A.S. Yatzenko, *Optical Spectra of Atmospheric Gases* (Novosibirsk, Nauka, 2000, in Russian)
16. W.L. Wiese, M.W. Smith, B.M. Glennon, *Atomic Transition Probabilities–H through Ne*, vol. 1. Natl. Stand. Ref. Data Ser., Natl. Bur. Stand. **4** (1966)
17. W.L. Wiese, M.W. Smith, B.M. Miles, *Atomic Transition Probabilities–Na through Ca*, vol. 2. Natl. Stand. Ref. Data Ser., Natl. Bur. Stand. **22** (1969)
18. W.L. Wiese, B.M. Glennon, Atomic transition probabilities, in *American Institute of Physics Handbook*, ed. by D.E. Gray (New York, McGrow Hill, 1972), Chap. 7, p. 200263
19. W.L. Wiese, G.A. Martin, Transition probabilities, in *Wavelengths and Transition Probabilities for Atoms and Atomic Ions*. Natl. Stand. Ref. Data Ser., Natl. Bur. Stand. **68**, 359–406, Part II (1980)
20. https://physics.nist.gov/PhysRefData/ASD1
21. https://www.nist.gov/srd
22. https://www.nist.gov/pml/productsservices/physical-reference-data
23. https://www.nist.gov/pml/atomic-spectra-database

V. Krainov and B. M. Smirnov, *Atomic and Molecular Radiative Processes*,
Springer Series on Atomic, Optical, and Plasma Physics 108,
https://doi.org/10.1007/978-3-030-21955-0

24. Y.B. Zel'dovich, Y.P. Raizer, *Physics of Shock Waves and High-Temperature Hydrodynamic Phenomena* (Academic Press, New York, 1966)
25. C. Kittel, H. Kroemer, *Thermal Physics* (Wiley, New York, 1980)
26. M. Capitelli, C.M. Ferreira, B.F. Gordiets, A.I. Osipov, *Plasma Kinetics in Atmospheric Gases* (Springer, Berlin, 2000)
27. K.Y. Kondrat'ev, *Radiation in the Atmosphere* (Academic Press, New York, 1969)
28. E.J. McCartney, *Absorption and Emission by Atmospheric Gases* (Wiley, New York, 1983)
29. M. Iqbal, *An Introduction to Solar Radiation* (Acad. Press, New York, 1983)
30. G.W. Petry, *A First Course in Atmospheric Radiation* (Sunlog Publ, Madison, 2006)
31. W. Zdunkowski, T. Trautmann, A. Bott, *Radiation in the Atmosphere* (Cambridge University Press, Cambridge, 2007)
32. H.R. Griem, *Spectral Line Broadening by Plasmas* (Academic Press, New York, 1974)
33. I.I. Sobelman, L.A. Vainstein, E.A. Ukov, *Excitation of Atoms and Broadening of Spectral Lines* (Springer, Berlin, 1998)
34. E. Collet, *Polarized Light: Fundamentals and Applications* (Dekker, New York, 1993)
35. D.S. Klinger, J.W. Lewis, C.E. Randull, *Polarized Light in Optics and Spectroscopy* (Academic Press, Boston, 1997)
36. S. Huard, G. Vacca, *Polarization of Light* (Wiley, Chichester, 1997)
37. V.S. Lebedev, I.L. Beigman, *Physics of Highly Excited Atoms and Ions* (Springer, Berlin, 1988)
38. H.I. Metcalf, P. van der Straaten, *Laser Cooling and Trapping* (Springer, Berlin, 1999)
39. V.P. Krainov, B.M. Smirnov, *Radiative Processes in Atomic Physics* (Moscow, High Education, 1983, in Russian)
40. V.P. Krainov, B.M. Smirnov, *Quantum Theory of Radiation of Atomic Physics* (Dolgoprudny, Intellect, 2015, in Russian)
41. M. Griggs, J. Chem. Phys. **49**, 857 (1958)
42. Y.B. Zel'dovich, Y.P. Raizer, *Physics of Shock Waves and High-Temperature Hydrodynamic Phenomena* (Academic Press, New York, 1966)
43. B.M. Smirnov, *Physics of Weakly Ionized Gases* (Mir, Moscow, 1981)
44. J. Dalibard, C. Cohen-Tannoudji, Sisyphys cooling. J. Opt. Soc. Am. B **6**, 2023 (1989)
45. P.J. Ungar, D.S. Weiss, E. Riis, S. Chu, Sisyphys cooling. JOSA **B6**, 2058 (1989)
46. L.V. Zasova, N.I. Ignatiev, I.V. Khatuntsev, V.M. Linkin, Planet Space Sci. **55**, 1712 (2007)

Index

V. Krainov and B. M. Smirnov, *Atomic and Molecular Radiative Processes*,
Springer Series on Atomic, Optical, and Plasma Physics 108,
https://doi.org/10.1007/978-3-030-21955-0

MIX
Papier aus verantwortungsvollen Quellen
Paper from responsible sources
FSC® C105338

If you have any concerns about our products,
you can contact us on
ProductSafety@springernature.com

In case Publisher is established outside the EU,
the EU authorized representative is:
Springer Nature Customer Service Center GmbH
Europaplatz 3, 69115 Heidelberg, Germany

Printed by Libri Plureos GmbH
in Hamburg, Germany